N. N. Bogoljubov, Ju. A. Mitropoliskii and
A. M. Samoilenko

METHODS OF ACCELERATED CONVERGENCE IN NONLINEAR MECHANICS

Translated from the Russian by V. KUMAR

Translation Editor I. N. SNEDDON

Hindustan Publishing Corporation (India) Delhi
1976
Springer-Verlag Berlin Heidelberg New York

METHODS OF ACCELERATED CONVERGENCE
IN NONLINEAR MECHANICS

Title of the Russian Original Edition : *Metod Uskorennoi skhodimosti v nelineinoi mekhanike,* which does not include Appendices I to XV contributed by the authors to the English edition only.

Distributed in Japan and all non-Asian countries by
Springer-Verlag Berlin Heidelberg New York

ISBN 3-540-07106-7 Springer-Verlag Berlin Heidelberg New York
ISBN 0-387-07106-7 Springer-Verlag New York Heidelberg Berlin

Printed in India by
Hindustan Publishing Corporation (India), Delhi 110007 (India)

PREFACE TO THE RUSSIAN EDITION

This monograph deals with some of the latest results in nonlinear mechanics, obtained recently by the use of a modernized version of Bogoljubov's method of successive changes of variables which ensures rapid convergence.

This method visualised as early as 1934 by Krylov and Bogoljubov provides an effective tool for solving many interesting problems of nonlinear mechanics. It led, in particular, to the solution of the problem of the existence of a quasi-periodic regime, with the restriction that approximate solutions obtained in the general case involved divergent series.

Recently, making use of the research of Kolmogorov and Arno'ld, Bogoljubov has modernised the method of successive substitutions in such a way that the convergence of the corresponding expansions is ensured.

This book consists of a short Introduction and seven chapters.

The first chapter presents the results obtained by Bogoljubov in 1963 on the extension of the method of successive substitutions and the study of quasi-periodic solutions applied to non-conservative systems (*inter alia* making explicit the dependence of these solutions on the parameter, indicating methods of obtaining asymptotic and convergent series for them, etc.).

The second chapter elaborates the results due to Mitropoliskii. Here, with the aid of the method of successive substitutions, the general solutions of a system of nonlinear equations are constructed and the behaviour of these solutions in the vicinity of quasi-periodic solutions is considered (discussing in the course of this the questions of their stability, their attraction to a particular quasi-periodic solution, etc.). The problem of reducing a nonlinear system of equations to a linear system with constant coefficients is solved.

The third chapter gives a resumé and exposition of the technique for the smoothing of functions, offering the possibility of applying a series of results obtained in the first two chapters for systems of differential equations with analytic right-hand sides to those with differentiable right-hand sides.

The fourth chapter deals with the results derived by Mitropoliskii and Samoilenko in their study of the behaviour of trajectories on the n-dimensional torus for the case when the torus is not analytic, a situation confronted in many problems of nonlinear mechanics.

The fifth chapter is devoted to the question of the reducibility of linear systems with quasi-periodic coefficients. It draws upon the results obtained jointly by Mitropoliskii and Samoilenko concerning the construction and examination of the fundamental matrix of linear systems with quasi-periodic coefficients, the investigation of the measure of reducible systems, etc. Systems of equations both with analytic and with differentiable right-hand sides are discussed.

The sixth and seventh chapters treat Samoilenko's results on the behaviour of integral curves of systems of nonlinear equations in the neighbourhood of smooth toroidal and compact invariant manifolds.

Thanks are due to V. K. Lisnichenko and T. A. Tishko for their help in typing the manuscript.

PREFACE TO THE ENGLISH EDITION

One of the methods of nonlinear mechanics conceived by Krylov and Bogoljubov as early as 1934, consisting of the successive substitutions of variables, provides an effective tool for solving many problems of nonlinear mechanics, principally those involving the study of : (i) equations with quasi-periodic coefficients, (ii) almost-periodic and quasi-periodic regimes and (iii) problems of stability of motion and those connected with the reducibility of systems of equations.

This method has recently been considerably modernized and significantly extended, and has come to be widely applied to the solution of diverse problems of nonlinear mechanics.

The present English translation of our Russian monograph will make the methods and ideas presented in this work accessible to a wider circle of research workers interested in problems in the theory of nonlinear mechanics and of differential equations containing a small parameter.

The authors have contributed several short Appendices to the English translation which bring the work up to date through the inclusion of certain results obtained by the authors and others in further extending the ideas expounded in the monograph and their application to the solution of certain new and interesting problems.

We shall be very happy if the methods and ideas elaborated in the monograph are of interest and use to engineers and scientists studying the theory of nonlinear oscillations as well as to mathematicians interested in ordinary differential equations containing a small parameter.

We take this opportunity of expressing our sincere thanks to the publisher and the translator for pains taken by them in translating and publishing our monograph.

Kiev,
12 August 1972

JU. A. MITROPOLISKII

TABLE OF CONTENTS

Chapter 0

INTRODUCTION

The need to study oscillatory systems is encountered while solving various physical and engineering problems. This study had its origins in celestial mechanics and led to such high points of achievement as : (i) the local theory of periodic solutions of Poincaré-Lyapunov, (ii) the qualitative theory of dynamical systems on a plane, due to Poincaré-Bendizon, and (iii) Birkhoff's theory of dynamical systems.

The rapid and extensive developments in the field of automation and communication engineering evoked the demand for evolving new methods for investigating oscillatory systems. The contributions made by N. M. Krylov and N. N. Bogoljubov towards the solution of the resulting problems are of basic importance. The methods developed and rigorously proved by them [30-33] provide a powerful and practical tool of mathematical investigation. Of the numerous variants of these methods developed for very diverse classes of nonlinear systems, the particular method of successive substitutions of variables discovered by Krylov and Bogoljubov as early as 1934 is found to be most convenient in many cases.

To bring out certain important aspects of this method [30], we consider a system of differential equations in the standard form

$$\frac{dx}{dt} = \varepsilon X(t, x, \varepsilon), \tag{1}$$

where $x = (x_1, \ldots, x_n)$ and $X = (X_1, \ldots, X_n)$ are points of n-dimensional Euclidean space E_n, t is time and ε is a small positive parameter. The m-th approximation for eqn. (1) is given by

$$x^{(m)} = \xi + \varepsilon F^{(1)}(t, \xi) + \ldots + \varepsilon^m F^{(m)}(t, \xi), \tag{2}$$

where the variable ξ represents a solution of the system

$$\frac{d\xi}{dt} = \varepsilon P^{(1)}(\xi) + \varepsilon^2 P^{(2)}(\xi) + \ldots + \varepsilon^m P^{(m)}(\xi). \tag{3}$$

The functions $F^{(1)}(t, \xi), \ldots, F^{(m)}(t, \xi)$ and $P^{(1)}(\xi), \ldots, P^{(m)}(\xi)$ are here chosen (proceeding from the known expressions for $X(t, \xi, \varepsilon)$) such that the series (2) satisfies eqn. (1) to within terms of order ε^{m+1}.

Now, if, after determining the functions $F^{(1)}(t, \xi), \ldots, F^{(m)}(t, \xi)$, the expression (2) is regarded not as an approximate solution of the system (1) but as some form of change of variables transforming the unknown x into a new unknown ξ, then eqn. (1) reduces to the equation

$$\frac{d\xi}{dt} = \varepsilon P^{(1)}(\xi) + \varepsilon^2 P^{(2)}(\xi) + \ldots + \varepsilon^m P^{(m)}(\xi) + \varepsilon^{m+1} R(t, \xi, \varepsilon), \tag{4}$$

which consists of an 'integrable' part and the perturbation $\varepsilon^{m+1} R(t, \xi, \varepsilon)$, a term of the order of ε^{m+1}. Moreover, if the variable ξ satisfies Eqn. (4), the expression (2) becomes an exact solution of eqn. (1).

In attempting to develop this method further, it is natural to consider it in the limit, i.e., to let m tend to ∞ in the expressions (2)—(4). If the series (2) is found to be convergent, the system of equations (1) reduces to the 'integrable' system

$$\frac{d\xi}{dt} = \varepsilon P(\xi, \varepsilon). \tag{5}$$

However, it soon becomes clear that a development along these lines is not possible; for in formula (2) with functions $X(t, x, \varepsilon)$ quasi-periodic in t, there occur, as a rule, small divisors and the series (2) becomes divergent.

The idea of reducing the 'non-integrable' system (1) to the integral form (5), though tempting, was, for some time, impossible to put into practice. A break-through came only recently; as a consequence of the works of A. N. Kolmogorov [27-29] and V. I. Arnold [2-4], N. N. Bogoljubov [11] was able to develop a new method of successive substitutions of variables, and other workers were able to give a concrete shape to the reduction concept.

We shall deal briefly with the basic ideas leading to the results obtained by Kolmogorov and Arnold. Consider the conservative dynamical systems defined by the canonical equations

$$\begin{aligned} \frac{dp}{dt} &= -\frac{\partial H}{\partial q}, \\ \frac{dq}{dt} &= \frac{\partial H}{\partial p} \qquad (p = (p_1, \ldots, p_n); \quad q = (q_1, \ldots, q_n)), \end{aligned} \tag{6}$$

where the Hamiltonian function $H(p, q, \varepsilon)$ is analytic and periodic in q with period 2π. Assume that the system (6) differs from an integrable system by a small perturbation, i.e. its Hamiltonian can be represented in the form

$$H(p, q, \varepsilon) = H_0(p) + \varepsilon H_1(p, q) + \ldots. \tag{7}$$

Substituting the expression (7) in eqn. (6), we obtain the system

$$\begin{aligned} \frac{dp}{dt} &= -\varepsilon \frac{\partial H_1}{\partial q} + \ldots, \\ \frac{dq}{dt} &= \omega(p) + \varepsilon \frac{\partial H_1}{\partial p} + \ldots \qquad \left(\omega(p) = \frac{\partial H_0}{\partial p}\right). \end{aligned} \tag{8}$$

In the system of equations (8), we introduce the canonical transformation defined by the formulae

$$p = p' + \varepsilon \frac{\partial S(p', q)}{\partial q},$$
$$q' = q + \varepsilon \frac{\partial S(p', q)}{\partial p'}, \tag{9}$$

which reduces $H(p, q, \varepsilon)$ to the form

$$H(p, q, \varepsilon) = H_0'(p', \varepsilon) + \varepsilon^2 H_1'(p', q') + \dots, \tag{10}$$

and, consequently, eqns. (8) to the system

$$\frac{dp'}{dt} = -\varepsilon^2 \frac{\partial H_1'}{\partial q'} + \dots,$$
$$\frac{dq'}{dt} = \omega'(p') + \varepsilon^2 \frac{\partial H_1'}{\partial p'} + \dots. \tag{11}$$

Performing again the same type of transformation in the system of equations (11) as the substitution (9), we arrive at the equations

$$\frac{dp''}{dt} = -\varepsilon^4 \frac{\partial H_1''}{\partial q''} + \dots,$$
$$\frac{dq''}{dt} = \omega''(p'') + \varepsilon^4 \frac{\partial H_1''}{\partial p''} + \dots, \tag{12}$$

and so on. The order of the 'non-integrable' members in the transformed equations is here proportional respectively to $\varepsilon^2, \varepsilon^4, \varepsilon^8, \dots, \varepsilon^{2s}, \dots$.

The method of repeated transformations presented here contains a crucial new element: to carry out the approximation process the transformation (9) is refined not by adding higher order terms, as was done in [30], but by repeated applications of the same transformation.

The resultant 'rapid convergence' process is analogous to Newton's method of tangents. It suppresses the effect of small denominators appearing in the substitution of variable formulae (9), and for a 'majority' of initial values of p the iterations of such substitutions converge.

Combining the method of rapid convergence with the method of integral manifolds and, in addition, drawing upon various specific characteristics peculiar to nonlinear systems, N. N. Bogoljubov succeeded in extending considerably the range of applicability of the method, and solved the problem of the existence of quasi-periodic solutions for the general case for $n > 2$. Later, this method was further extended by the authors of the present monograph and used for solving a series of interesting and important problems of nonlinear mechanics (the reducibility of a nonlinear system to a linear system with constant coefficients, the reducibility of a system of linear differential equations with quasi-periodic coefficients, the behaviour of integral curves in the neighbourhood of analytic and smooth manifolds, etc.). These problems will be treated in the chapters that follow. Moreover, in addition to the solution and detailed analysis of the corresponding differential equations, the usefulness of the method of 'accelerated convergence' will be clearly shown.

Chapter 1

QUASI-PERIODIC SOLUTIONS IN PROBLEMS OF NONLINEAR MECHANICS

§ 1. Statement of the Problem. The Existence of an Invariant Manifold

It is well known [10, 30] that in problems of nonlinear mechanics it is possible in many cases to establish the existence of integral (invariant) manifolds having the property of asymptotic attraction of nearby trajectories.

Let a dynamical system be characterized by the equations

$$\frac{dx}{dt} = X(x, \varepsilon), \tag{1.1}$$

where $x = (x_1, \ldots, x_n)$, $X = (X_1, \ldots, X_n)$ are vectors of an n-dimensional Euclidean space, and ε is a small positive parameter.

Under specific conditions, it is possible to establish the existence of the invariant toroidal manifold

$$\begin{aligned} x &= \Phi(\varphi), \\ \varphi &= (\varphi_1, \ldots, \varphi_m), \end{aligned} \tag{1.2}$$

for the system of equations (1.1)

In this case the original system (1.1) reduces to the following equation on the torus

$$\frac{d\varphi}{dt} = \nu + f(\varphi, \varepsilon), \tag{1.3}$$

where $\varphi = (\varphi_1, \ldots, \varphi_m)$, $\nu = (\nu_1, \ldots, \nu_m)$ and $f(\varphi, \varepsilon)$ is a periodic function probably sufficiently small [because of ε appearing in system (1.1)], i.e. a variation of φ effects an almost uniform rotation with constant angular velocity ν.

Under specific conditions, the manifold (1.2) has the property of asymptotic attraction of the trajectories of any solution of eqns. (1.1) not lying on the torus. For certain differential equations, such properties are determined in [30]. This problem has been further developed in [10, 12].

In problems of nonlinear mechanics, our interest is not only to obtain a stable toroidal manifold but also to investigate the behaviour of integral curves on this manifold.

The earlier investigations of this problem [30] were carried out with the aid of the results of Poincaré and Denjoy on the transformation of a circle into itself. However, the Poincaré-Denjoy theory is related to the one-dimensional case, where the original system of differential equations (1.1) reduces to the following two equations :

$$\frac{d\varphi}{dt} = \nu + f(\varphi + \theta), \qquad \frac{d\theta}{dt} = \omega. \tag{1.4}$$

According to this theory, the behaviour of the solutions on a two-dimensional torus is characterized by the rotation number Ω : (i) if Ω is irrational, then the solution on the torus is quasi-periodic ; (ii) if Ω is rational, then periodic solutions exist, and all the remaining solutions approximate them with increasing time.

If the original system of equations (1.1) is reducible to the special form (1.4), then it is possible to prove [30] the existence of a quasi-periodic (with the two frequencies ω_1, ω_2) solution, and to establish its stability, etc.

It has not yet been possible to investigate within the framework of the Poincaré-Denjoy theory the general case of manifolds of higher dimensions, i.e. when eqns. (1.1) on the torus are reducible to equations of the form (1.3).

A combination of the method of integral manifolds with that of accelerated convergence has unfolded new vistas for investigating considerably more general cases because of the removal of the restrictions on the dimension of the invariant manifolds.

In this chapter we deal with the following problems : the existence of quasi-periodic solutions, the investigation of their stability, their analyticity in relation to the parameter, and the derivation of power series expansions whose convergence is obvious [12].

Before taking up the formulation and rigorous proof of the fundamental theorems, let us analyze intuitively the behaviour of the solutions of the system of equations

$$\frac{d\varphi}{dt} = \nu + f(\varphi), \qquad \varphi = (\varphi_1, \ldots, \varphi_m), \quad f = (f_1, \ldots, f_m), \tag{1.5}$$

where $f(\varphi)$ is a small periodic function of φ representable in the form of the m-multiple Fourier series

$$f(\varphi) = \sum_{(k)} f_k e^{i(k,\varphi)}, \quad (k, \varphi) = k_1\varphi_1 + \ldots + k_m\varphi_m. \tag{1.6}$$

Obviously, the motion described by the system of equations (1.5) represents uniform rotation, on which certain small oscillatory terms are superimposed.

On averaging the right-hand sides of the system of equations (1.5) in φ, by the well-known principles of nonlinear mechanics, we obtain in the zero approximation the equations

$$\frac{d\bar{\varphi}}{dt} = \nu + \bar{f}(\varphi), \tag{1.7}$$

whence it is seen that in this approximation the rotation takes place with a somewhat changed frequency

$$\omega = \nu + \bar{f}. \tag{1.8}$$

To obtain solutions of the system of eqns. (1.5) in the first approximation, we must take into consideration the terms giving rise to vibrations. We obtain the equation

$$\varphi = \omega t + \varphi_0 + \sum_{(|k| \neq 0)} \frac{f_k e^{i(k, \omega t + \varphi_0)}}{i(k, \omega)}, \quad (k, \omega) = k_1\omega_1 + \ldots + k_m\omega_m, \tag{1.9}$$

where we have assumed that the ω_i are linearly independent.

Since the right-hand side of eqn. (1.9) consists of small summands, it follows by applying perturbation theory that higher approximations can also be obtained. However, it is necessary to note that the sums (k, ω), the so-called small divisors, will appear in the denominators and, consequently, it will not be possible to express ω in powers of ε. It becomes obvious from these purely intuitive considerations that it will be more expedient if, in place of finding ω in terms of ν and f, we regard it as given and determine $\Delta = \nu - \omega$ as a function of ω:

$$\Delta = \Delta(\omega). \tag{1.10}$$

In the first approximation, we obtain

$$\Delta = -\bar{f}. \tag{1.11}$$

It should be noted that the appropriateness of expressing the 'frequencies of zero-approximation' appearing in the equation in terms of exact frequencies has already been indicated in [31].

Let us first examine the convergence of the series appearing in the expression (1.9) for the first approximation. Even if f_k decreases so rapidly that $|f_k| \leqslant C\rho^{|k|}$ where $\rho < 1$, it is possible to choose ω to be such that the series (1.9) diverges. Consequently, it is necessary to impose certain conditions on ω, for example,

$$|(k, \omega)| \geqslant K|k|^{-(m+1)}, \tag{1.12}$$

where m is the dimension of the space, $|k| = |k_1| + |k_2| + \ldots + |k_m|$ and $k_1, \ldots, k_m$ are any integers (positive or negative).

It is well known that if we consider a sphere in the space $\omega = (\omega_1, \ldots, \omega_m)$, then the relative measure of the set of those ω's which do not satisfy the condition (1.12) will tend to zero together with K. Thus, for sufficiently small K, *most* of the ω's will satisfy the inequality (1.12).

In investigating the problem of successively constructing the higher approximations of the system of eqns. (1.5), we encounter serious difficulties described below.

Let $f(\varphi)$ be an analytic function satisfying on the domain

$$|\operatorname{Im} \varphi| \leqslant \rho \tag{1.13}$$

the inequality $|f(\varphi)| \leqslant M$.

Let us introduce the notation

$$\tilde{f}(\varphi) = \sum_{(k \neq 0)} \frac{f_k e^{i(k,\varphi)}}{i(k, \omega)}. \tag{1.14}$$

Then, as is well-known, on the domain

$$|\operatorname{Im} \varphi| \leqslant \rho - 2\delta \tag{1.15}$$

the inequality

$$|\tilde{f}(\varphi)| \leqslant \frac{MC}{\delta^{2m+1}} \tag{1.16}$$

holds.

We note that the usual iteration process is characterized by the estimates of the form

$$M_{s+1} \leqslant C_s M_s. \tag{1.17}$$

If $C_s < 1$, the process converges as a geometric progression. However, if $C_s \to \infty$ as $s \to \infty$, then starting from the estimates (1.17) it is not possible to prove the convergence of the series obtained under successive constructions of higher approximations.

Consider the expression (1.16). Applying the iteration process successively from ρ to $\rho - 2\delta$, from $\rho - 2\delta$ to $\rho - 2\delta - 2\delta_1, \ldots$, to $\rho - 2\delta - 2\delta_1 - \ldots - 2\delta_i - \ldots$, we must decrease δ_i. This, evidently, according to the inequality (1.16) must give rise to terms that increase at the expense of the decrease of the denominators δ_i. This is also the well-known difficulty connected with small divisors, which was encountered in celestial mechanics in the last century. This difficulty was circumvented in [2] in an exceedingly simple and novel manner. In order to illustrate this by means of the sample system (1.5), we revert to the expression (1.9) for the first approximation in the form

$$\varphi = \omega t + \varphi_0 + \tilde{f}(\omega t + \varphi_0). \tag{1.18}$$

We treat this not as the first approximation of the solution of eqns. (1.5) but (according to our assertions in the Introduction) as a substitution of the variable

$$\varphi = \varphi^{(1)} + \tilde{f}(\varphi^{(1)}) \tag{1.19}$$

in the equation

$$\frac{d\varphi}{dt} = \omega + \Delta + f(\varphi). \tag{1.20}$$

Here we have already introduced the correction term $\Delta = \nu - \omega$.

With the notation (1.14), we have the relation

$$\frac{\partial \tilde{f}}{\partial \varphi^{(1)}} \omega = f(\varphi^{(1)}) - \bar{f}(\varphi). \tag{1.21}$$

Substituting (1.19) in eqn. (1.20), we get

$$\frac{d\varphi^{(1)}}{dt} + \frac{\partial \tilde{f}}{\partial \varphi^{(1)}} \frac{d\varphi^{(1)}}{dt} = \omega + \Delta + f(\varphi^{(1)} + \tilde{f}(\varphi^{(1)})). \tag{1.22}$$

Using the relation (1.21), we obtain

$$\left(E + \frac{\partial \tilde{f}}{\partial \varphi^{(1)}}\right)\left(\frac{d\varphi^{(1)}}{dt} - \omega\right) = \Delta + \bar{f}(\varphi) + f(\varphi^{(1)} + \tilde{f}) - f(\varphi^{(1)}), \tag{1.23}$$

where E is the unit matrix.

Dividing the expression (1.23) by $\left(E + \frac{\partial \tilde{f}}{\partial \varphi^{(1)}}\right)$, we obtain

$$\frac{d\varphi^{(1)}}{dt} = \omega + \Delta + \tilde{f} + \left\{\left(E + \frac{\partial \tilde{f}}{\partial \varphi^{(1)}}\right)^{-1} - E\right\}(\Delta + \bar{f}) + \\ + \left(E + \frac{\partial \tilde{f}}{\partial \varphi^{(1)}}\right)^{-1} \{f(\varphi^{(1)} + \tilde{f}) - f(\varphi^{(1)})\}. \tag{1.24}$$

Introducing the deviation $\Delta^{(1)}$ from the frequency by the equation

$$\Delta^{(1)} = \Delta + \bar{f}(\varphi), \tag{1.25}$$

we can rewrite eqn. (1.24) in the form

$$\frac{d\varphi^{(1)}}{dt} = \omega + \Delta^{(1)} + f^{(1)}(\varphi^{(1)}, \Delta^{(1)}), \tag{1.26}$$

where

$$f^{(1)}(\varphi^{(1)}, \Delta^{(1)}) = \left\{\left(E + \frac{\partial \tilde{f}(\varphi^{(1)}, \Delta^{(1)})}{\partial \varphi^{(1)}}\right)^{-1} - E\right\}(\Delta + \bar{f}(\varphi)) + \\ + \left(E + \frac{\partial \tilde{f}(\varphi^{(1)}, \Delta^{(1)})}{\partial \varphi^{(1)}}\right)^{-1} \{f(\varphi^{(1)} + \tilde{f}) - f(\varphi^{(1)})\}. \tag{1.27}$$

Let us now determine the order of the terms appearing in the expression (1.27): If $f(\varphi)$ and Δ are first-order terms (of order ε), then after the first substitution, $f^{(1)}(\varphi^{(1)}, \Delta^{(1)})$ and $\Delta^{(1)}$ will become second-order terms (of order ε^2).

Making transformations of the same kind in eqns. (1.26), we obtain equations of the form

$$\frac{d\varphi^{(2)}}{dt} = \omega + \Delta^{(2)} + f^{(2)}(\varphi^{(2)}, \Delta^{(2)}), \tag{1.28}$$

and so on. In addition, the order of the functions $f^{(i)}(\varphi^{(i)}, \Delta^{(i)})$ and of the corrections to the frequency $\Delta^{(i)}$ $(i = 1, 2, \ldots, s)$ will be proportional respectively to $\varepsilon^2, \varepsilon^4, \varepsilon^8, \ldots, \varepsilon^{2s}$.

To investigate the convergence of the process under consideration, we assume that the inequality

$$|f(\varphi)| \leqslant M$$

holds for the initial function $f(\varphi)$ on the domain

$$|\operatorname{Im} \varphi| \leqslant \rho.$$

Then, for the next function $f^{(1)}(\varphi^{(1)}, \Delta^{(1)})$ on the domain

$$|\operatorname{Im} \varphi^{(1)}| \leqslant \rho - 2\delta, \quad |\Delta^{(1)}| \leqslant M$$

we have the inequality

$$|f^{(1)}(\varphi^{(1)}, \Delta^{(1)})| \leqslant M_1,$$

where

$$M_1 = P\frac{M^2}{\delta^{2n+2}}, \tag{1.29}$$

and

$$P = \frac{C}{K}, \quad C = \text{const.} \tag{1.30}$$

In constructing the iteration process, it is necessary to make a suitable choice of ρ and δ. According to the condition (1.13), the initial domain is characterized by the number ρ. After the first iteration, we obtain

$$\rho_1 = \rho - 2\delta; \tag{1.31}$$

after the second one,

$$\rho_2 = \rho_1 - 2\delta_1 = \rho - 2\delta - 2\delta_1, \tag{1.32}$$

and so on.

Obviously, it is necessary that all the ρ_i be bounded below; for example, $\rho_i \geqslant \rho/2$. Thus, we set

$$\delta = \gamma, \quad \delta_1 = \gamma^2, \ldots, \delta_i = \gamma^{i+1}, \tag{1.33}$$

and choose γ such that it satisfies the equation

$$2(\gamma + \gamma^2 + \ldots + \gamma^i + \ldots) = \frac{\rho}{2},$$

i.e. it is defined by the expression

$$\rho = \frac{\rho}{4+\rho}. \tag{1.34}$$

Then, as a result of successive substitutions of the type (1.19), for all the new variables ($\varphi^{(1)}, \varphi^{(2)}, \ldots, \varphi^{(s+1)}$) we will successively obtain differential equations, in which the deviations from uniform rotation are given respectively by

$$\begin{aligned} M_1 &= P\,\frac{M^2}{\delta^{2m+2}} \\ &\vdots \\ M_{s+1} &= P\,\frac{M_s^2}{\gamma^{(2m+2)(s+1)}}, \end{aligned} \tag{1.35}$$

or,

$$M_{s+1} = PN^{s+1} M_s^2, \tag{1.36}$$

where

$$N = \frac{1}{\gamma^{2m+2}}.$$

If $M_0 = M$ is sufficiently small, then the process converges rapidly. Thus, setting $PN^2M \leqslant r_0 < 1$, we obtain from the iteration formulas (1.35)

$$M_s \leqslant \frac{r_0^{2s}}{N^{s+2}}, \tag{1.37}$$

which also ensures rapid convergence of the process considered.

It is appropriate here to emphasize the extraordinary stability of the iteration process involving the quadratic convergence. Thus, if instead of the relation (1.36), we even had a relation of the form

$$M_{s+1} = N^{(1+\gamma)^s} M_s^2, \tag{1.38}$$

we should have convergence for sufficiently small M_0, only for $0 < \gamma < 1$. In fact, by raising eqn. (1.38) to the $(1-\gamma)$-th power we obtain

$$M_{s+1}^{1-\gamma} = N^{(1+\gamma)^s(1-\gamma)} (M_s^{(1-\gamma)})^2. \tag{1.39}$$

Set

$$M_s^{1-\gamma} = N^{-(1+\gamma)^s} \lambda_s. \tag{1.40}$$

Then, substituting this value into (1.39), we get

$$N^{-(1+\gamma)^s(1+\gamma)}\lambda_{s+1} = N^{(1+\gamma)^s(1-\gamma)-2(1+\gamma)^s}\lambda_s^2 = N^{-(1+\gamma)(1+\gamma)^s}\lambda_s^2. \tag{1.41}$$

Note that

$$\lambda_m = \lambda_0^{2m}. \tag{1.42}$$

Putting $s = 0$ in the relation (1.40), we find

$$\lambda_0 = M_0^{1-\gamma} N, \tag{1.43}$$

which implies

$$\lambda_0^{\frac{1}{1-\gamma}} = M_0 N^{\frac{1}{1-\gamma}}. \tag{1.44}$$

Making use of eqn. (1.42), we obtain from (1.40) the equation

$$M_s = \frac{1}{N^{(1+\gamma)^s \cdot \frac{1}{1-\gamma}}} \lambda_s^{\frac{1}{1-\gamma}} = \frac{1}{N^{(1+\gamma)^s \cdot \frac{1}{1-\gamma}}} \lambda_0^{\frac{1}{1-\gamma} 2^s}. \tag{1.45}$$

Furthermore, keeping in view eqn. (1.44), we finally get

$$M_s = \frac{1}{N^{(1+\gamma)^s \frac{1}{1-\gamma}}} \left(M_0 N^{\frac{1}{1-\gamma}} \right)^{2^s}, \tag{1.46}$$

from which we deduce the validity of the above assertion.

Thus, passing to the limit in the iteration process under consideration, we get the substitution of the variable

$$\varphi = \theta + \Phi(\theta), \tag{1.47}$$

and a value of the correction term Δ which yields in place of eqn. (1.20), the equation

$$\frac{d\theta}{dt} = \omega, \tag{1.48}$$

which corresponds to uniform rotation.

In particular, if $f(\varphi) = \varepsilon f(\varphi)$, the successive iterations lead to analytic functions on the sufficiently small domain $|\varepsilon| < \varepsilon_0$, and since the iteration process is uniformly convergent, the limiting functions are analytic on this domain. This means that the functions

$$\Delta = \Delta(\varepsilon), \qquad \Phi = \Phi(\varphi, \varepsilon) \tag{1.49}$$

can be expanded as convergent series in powers of ε. The direct investigation of these series is, howerver, beset with considerable difficulties.

In the reasoning used above, it is essential that $f(\varphi)$ be an analytical function of φ on a certain complex domain

$$|\operatorname{Im} \varphi| < \rho.$$

It may be noted that under suitable conditions it is possible to prove by the methods usually applied in nonlinear mechanics that the functions defining the invariant torus $x = \Phi(\varphi)$ possess sufficient numbers of derivatives. However, these methods are inadequate to establish the analytic character of this torus.

This preliminary discussion of the system of equations presented above reveals the difficulties met with in constructing solutions and also shows the ways of overcoming them.

We now proceed to formulate the fundamental problem to be dealt with in this chapter.

In investigating the system (1.1) we usually find it convenient to introduce new variables h and φ in place of the variables $x = (x_1, \ldots, x_m)$, so that eqns. (1.1) reduce to the equations of the form

$$\begin{aligned} \frac{dh}{dt} &= Hh + F(h, \varphi), \\ \frac{d\varphi}{dt} &= \nu + f(h, \varphi)\,; \end{aligned} \tag{1.50}$$

here $h = (h_1, \ldots, h_n)$, $\varphi = (\varphi_1, \ldots, \varphi_m)$ (the sum $n + m$ of the dimensions of the vectors h and φ being equal to the dimension of the vector x, is denoted by $n + m$ for the sake of convenience); the real parts of the eigenvalues of the square $n \times n$ matrix H are all negative:

$$|e^{Ht}| \leqslant Pe^{-\alpha t}, \qquad t \geqslant 0, \tag{1.51}$$

where $P > 0$, $\alpha > 0$ are constants; the vector functions $F(h, \varphi)$ and $f(h, \varphi)$ are small for sufficiently small h and are regular.

We remark that even if the functions on the right-hand sides of the syetem of eqns. (1.50) are analytic and are arbitrarily small, it is not possible to prove directly the existence of an analytic torus on the complex domain for these equations. This will be made clear by a quite simple example.

Let us consider the system of three differential equations

$$\begin{aligned} \frac{dh}{dt} &= -\alpha h + f(\varphi) = -\alpha h + \sum_{(k)} \rho^{(k)} e^{i(k,\varphi)}, \\ \frac{d\varphi}{dt} &= \nu + \varepsilon\gamma, \quad (h = h_1, \quad \varphi = (\varphi_1, \varphi_2)). \end{aligned} \tag{1.52}$$

For the system (1.52), it is easy to find the invariant manifold

$$h = S(\varphi), \tag{1.53}$$

where

$$S(\varphi) = \sum_{(k)} \rho^{(k)} \frac{e^{i(k,\varphi)}}{ik(\nu + \varepsilon\gamma) + \alpha}, \qquad (k = (k_1, k_2)). \tag{1.54}$$

It is obvious that, when ε and γ are real, the torus (1.53) always exists.

If, however, ε or γ is complex, then we can always handle them such that

$$k(\nu + \operatorname{Re} \varepsilon\gamma) = 0,$$

and

$$k(\operatorname{Im} \varepsilon\gamma) + \alpha = 0.$$

In this case, we can obtain for the function $S(\varphi)$ a set of poles for arbitrarily small $|\varepsilon|$. Consequently, it is not possible to establish directly the existence of the invariant torus which will be analytic not only in the angle φ but also in the parameter ε in the neighbourhood of the value $\varepsilon = 0$.

In what follows we shall be interested not only in the question of the existence of quasi-periodic solutions of the system of eqns. (1.50) but also in establishing their analytic properties in relation to the parameter, and thus enabling us to obtain power series expansions for them whose convergence is obvious. To this end, we shall combine the customary method of proof of the existence of integral manifolds with the one ensuring accelerated convergence.

First of all, we transform the system of eqns. (1.50). We introduce the correction for the frequency, Δ, by the formula $\Delta = \nu - \omega$, where ω are natural frequencies of the system. Then, together with the system of eqns. (1.50), we consider the system

$$\begin{aligned} \frac{dh}{dt} &= Hh + F(h, \varphi, \Delta), \\ \frac{d\varphi}{dt} &= \omega + \Delta + f(h, \varphi, \Delta), \end{aligned} \tag{1.55}$$

where $F(h, \varphi, \Delta), f(h, \varphi, \Delta)$ are functions regular on the domain

$$|h| < \eta, \quad |\operatorname{Im} \varphi| < \rho, \quad |\Delta| < \sigma. \tag{1.56}$$

The problem consists in choosing Δ such that by means of some substitution of variables, eqns. (1.55) reduce to the system

$$\begin{aligned} \frac{dh}{dt} &= Hh + F(h, \varphi), \\ \frac{d\varphi}{dt} &= \omega, \end{aligned} \tag{1.57}$$

where ω are real.

For the system of equations (1.57), there always exists an invariant analytic torus, as already remarked, since the vector function $F(h, \varphi)$ is analytic and ω is real. We shall prove the existence of this torus, and hence the existence of quasi-periodic solution for this system.

We now consider certain auxiliary definitions.

As is well known, it is customary to define the norm in the form

$$|h| = \sup_{k=1,\ldots,n} |h_k|,$$

where $h = h_1, \ldots, h_n$ are complex terms.

We introduce the new norm

$$\|h\| = \sup_{\substack{k=1,\ldots,n \\ 0 \leqslant t < \infty}} |e^{Ht} h_k| e^{\alpha t}, \tag{1.58}$$

where H represents the matrix considered in the system of equations (1.57) and satisfies the condition

$$|e^{Ht}| \leqslant Pe^{-\alpha t} \tag{1.59}$$

for $t \geqslant 0$, $\alpha > 0$, $P = \text{const.} \geqslant 1$.

The new norm obviously satisfies the inequality

$$|h| \leqslant \|h\| \leqslant P|h|. \tag{1.60}$$

It is easy to see that the norm $\|h\|$ can be accepted as the customary norm on a linear n-dimensional complex space.

We now proceed to prove the existence of the quasi-periodic solutions for the system (1.57). We have the theorem

Theorem 1. *Let the system of eqns.* (1.57) *satisfy the following conditions:*

(i) *The functions $F(h, \varphi)$ satisfy the inequalities*

$$\|F(h, \varphi)\| \leqslant N, \quad n \left\| \frac{\partial F(h, \varphi)}{\partial h_q} \right\| \leqslant L \tag{1.61}$$

on the domain

$$\|h\| \leqslant \eta, \quad |\operatorname{Im} \varphi| \leqslant \rho;$$

here η, ρ, N, L are certain constants, and

$$\frac{N}{\alpha} \leqslant \eta, \quad L \leqslant \frac{\alpha}{2}. \tag{1.62}$$

(ii) *All the eigenvalues of the matrix H have negative real parts and, consequently, the inequality*

$$|e^{Ht}| \leqslant Pe^{-\alpha t} \tag{1.63}$$

holds, where $P = \text{const.} \geqslant 1$, $t \geqslant 0$, $\alpha > 0$.

Then the system of equations (1.57) *admits the existence of the unique integral manifold*

$$h = S(\varphi), \tag{1.64}$$

where $S(\varphi)$ *is a periodic function in* φ *of period* 2π *satisfying the inequality*

$$\| S(\varphi) \| \leqslant \frac{N}{\alpha} \quad \text{for} \quad | \operatorname{Im} \varphi | \leqslant \rho. \tag{1.65}$$

If, for any solution not lying on the manifold, the initial values h_0, φ_0 *satisfy the conditions*

$$\| h_0 \| \leqslant \eta, \quad | \operatorname{Im} \varphi_0 | \leqslant \rho,$$

for $t = t_0$, *then for any* $t \geqslant t_0$ *the inequality*

$$\| h_t - S(\varphi_t) \| \leqslant 2\eta e^{-\alpha t/2}, \qquad t \geqslant 0, \tag{1.66}$$

remains valid and, moreover,

$$| h_t | \leqslant \eta, \qquad t \geqslant 0.$$

Proof. Consider the solution of the integral equation

$$h_t = \int_{-\infty}^{t} e^{H(t-\tau)} F(h_\tau, \varphi + \omega\tau) \, d\tau, \tag{1.67}$$

satisfying the differential eqns. (1.57).

Let

$$h_t = S(\varphi + \omega t). \tag{1.68}$$

Then

$$S(\varphi + \omega t) \equiv \int_{-\infty}^{t} e^{H(t-\tau)} F\{S(\varphi + \omega\tau), \varphi + \omega\tau\} \, d\tau. \tag{1.69}$$

Inserting in the expression (1.69), the substitution of variables

$$\varphi \to \varphi - \omega t, \quad \tau = t - z,$$

we reduce the problem of the existence of the required integral invariant manifold to that of investigating the equation

$$S(\varphi) = \int_{0}^{\infty} e^{Hz} F\{S(\varphi - \omega z), \varphi - \omega z\} \, dz. \tag{1.70}$$

We seek to solve Eqn. (1.70) by means of the usual iteration process:

$$\begin{aligned} S_0(\varphi) &= 0, \\ S_1(\varphi) &= \int_0^\infty e^{Hz} F\{0, \varphi - \omega z\}\, dz, \\ &\vdots \\ S_{s+1}(\varphi) &= \int_0^\infty e^{Hz} F\{S_s(\varphi - \omega z), \varphi - \omega z\}\, dz. \end{aligned} \tag{1.71}$$

Keeping in view the hypotheses of the theorem, we can set

$$\| S_1(\varphi) \| \leqslant \left| \int_0^\infty e^{-\alpha z}\, dz \right|, \qquad N = \frac{N}{\alpha} \leqslant \eta. \tag{1.72}$$

Assume that the inequality

$$\| S_r(\varphi) \| \leqslant \frac{N}{\alpha} \tag{1.73}$$

holds for all $S_r(\varphi)$ $(r = 1, 2, \ldots, s)$.

It is required to show that this inequality remains valid also for $r = s + 1$. Since

$$\| S_s(\varphi) \| \leqslant \eta, \quad | \operatorname{Im} (\varphi - \omega z) | = | \operatorname{Im} \varphi | \leqslant \rho, \tag{1.74}$$

it follows that

$$\| S_{s+1}(\varphi) \| \leqslant N \int_0^\infty e^{-\alpha z}\, dz = \frac{N}{\alpha}. \tag{1.75}$$

In order to establish that the sequences S_s are convergent, consider the difference

$$\begin{aligned} S_{s+1}(\varphi) - S_s(\varphi) = \int_0^\infty e^{Hz} \{F[S_s(\varphi - \omega z), \ \varphi - \omega z] - F \\ \times [S_{s-1}(\varphi - \omega z), \ \varphi - \omega z]\}\, dz, \quad s = 1, 2, \ldots. \end{aligned} \tag{1.76}$$

Since $\left\| \dfrac{\partial F(h, \varphi)}{\partial h_q} \right\| \leqslant L$, it follows that

$$\| S_{s+1}(\varphi) - S_s(\varphi) \| \leqslant \int_0^\infty e^{-\alpha z} L \, \| S_s(\varphi - \omega z) - S_{s-1}(\varphi - \omega z) \|\, dz. \tag{1.77}$$

Let

$$\sup_{|\operatorname{Im} \chi| \leqslant \rho} \| S_s(\chi) - S_{s-1}(\chi) \| = D_s. \tag{1.78}$$

Then

$$D_1 \leqslant \frac{N}{\alpha},$$
$$\vdots \qquad \vdots \tag{1.79}$$
$$D_{s+1} \leqslant \frac{L}{\alpha} D_s \leqslant \frac{1}{2} D_s.$$

Thus

$$D_s < \frac{N}{\alpha}\left(\frac{1}{2}\right)^{s-1}. \tag{1.80}$$

From this we deduce the uniform convergence of the sequence $S_s(\varphi)$:

$$S_s(\varphi) \to S(\varphi) \text{ in the domain } |\operatorname{Im}\varphi| \leqslant \rho. \tag{1.81}$$

We now prove the property of the integral manifold of the trajectories attracting any solution not lying on them, for which $| h_0 | \leqslant \eta$, $| \operatorname{Im} \varphi_0 | < \rho$.

Substituting $h = S(\varphi + \omega t)$ in eqn. (1.57), we obtain

$$\frac{dS(\varphi + \omega t)}{dt} = HS(\varphi + \omega t) + F\{S(\varphi + \omega t), \varphi + \omega t\}. \tag{1.82}$$

Let $h = h_0$ satisfy the inequality

$$\| h \| \leqslant \eta.$$

Then, from the expression

$$h_t = \int_0^t e^{H(t-\tau)} F(h_\tau, \varphi + \omega\tau)\, d\tau + e^{Ht} h_0, \tag{1.83}$$

it is easy to see that

$$| h_t | \leqslant \eta, \qquad t \geqslant 0.$$

In fact,

$$\| h_t \| \leqslant e^{-\alpha t} \| h_0 \| + N \int_0^t e^{-\alpha(t-\tau)}\, d\tau, \tag{1.84}$$

or,

$$\| h_t \| \leqslant e^{-\alpha t} \| h_0 \| + \frac{1 - e^{-\alpha t}}{\alpha} N \leqslant e^{-\alpha t}\eta + \eta(1 - e^{-\alpha t}) \leqslant \eta.$$

Further, we have

$$\frac{dh_t}{dt} = Hh_t + F(h_t, \varphi + \omega t). \tag{1.85}$$

Subtracting the eqns. (1.82) from the system of eqns. (1.85), we obtain

$$\frac{d(h_t - S(\varphi + \omega t))}{dt} = H[h_t - S(\varphi + \omega t)] + F(h_t, \varphi + \omega t) - F(S(\varphi + \omega t), \varphi + \omega t)). \tag{1.86}$$

Solving the system (1.86), we find that

$$h_t - S(\varphi + \omega t) = e^{Ht}(h - S(\varphi)) + \int_0^t e^{H(t-\tau)} \{F(h_\tau, \varphi + \omega\tau) - F(S(\varphi + \omega\tau), \varphi + \omega\tau)\} d\tau.$$

From this, after majorising the right-hand side, we obtain

$$\| h_t - S(\varphi + \omega t) \| \leqslant e^{-\alpha t} 2\eta + \int_0^t e^{-\alpha(t-\tau)} L | h_\tau - S(\varphi + \omega\tau) | d\tau, \tag{1.87}$$

or,

$$\| h_t - S(\varphi + \omega t) \| \leqslant G(t), \tag{1.88}$$

where

$$G(t) = e^{-\alpha t} 2\eta + L \int_0^t e^{-\alpha(t-\tau)} G(\tau)\, d\tau, \tag{1.89}$$

$$G(0) = 2\eta.$$

Differentiating the expression (1.89) with respect to t as the parameter, we get

$$\frac{dG}{dt} = -\alpha G + LG = -(\alpha - L)G. \tag{1.90}$$

Noting that $\alpha - L \geqslant \alpha - \frac{\alpha}{2} = \frac{\alpha}{2}$, $G(0) = 2\eta$, we deduce from eqn. (1.90) that

$$G(t) = 2\eta e^{-(\alpha - L)t} \leqslant 2\eta e^{-\frac{\alpha}{2}t}. \tag{1.91}$$

Comparing the inequalities (1.88) and (1.91), we finally obtain

$$\| h_t - S(\varphi + \omega t) \| \leqslant 2\eta e^{-\frac{\alpha}{2}t}, \quad t \geqslant 0, \tag{1.92}$$

which completes the proof of the theorem.

Remarks. 1. If H is a real matrix and $F(h, \varphi)$ is a real function for real h, φ, then the manifold $S(\varphi)$ is also real for real φ.

2. If $F(h, \varphi)$ is a function analytic in μ on the domain ϑ and satisfies the hypotheses of the theorem on this domain with fixed ρ, η, then $S = S(\varphi, \mu)$ is also analytic with respect to μ on the domain ϑ.

§ 2. Auxiliary Theorems

In this section, we will derive some estimates needed for solving the already formulated problem of reducing the system of eqns. (1.55) to the system (1.57).

Assume that the analytic function $f(\varphi)$ is given on the domain $|\operatorname{Im} \varphi| \leqslant \rho$, where φ is the angular variable $\varphi = (\varphi_1, \ldots, \varphi_m)$, and that $f(\varphi)$ is bounded by the constant M:

$$|f(\varphi)| \leqslant M. \tag{2.1}$$

Then the Fourier expansion of $f(\varphi)$ is given by

$$f(\varphi) = \sum_{k} f_k e^{i(k,\varphi)}, \tag{2.2}$$

where

$$k = k_1 + k_2 + \ldots + k_m \quad \text{and} \quad (k, \varphi) = k_1\varphi_1 + \ldots + k_m\varphi_m.$$

Introducing the notation

$$\tilde{f}(\varphi) = \sum_{\substack{k \\ (|k| \neq 0)}} \frac{f_k}{i(k, \omega)} e^{i(k,\varphi)}, \tag{2.3}$$

we wish to establish certain properties of the function $\tilde{f}(\varphi)$. Obviously,

$$\left(\frac{\partial \tilde{f}(\varphi)}{\partial \varphi}, \omega\right) = f(\varphi) - \bar{f}(\varphi) = \sum_{\substack{k \\ (|k| \neq 0)}} f_k e^{i(k,\varphi)}. \tag{2.4}$$

Assume that the numbers $\omega = (\omega_1, \ldots, \omega_m)$ satisfy the conditions

$$|(k, \omega)| \geqslant K|k|^{-(m+1)} \; (|k| \neq 0, \; |k| = |k_1| + \ldots + |k_m|) \tag{2.5}$$

for all integral vectors $k = (k_1, \ldots, k_m)$, where K is a certain constant. Then the assertion made below in Lemma 1 holds for the function $\tilde{f}(\varphi)$.

Lemma 1. *Subject to the fulfilment of the conditions* (2.5), $\tilde{f}(\varphi)$ *is an analytic function on the domain* $|\operatorname{Im} \varphi| < \rho - 2\delta$, $0 < 2\delta < \rho$, $\delta < 1$; *in addition, the estimates*

$$|\tilde{f}(\varphi)| \leqslant \frac{M}{K}\left(\frac{1}{\delta}\right)^{2m+1}\left(\frac{m+1}{e}\right)^{m+1}(1 + e)^m, \tag{2.6}$$

$$\sum_{(k)} \left| \frac{\partial \tilde{f}(\varphi)}{\partial \varphi_k} \right| \leqslant \frac{M}{K} \left(\frac{1}{\delta} \right)^{2m+2} \left(\frac{m+2}{e} \right)^{m+2} (1+e)^m \tag{2.7}$$

hold on this domain.

Proof. As is known, the coefficients of the series (2.2) are defined by the formulae

$$f_k = \frac{1}{(2\pi)^m} \int_0^{2\pi} \dots \int_0^{2\pi} f(\theta) e^{-i(k,\theta)} d\theta_1 \dots d\theta_m =$$

$$= \frac{1}{(2\pi)^m} \int_0^{2\pi} \dots \int_0^{2\pi} f(\theta + i\varphi) e^{-i(k,\theta)} e^{(k,\varphi)} d\theta_1 \dots d\theta_m.$$

Setting $\varphi_j = -\rho \operatorname{sign} k$, we have $(k, \varphi) = -\rho \mid k \mid$ and recalling the inequatity (2.1), we obtain

$$| f_k | \leqslant M e^{-\rho |k|}.$$

It is easy to show that, subject to the fulfilment of the conditions (2.1) and (2.7), the inequalities

$$| \tilde{f}(\varphi) | \leqslant \frac{M}{K} \sum_{(k)} | k |^{m+1} e^{-2\delta |k|}, \tag{2.8}$$

$$\sum_{(j)} \left| \frac{\partial \tilde{f}(\varphi)}{\partial \varphi_j} \right| \leqslant \frac{M}{K} \sum_{(k)} | k |^{m+2} e^{-2\delta |k|} \tag{2.9}$$

remain valid in the domain $| \operatorname{Im} \varphi | \leqslant \rho - 2\delta$.

It is required to estimate the expression

$$\sum_{(k)} | k |^{\nu} e^{-2\delta |k|} \tag{2.10}$$

for $0 < \delta < 1$, $\nu > 1$. To do this we shall find the value of $z \geqslant 1$ yielding a maximum value for the expression:

$$\nu \ln z - \delta z. \tag{2.11}$$

Differentiating the expression (2.11) in z and equating the result to zero, we get

$$\frac{\nu}{z} = \delta, \qquad z = \frac{\nu}{\delta} > 1.$$

From this it follows that

$$\nu \ln z - \delta z \leqslant \nu \left(\ln \frac{\nu}{\delta} - 1 \right).$$

The expression obtained yields

$$z^\nu \leqslant e^{\delta z} e^{\nu\left(\ln \frac{\nu}{\delta} - 1\right)} = \left(\frac{\nu}{e}\right)^\nu \frac{e^{\delta z}}{\delta^\nu}. \tag{2.12}$$

Substituting the expression (2.12) in (2.10), we obtain

$$\sum_{(k)} e^{-2|k|\delta} |k|^\nu \leqslant \left(\frac{\nu}{e}\right)^\nu \frac{1}{\delta^\nu} \sum_{(k)} e^{-|k|\delta}$$

$$= \left(\frac{\nu}{e}\right)^\nu \frac{1}{\delta^\nu} \left(1 + 2 \sum_{1<q<\infty} e^{-2\delta}\right)^m$$

$$= \left(\frac{\nu}{e}\right)^\nu \frac{1}{\delta^\nu} \left(\frac{1 + e^{-\delta}}{1 - e^{-\delta}}\right)^m.$$

But $1 - e^{-\delta} > \delta e^{-\delta}$, so that

$$\frac{1 + e^{-\delta}}{1 - e^{-\delta}} \leqslant \frac{1}{\delta} (1 + e^{\delta}) < \frac{1 + e}{\delta},$$

and hence that

$$\sum_{(k)} e^{-2|k|\delta} |k|^\nu < \left(\frac{\nu}{e}\right)^\nu \frac{1}{\delta^{\nu+m}} (1 + e)^m. \tag{2.13}$$

Substituting the inequality (2.13) in the inequalities (2.8) and (2.9), for $\nu = m + 1$ and $\nu = m + 2$ we obtain the estimates (2.6) and (2.7) respectively.

Remark. If $f(\varphi)$ is a real function for real φ, then $\tilde{f}(\varphi)$ is also a real function for real φ.

If $f(\varphi) = f(\varphi, \mu)$ is a function analytic in μ on the domain ϑ and satisfies the condition (2.1) for $f(\varphi, \mu)$ when $\mu \in \vartheta$ with fixed M and ρ, then $\tilde{f}(\varphi, \mu)$ is also analytic with respect to μ on the domain ϑ.

Further, let h be any point of the set Σ_η of points h, for which $\| h \| \leqslant \eta$. Set $q = 1, \ldots, n$ and consider h', satisfying the relations

$$h'_k = h_k,\ k \neq q,\ | h'_q - h_q | \leqslant \frac{\eta - \eta_1}{p},\quad \eta > \eta_1,$$

where p is a constant. For these h', we have

$$| h' - h | \leqslant \frac{\eta - \eta_1}{p}, \qquad \| h' - h \| \leqslant \eta - \eta_1,$$

and

$$\| h' \| \leqslant \| h \| + \| h' - h \| < \eta.$$

Hence, if $f(h)$ is an analytic function of h in the domain Σ_η, we have further that

$$|f(h)| \leqslant N, \qquad \|h\| \leqslant \eta, \tag{2.14}$$

and that

$$\left|\frac{\partial f(h)}{\partial h_q}\right| \leqslant \frac{Np}{\eta - \eta_1}, \qquad \|h\| \leqslant \eta_1 - \eta. \tag{2.15}$$

We now consider the problem of reducing the system of eqns. (1.55) to the system (1.57).

It is required to find the transformation

$$\varphi = \theta + \Phi(h, \theta, \Delta) \tag{2.16}$$

of the angular variables φ to the angular variables θ, which would result in the represntation of eqns. (1.55) in the form

$$\begin{aligned} \frac{dh}{dt} &= Hh + F(h, \theta + \Phi, \Delta), \\ \frac{d\theta}{dt} &= \omega. \end{aligned} \tag{2.17}$$

The function $\Phi(h, \theta, \Delta)$ must, obviously, satisfy the equation

$$\left(E + \frac{\partial \Phi}{\partial \theta}\right)\omega + \frac{\partial \Phi}{\partial h}\{Hh + F(h, \theta + \Phi, \Delta)\} = \omega + \Delta + f(h, \theta + \Phi, \Delta). \tag{2.18}$$

A direct solution of eqn. (2.18) for Φ is somewhat difficult. Therefore, in order to make use of the fundamental idea of accelerated convergence, we choose Φ such that eqn. (2.18) is satisfied to within second order terms (if f is regarded to be of second order). Since Φ is a first order quantity, it follows that on disregarding the second order terms in eqn. (2.18) we obtain

$$\frac{\partial \Phi}{\partial \theta}\omega + \frac{\partial \Phi}{\partial h}\{Hh + F(h, \theta, \Delta)\} = f(h, \theta, \Delta) + C, \tag{2.19}$$

where C is a constant.

To solve eqn. (2.19), we must take C to be the quantity $\bar{f}(S(\theta), \theta, \Delta)$, where $S(\theta)$ defines the corresponding invariant toroidal manifold for the system

$$\begin{aligned} \frac{dh}{dt} &= Hh + F(h, \theta, \Delta), \\ \frac{d\theta}{dt} &= \omega. \end{aligned} \tag{2.20}$$

In what follows, it will be shown that a transformation of the type (2.16), where $\Phi(h, \theta, \Delta)$ is determined from a system of the form (2.19), indeed increases the order of the second term in the second equation of the system (1.55) from 1 to 2, and thus gives rises to the possibility of obtaining an iteration process with accelerated convergence.

Now, consider the system of partial differential equations

$$\left(\frac{\partial u}{\partial \varphi}, \omega\right) + \frac{\partial u}{\partial h}(Hh + F(h, \varphi)) = f(h, \varphi) - \overline{f}(S(\varphi), \varphi), \tag{2.21}$$

where $S(\varphi)$ is an invariant manifold defined by eqns. (1.57), $F(h, \varphi)$ and $f(h, \varphi)$ are vector functions periodic in φ with period 2π, H is a matrix satisfying the condition (1.51), and $\overline{f}(S(\varphi), \varphi)$ is a constant assuring the solvability of (2.21):

$$\overline{f}(S(\varphi), \varphi) = \frac{1}{(2\pi)^m}\int_0^{2\pi} \cdots \int_0^{2\pi} f(S(\varphi), \varphi)\, d\varphi_1 \ldots d\varphi_m.$$

We proceed to investigate the solutions of the system (2.21) that are periodic in φ. Theorem 2 below establishes the existence of such solutions and some of their properties.

Theorem 2. *For eqn.* (2.21), *let the following conditions hold:*

(i) *The functions* $F(h, \varphi)$ *and* $f(h, \varphi)$, *analytic in* h, φ *on the domain*

$$\| h \| \leqslant \eta, \quad | \operatorname{Im} \varphi | \leqslant \rho, \tag{2.22}$$

satisfy the inequalities

$$| f(h, \varphi) | \leqslant M, \quad \| F(h, \varphi) \| \leqslant N, \quad n\left\| \frac{\partial F(h, \varphi)}{\partial h_q} \right\| \leqslant L, \tag{2.23}$$

where η, ρ, M, N, L *are constants satisfying the inequalities*

$$L \leqslant \frac{\alpha}{2}, \quad \frac{N}{\alpha} \leqslant \eta\left(1 - \frac{\delta}{\rho}\right), \quad 0 \leqslant 2\delta < \rho, \quad \delta > 1. \tag{2.24}$$

(ii) *The frequencies* $\omega = (\omega_1, \ldots, \omega_m)$ *satisfy the condition*

$$| (k, \omega) | \geqslant K | k |^{-(m+1)}.$$

Then, eqn. (2.21) *has the solution*

$$u = \hat{f}(h, \varphi) = \tilde{f}(S(\varphi), \varphi) - \int_0^{\infty} \{ f(h_\tau, \varphi_\tau) - f(S(\varphi_\tau), \varphi_\tau) \}\, d\tau, \tag{2.25}$$

which is periodic in φ *with period* 2π, *is analytic in h and* φ *on the domain*

$$\| h \| \leqslant \eta \left(1 - \frac{2\delta}{\rho}\right), \qquad | \operatorname{Im} \varphi | \leqslant \rho \left(1 - \frac{2\delta}{\rho}\right), \tag{2.26}$$

and satisfies the inequalities

$$\left| \hat{f}(h, \varphi) \right| \leqslant M \left\{ 4Pn \frac{\rho}{\alpha\delta} + \left(\frac{m+1}{e}\right)^{m+1} \frac{(1+e)^m}{K\delta^{2m+1}} \right\},$$

$$\sum_{(q)} \left| \frac{\partial \hat{f}(h, \varphi)}{\partial \varphi_q} \right| \leqslant M \left\{ 2Pnm \frac{\rho}{\alpha\delta^2} + \left(\frac{m+2}{e}\right)^{m+2} \frac{(1+e)^m}{R\delta^{2m+1}} \right\}, \tag{2.27}$$

$$\left| \frac{\partial \hat{f}(h, \varphi)}{\partial h_q} \right| \leqslant M \frac{4P^2\rho^2 n}{\alpha\eta\delta^2}.$$

Proof. Rewrite the function $\hat{f}(h, \varphi)$ in the form

$$\hat{f}(h, \varphi) = v(h, \varphi) + \tilde{f}(S(\varphi), \varphi) = v(h, \varphi) + w(\varphi), \tag{2.28}$$

and show that it satisfies eqn. (2.21). By definition we have

$$w(\varphi) = \tilde{f}(S(\varphi), \varphi) = \sum_{|k| \neq 0} \frac{f_k}{i(k, \omega)} e^{i(k, \varphi)}.$$

Differentiating this, we find

$$\left(\frac{\partial w}{\partial \varphi}, \omega\right) = \sum_{|k| \neq 0} f_k e^{i(k, \varphi)}.$$

Thus, the function $w(\varphi)$ satisfies the equation

$$\frac{\partial w}{\partial h}(Hh + F) + \left(\frac{\partial w}{\partial \varphi}, \omega\right) = f(S(\varphi), \varphi) - \bar{f}(S(\varphi), \varphi).$$

It now remains to show that

$$\frac{\partial v(h, \varphi)}{\partial h}(Hh + F(h, \varphi)) + \left(\frac{\partial v(h, \varphi)}{\partial \varphi}, \omega\right) \equiv f(h, \varphi) - f(S(\varphi), \varphi). \tag{2.29}$$

Let

$$\varphi_t = \varphi + \omega t,$$

and let $h_t = h(t, h)$ be a solution of the equation

$$\frac{dh}{dt} = Hh + F(h, \varphi_t), \qquad t \geqslant 0$$

under the initial condition

$$h(0, h) = h.$$

On carrying out the substitution of variables

$$h \to h_t, \qquad \varphi \to \varphi_t,$$

we find that eqn. (2.29) takes the form

$$\frac{dv(h_t, \varphi_t)}{dt} = f(h_t, \varphi_t) - f(S(\varphi_t), \varphi_t). \tag{2.30}$$

The solution of eqn. (2.30) emerges as the expression

$$v(h_t, \varphi_t) = -\int_0^\infty [f(h_{t+\tau}, \varphi_{t+\tau}) - f(S(\varphi_{t+\tau}), \varphi_{t+\tau})]\, d\tau =$$

$$= \int_0^\infty [f(h_z, \varphi_z) - f(S(\varphi_z), \varphi_z)]\, dz,$$

which satisfies the system (2.29) for $t = 0$. Consequently, the function $\hat{f}(h, \varphi)$ is a solution of eqn. (2.21).

We now proceed to establish the estimates (2.27). Consider the domain

$$D\left\{\| h \| \leqslant \eta\left(1 - \frac{\delta}{\rho}\right), \quad |\operatorname{Im} \varphi| \leqslant \rho\right\}.$$

From the relations (2.23) and (2.15), we deduce the inequalities

$$\left|\frac{\partial f(h, \varphi)}{\partial h_q}\right| \leqslant \frac{M\rho}{\eta\delta} \quad (q = 1, 2, \ldots, n)$$

on the D-domain. Hence, taking into account the bound (1.13) and the inequalities

$$\| h_t \| \leqslant \eta\left(1 - \frac{\delta}{\rho}\right), \quad \| S(\varphi_t) \| \leqslant \eta\left(1 - \frac{\delta}{\rho}\right) \qquad \text{when } t \geqslant 0,$$

we obtain

$$| f(h_\tau, \varphi_\tau) - f(S(\varphi_\tau), \varphi_\tau) | \leqslant \frac{PM\rho n}{m\delta} \left| h_\tau - S(\varphi_\tau) \right| \leqslant$$

$$\leqslant \frac{2P\rho n}{\delta} e^{-\frac{\alpha}{2}\tau}, \qquad t \geqslant 0.$$

Consequently, on the D-domain we have

$$|v(h, \varphi)| \leqslant 2 \frac{PM\rho n}{\delta} \int_0^\infty e^{-\frac{\alpha}{2}\tau} d\tau = \frac{4PM\rho n}{\alpha\delta}, \tag{2.31}$$

and hence

$$\left|\frac{\partial v(h, \varphi)}{\partial h_q}\right| \leqslant \frac{4MP^2\rho^2 n}{\alpha\eta\delta^2}; \quad \left|\frac{\partial v(h, \varphi)}{\partial \varphi_k}\right| \leqslant \frac{2PM\rho n}{\alpha\delta^2} \tag{2.32}$$

on the still narrower domain (2.26).

On the other hand, by the estimates (2.6) and (2.7), we obtain on the domain (2.26) the inequalities

$$|w(\varphi)| \leqslant \frac{M}{K}\left(\frac{1}{\delta}\right)^{2m+1}\left(\frac{m+1}{e}\right)^{m+1}(1+e)^m,$$

$$\sum_{(k)}\left|\frac{\partial w(\varphi)}{\partial \varphi_k}\right| \leqslant \frac{M}{K}\left(\frac{1}{\delta}\right)^{2m+2}\left(\frac{m+2}{e}\right)^{m+2}(1+e)^m,$$

which together with inequalities (2.31) and (2.32) yield the desired estimates (2.27).

Remarks. 1. If H is a real matrix, and $F(h, \varphi)$ and $f(h, \varphi)$ are real functions for real h, φ, then the function $\hat{f}(h, \varphi)$ is also real for real h, φ.

2. If $F = F(h, \varphi, \mu)$ and $f = f(h, \varphi, \mu)$ are analytic functions of μ on the domain ϑ, and if on this domain all these estimates hold uniformly, then $\hat{f}(h, \varphi, \mu)$ is also an analytic function of μ on ϑ.

§ 3. Lemma on Iterations

We revert to the system of equations

$$\begin{aligned} \frac{dh}{dt} &= Hh + F(h, \varphi, \Delta), \\ \frac{d\varphi}{dt} &= \omega + \Delta + f(h, \varphi, \Delta), \end{aligned} \tag{3.1}$$

where $F(h, \varphi, \Delta)$, $f(h, \varphi, \Delta)$ are analytic functions on the domain

$$\|h\| \leqslant \eta, \quad |\operatorname{Im} \varphi| < \rho, \quad |\Delta| < \sigma, \tag{3.2}$$

and periodic in φ with period 2π, while $\omega = (\omega_1, \ldots, \omega_m)$ are real frequencies satisfying the conditions (2.5).

The problem consists in choosing $\Delta = (\Delta_1, \ldots, \Delta_m)$ so that there exists a change of variables reducing eqns. (3.1) to the form

$$\frac{dh}{dt} = Hh + F(h, \theta, \Delta),$$
$$\frac{d\theta}{dt} = \omega. \tag{3.3}$$

In determining Δ and finding the required transformation $h, \varphi \to h, \theta$, the following lemma plays an important role.

Lemma 2. *Suppose that in eqns.* (3.1) *the functions* $F(h, \varphi, \Delta)$ *and* $f(h, \varphi, \Delta)$ *are sufficiently small for small* h, $\operatorname{Im} \varphi$, δ *and satisfy the inequalities*

$$\| F(h, \varphi, \Delta) \| \leqslant N, \quad | f(h, \varphi, \Delta) | \leqslant M,$$
$$n \left\| \frac{\partial F(h, \varphi, \Delta)}{\partial h_q} \right\| \leqslant L, \quad (q = 1, \ldots, n), \tag{3.4}$$

with h, φ, Δ *varying in the domain* (3.2) *and the constants* N, L, M *being related to* η, ρ, σ *by the inequalities*

$$\frac{N}{\alpha} \leqslant \eta\left(1 - \frac{\delta}{\rho}\right), \quad L \leqslant \frac{\alpha}{2}, \quad M \leqslant \frac{\sigma}{4\pi},$$
$$0 < \delta < \frac{\rho}{2}, \quad \delta < 1, \quad \frac{MQ(\rho, \delta)}{\alpha\delta^{2m+4}} \leqslant 1, \tag{3.5}$$

where

$$Q(\rho, \delta) = Q\left(\rho, \delta, P, \frac{\alpha}{K}\right)$$
$$= 4Pnm\rho\delta^{2m+2} + 2\left(\frac{m+2}{e}\right)^{m+2} \frac{\alpha}{K} \times$$
$$\times (1+e)^m \delta^2 + \left(\frac{m\delta^2}{1 - \dfrac{1}{2m}} + \frac{4P^2\rho^2 n^2 m}{1 - \dfrac{1}{2m}}\right) \times$$
$$\times \left[4Pn\rho\delta^{2m} + \frac{\alpha}{K}\left(\frac{m+1}{e}\right)^{m+1} (1+e)^m\right].$$

Then for sufficiently small η, σ, *we have the analytic transformation*

$$\varphi = \theta + u(h, \theta, \Delta^{(1)}),$$
$$\Delta = \Delta(\Delta^{(1)}), \tag{3.6}$$

periodic in θ *with period* 2π, *which reduces the system of eqns.* (3.1) *to the form*

$$\frac{dh}{dt} = Hh + F_1(h, \theta, \Delta^{(1)}),$$
$$\frac{d\theta}{dt} = \omega + \Delta^{(1)} + f_1(h, \theta, \Delta^{(1)}), \tag{3.7}$$

where

$$F_1(h, \theta, \Delta^{(1)}) = F(h, \theta + u(h, \theta, \Delta^{(1)}), \quad \Delta(\Delta^{(1)})),$$
$$f_1(h, \theta, \Delta^{(1)}) = f(h, \theta + u(h, \theta, \Delta^{(1)}), \quad \Delta(\Delta^{(1)})); \tag{3.8}$$

in addition, the functions $\Delta(\Delta^{(1)}), f_1(h, \theta, \Delta^{(1)}), F_1(h, \theta, \Delta^{(1)})$ *and* $u(h, \theta, \Delta^{(1)})$ *are also analytic on the domain*

$$\| h \| \leqslant \eta\left(1 - \frac{2\delta}{\rho}\right), \quad |\operatorname{Im} \theta| \leqslant \rho\left(1 - \frac{2\delta}{\rho}\right), \quad |\Delta^{(1)}| \leqslant M, \tag{3.9}$$

and satisfy the inequalities

$$|\Delta(\Delta^{(1)}) - \Delta^{(1)}| \leqslant M, \quad \sum_{(q)} \left| \frac{\partial \Delta(\Delta^{(1)})}{\partial \Delta_q^{(1)}} \right| \leqslant 1 + \frac{4Mm}{\sigma} \leqslant 2,$$

$$|u(h, \theta, \Delta^{(1)})| \leqslant \frac{MA(\rho, \delta)}{\alpha \delta^{2m+1}} \leqslant \frac{\delta}{m},$$

$$\sum_{(q)} \left| \frac{\partial u(h, \theta, \Delta^{(1)})}{\partial \theta_q} \right| \leqslant \frac{MQ(\rho, \delta)}{2\alpha\delta^{2m+4}} \leqslant \frac{1}{2}, \tag{3.10}$$

$$\sum_{q} \left| \frac{\partial u(h, \theta, \Delta^{(1)})}{\partial \Delta_q^{(1)}} \right| \leqslant \frac{M 2n \left(1 - \frac{1}{2m}\right)^{-1} A(\rho, \delta)}{\alpha \delta^{2m+1} \sigma},$$

$$\left| \frac{\partial u(h, \theta, \Delta^{(1)})}{\partial h_q} \right| \leqslant M \frac{4P^2\rho^2 n}{\alpha \eta \delta^2},$$

$$\| F_1(h, \theta, \Delta^{(1)}) \| \leqslant N, \quad n \left\| \frac{\partial F_1(h, \theta, \Delta^{(1)})}{\partial h_q} \right\| \leqslant L_1,$$
$$\| f_1(h, \theta, \Delta^{(1)}) \| \leqslant M_1, \tag{3.11}$$

where

$$A(\rho, \delta) = A\left(\rho, \delta, P, \frac{\alpha}{K}\right) = 4P_0\rho\delta^{2m} + \frac{\alpha}{K}\left(\frac{m+1}{e}\right)^{m+1}(1+e)^m,$$

$$L_1 = L + \frac{2MP^2n^2m\rho^2}{\delta^3\left(1 - \frac{1}{2m}\right)}, \tag{3.12}$$

$$M_1 = \frac{M^2 Q(\rho, \delta)}{\delta^{2m+4\alpha}}.$$

Proof. We shall first reduce the system (3.1) to a form in which the function $f_1(h, \theta, \Delta^{(1)})$ appears as a second order term. For this we choose the transformation of the angular variable φ in the form

$$\varphi = \theta + \hat{f}(h, \theta, \Delta). \tag{3.13}$$

Substituting (3.13) in the system of eqns. (3.1), we obtain

$$\left(E + \frac{\partial \hat{f}}{\partial \theta}\right)\frac{d\theta}{dt} + \frac{\partial \hat{f}}{\partial h}\frac{dh}{dt} = \omega + \Delta + f(h, \theta + \hat{f}, \Delta). \tag{3.14}$$

Subtracting the system

$$\frac{\partial u(h, \varphi, \Delta)}{\partial h}(Hh + F(h, \varphi, \Delta)) + \frac{\partial u(h, \varphi, \Delta)}{\partial \varphi}\,\omega =$$
$$= f(h, \varphi, \Delta) - \bar{f}(S(\varphi, \Delta), \varphi, \Delta)$$

from eqns. (3.14), we get

$$\left(E + \frac{\partial \hat{f}}{\partial \theta}\right)\left(\frac{d\theta}{dt} - \omega\right) + \frac{\partial \hat{f}}{\partial h}\left(\frac{dh}{dt} - Hh - F(h, \theta, \Delta)\right) =$$
$$= \Delta + f(h, \theta + \hat{f}, \Delta) + \bar{f}(S(\varphi, \Delta), \varphi, \Delta) - f(h, \theta, \Delta). \tag{3.15}$$

In the expression (3.15) we insert for dh/dt its value

$$\frac{dh}{dt} = Hh + F(h, \varphi, \Delta),$$

to obtain

$$\left(E + \frac{\partial \hat{f}}{\partial \theta}\right)\left(\frac{d\theta}{dt} - \omega\right) = \Delta + f(h, \theta + \hat{f}, \Delta) + \bar{f}(S(\varphi, \Delta), \varphi, \Delta) - f(h, \theta, \Delta)$$
$$- \frac{\partial \hat{f}}{\partial h}[F(h, \theta + \hat{f}, \Delta) - F(h, \theta, \Delta)]. \tag{3.16}$$

We now introduce a new $\Delta^{(1)}$ by the formula

$$\Delta^{(1)} = \Delta + \bar{f}(S(\varphi, \Delta), \varphi, \Delta). \tag{3.17}$$

Then solving the system (3.16) for $d\theta/dt$ and making use of the equations in h, we finally obtain

$$\begin{aligned} \frac{dh}{dt} &= Hh + F(h, \theta + \hat{f}, \Delta), \\ \frac{d\theta}{dt} &= \omega + \Delta^{(1)} + E_1(h, \theta, \Delta), \end{aligned} \tag{3.18}$$

where

$$E_1(h, \theta, \Delta) = \left\{\left(E + \frac{\partial \hat{f}}{\partial \theta}\right)^{-1} - E\right\} \Delta^{(1)} + \left(E + \frac{\partial \hat{f}}{\partial \theta}\right)^{-1} \times$$
$$\times \left\{[f(h, \theta + \hat{f}, \Delta) - f(h, \theta, \Delta)]\right.$$
$$\left. - \frac{\partial \hat{f}}{\partial h} [F(h, \theta + \hat{f}, \Delta) - F(h, \theta, \Delta)]\right\}. \tag{3.19}$$

Before verifying that the estimates (3.10) and (3.11) are satisfied, we analyze formally the orders of the terms occurring in the right-hand side of the expression (3.19).

Treating f and Δ as first order terms, it is seen that $\hat{f}$ and $\Delta^{(1)}$ are also of first order. Consequently, it follows from the expression (3.19) that $E_1(h, \theta + \hat{f}, \Delta)$ is a second order term.

Now, consider the relationship between Δ and $\Delta^{(1)}$.

Introduce the notation

$$\bar{f}(S(\varphi, \Delta), \varphi, \Delta) = a(\Delta).$$

By the hypotheses of Lemma 2, $a(\Delta)$ is an analytic function of Δ on the domain $|\Delta| \leqslant \sigma$ and satisfies the inequality $|a(\Delta)| \leqslant M$. Hence, for $a(\Delta)$ the estimates

$$\sum_{(q)} \left| \frac{\partial a(\Delta)}{\partial \Delta_q} \right| \leqslant \frac{2Mm}{\sigma} \leqslant \frac{1}{2} \quad \text{when } |\Delta| \leqslant \frac{\sigma}{2} \tag{3.20}$$

hold and, consequently,

$$|a(\Delta') - a(\Delta'')| \leqslant \frac{1}{2} |\Delta' - \Delta''| \quad \text{when } |\Delta'|, |\Delta''| \leqslant \frac{\sigma}{2}.$$

From this it follows that the equation

$$\Delta + a(\Delta) = \Delta^{(1)} \tag{3.21}$$

can be solved for $\Delta : \Delta = \Delta(\Delta^{(1)})$, and $\Delta(\Delta^{(1)})$ represents an analytic function of $\Delta^{(1)}$ on the domain $|\Delta^{(1)}| \leqslant M$. Furthermore, on this domain, we have

$$|\Delta(\Delta^{(1)})| \leqslant 2M \leqslant \frac{\sigma}{2m} < \frac{\sigma}{2}.$$

We derive some more estimates of Δ. Differentiating eqn. (3.21), we get

$$\frac{\partial \Delta_k}{\partial \Delta_q^{(1)}} + \sum_{(r)} \frac{\partial a_k(\Delta)}{\partial (\Delta)_r} \frac{\partial \Delta_r}{\partial \Delta_q^{(1)}} = \delta_{kq} = \begin{cases} 1, & k = q, \\ 0, & k \neq q. \end{cases}$$

Keeping in view the expression (3.20), we see that this implies

$$\left|\frac{\partial \Delta_k}{\partial \Delta_q^{(1)}}\right| \leqslant \delta_{kq} + \frac{2M}{\sigma} \sum_{(r)} \left|\frac{\partial \Delta_r}{\partial \Delta_q^{(1)}}\right|, \tag{3.22}$$

or,

$$\sum_{(r)} \left|\frac{\partial \Delta_r}{\partial \Delta_q^{(1)}}\right| \leqslant 1 + \frac{2Mm}{\sigma} \sum_{(r)} \left|\frac{\partial \Delta_r}{\partial \Delta_q^{(1)}}\right|$$

by summation over r. Consequently,

$$\sum_{(q)} \left|\frac{\partial \Delta_k}{\partial \Delta_q^{(1)}}\right| \leqslant \frac{1}{1 - \dfrac{2M}{\sigma} m} \leqslant 2.$$

Hence the inequality (3.22) can be written in the form

$$\left|\frac{\partial \Delta_k}{\partial \Delta_q^{(1)}}\right| \leqslant \delta_{kq} + \frac{4M}{\sigma}. \tag{3.23}$$

Summing (3.23) over q, we obtain

$$\sum_{(q)} \left|\frac{\partial \Delta_k}{\partial \Delta_q^{(1)}}\right| \leqslant 1 + \frac{4M}{\sigma} m \leqslant 2.$$

It is also evident that

$$|\Delta(\Delta^{(1)}) - \Delta^{(1)}| \leqslant M.$$

We now estimate the function $\hat{f}(h, \theta, \Delta)$ appearing in the change of variables (3.13).

By virtue of the estimates of Theorem 2, we arrive at the inequalities

$$|\hat{f}(h, \theta, \Delta)| \leqslant \frac{MA(\rho, \delta)}{\alpha \delta^{2m+1}};$$

$$\sum_{(q)} \left|\frac{\partial \hat{f}(h, \theta, \Delta)}{\partial \theta_q}\right| \leqslant \frac{M}{\alpha \delta^{2m+2}} \left\{2Pnm\rho\delta^{2m} + \left(\frac{m+2}{e}\right)^{m+2} \frac{\alpha}{K} (1+e)^m\right\}$$

$$< \frac{MQ(\rho, \delta)}{2\alpha \delta^{2m+4}} \leqslant \frac{1}{2};$$

$$\left|\frac{\partial \hat{f}(h, \theta, \Delta)}{\partial h_q}\right| \leqslant \frac{M}{\alpha \delta^2} \cdot \frac{4P^2\rho^2 n}{\eta}$$

for

$$\| h \| \leqslant \eta\left(1 - \frac{2\delta}{\rho}\right), \qquad |\operatorname{Im}\theta| \leqslant \rho\left(1 - \frac{2\delta}{\rho}\right), \qquad |\Delta| \leqslant \sigma.$$

Note that

$$\frac{MA(\rho, \delta)}{\alpha\delta^{2m+1}} = \frac{M}{\alpha}\frac{4Pn\rho}{\delta} + \frac{\alpha}{K}\cdot\frac{\left(\frac{m+1}{e}\right)^{m+1}(1+e)^m}{\delta^{2m+1}} \leqslant$$

$$\leqslant 2\frac{\delta}{m}\frac{M}{\alpha}\left(\frac{2Pnm\rho}{\delta^2} + \left(\frac{m+2}{e}\right)^{m+2}\frac{\alpha}{K}(1+e)^m\right) \leqslant$$

$$\leqslant 2\frac{\delta}{m}\frac{M}{\alpha}\frac{Q(\rho, \delta)}{\delta^{2m+4}} \leqslant \frac{\delta}{m}.$$

Evidently,

$$\left|\frac{\partial\hat{f}(h, \theta, \Delta)}{\partial\Delta_k}\right| \leqslant \frac{\left(1 - \frac{1}{2m}\right)^{-1} MA(\rho, \delta)}{\alpha\delta^{2m+1}\sigma}$$

for $|\Delta| \leqslant (\sigma/2m)$.

Consequently, setting $\hat{f}(h, \theta, \Delta(\Delta^{(1)})) = u(h, \theta, \Delta^{(1)})$ we see that the inequalities (3.10) are certainly satisfied. Writing $F_1(h, \theta, \Delta^{(1)}) = F(h, \theta + \hat{f}, \Delta)$ we observe that the inequality $\| F_1 \| \leqslant N$ remains valid on the domain (3.9).

Further, we have

$$n\left\|\frac{\partial F_1}{\partial h_k}\right\| = n\left\| F'_{h_k}(h, \theta + \hat{f}, \Delta) + \sum_{(q)} F'_{\varphi_q}(h, \theta + \hat{f}, \Delta)\frac{\partial\hat{f}_q}{\partial h_k}\right\| \leqslant$$

$$\leqslant L + n\frac{Nm}{2\left(1 - \frac{1}{2m}\right)\delta}\cdot\frac{M}{\alpha\delta^2}\frac{4P^2\rho^2 n}{\eta} \leqslant$$

$$\leqslant L + \frac{2MP^2nm\rho^2}{\delta^3\left(1 - \frac{1}{2m}\right)}.$$

We now write

$$L_1 = L + \frac{2MP^2n^2m\rho^2}{\delta^3\left(1 - \frac{1}{2m}\right)}.$$

It remains to find M_1, i.e. to obtain the estimate of $E(h, \theta, \Delta)$.

We observe that, subject to the fulfilment of the condition $\sum_{(q)} |a_{kq}| \leqslant d \leqslant \frac{1}{2}$, the system

$$y_k + \sum_{(q)} a_{kq}y_q = z_k,$$

or, in the matrix notation

$$y = (E + a)^{-1} z$$

leads to the estimate

$$|(E + a)^{-1} - E| \leqslant \frac{d}{1 - d} \leqslant 2d. \tag{3.24}$$

Furthermore, keeping in mind (3.24), we see that the inequality

$$\sum_{(q)} \left| \frac{\partial \hat{f}}{\partial \theta_q} \right| \leqslant \frac{M}{\alpha \delta^{2m+2}} \left\{ 2Pnm\rho\delta^{2m} + \left(\frac{m+2}{e} \right)^{m+2} \frac{\alpha}{K} (1 + e)^m \right\} \leqslant \frac{1}{2}$$

implies the inequalities

$$\left| \left(E + \frac{\partial \hat{f}}{\partial \theta} \right)^{-1} \right| \leqslant 2,$$

and

$$\left| \left(E + \frac{\partial \hat{f}}{\partial \theta} \right)^{-1} - E \right| \leqslant \frac{2M}{\alpha \delta^{2m+2}} \left\{ 2Pnm\delta^{2m} + \left(\frac{m+2}{e} \right)^{m+2} \frac{\alpha}{K} (1 + e)^m \right\}.$$

Thus,

$$\begin{aligned} |E_1(h, \theta, \Delta)| &\leqslant \frac{2M^2}{\alpha \delta^{2m+2}} \left\{ 2Pnm\rho\delta^{2m} + \left(\frac{m+2}{e} \right)^{m+2} \frac{\alpha}{K} (1 + e)^m \right\} + \\ &+ 2 |f(h, \theta + \hat{f}, \Delta) - f(h, \theta, \Delta)| + 2n \frac{M}{\alpha \delta^2} \frac{4P^2\rho^2 n}{\eta} \\ &\times |F(h, \theta + \hat{f}, \Delta) - F(h, \theta, \Delta)|. \end{aligned} \tag{3.25}$$

However,

$$\begin{aligned} |f(h, \theta + \hat{f}, \Delta) - f(h, \theta, \Delta)| &\leqslant \frac{Mm}{2\left(1 - \frac{1}{2m}\right)\delta} |\hat{f}| \leqslant \\ &\leqslant \frac{M^2 m}{2\left(1 - \frac{1}{2m}\right)\alpha} \frac{A(\rho, \delta)}{\delta^{2m+2}}, \end{aligned} \tag{3.26}$$

because

$$|\operatorname{Im}(\theta + \hat{f})| \leqslant \rho - 2\delta + \frac{\delta}{m} = \rho - 2\left(1 - \frac{1}{2m}\right)\delta,$$

$$\begin{aligned} |F(h, \theta + \hat{f}, \Delta) - F(h, \theta, \Delta)| &\leqslant \frac{Nm}{2\left(1 - \frac{1}{2m}\right)\delta} |\hat{f}| \leqslant \\ &\leqslant \frac{Nm}{2\left(1 - \frac{1}{2m}\right)} \frac{MA(\rho, \delta)}{\delta^{2m+2}}. \end{aligned} \tag{3.27}$$

Inserting inequalities (3.26) and (3.27) in the right-hand side of the expression (3.25), we get

$$E_1(h, \theta, \Delta)| \leqslant \frac{M^2}{\alpha\delta^{2m+4}} \left\{4Pnm\rho\delta^{2m+2} + 2\left(\frac{m+2}{e}\right)^{m+2} \frac{\alpha}{K}(1+e)^m\delta^2 + \right.$$

$$\left. + \frac{m}{1-\frac{1}{2m}} A(\rho, \delta)\,\delta^2 + \frac{m}{1-\frac{1}{2m}} \frac{4P^2\rho^2n^2N}{\alpha\eta} A(\rho, \delta)\right\} \leqslant$$

$$\leqslant \frac{M^2}{\alpha\delta^{2m+4}} \left\{ 4Pnm\rho\delta^{2m+2} + 2\left(\frac{m+2}{e}\right)^{m+2} \frac{\alpha}{K}(1+e)^m\delta^2 + \right.$$

$$+ \left[4Pn\rho\delta^{2m} + \frac{\alpha}{K}\left(\frac{m+1}{e}\right)^{m+1} (1+e)^m \right] \times$$

$$\left. \times \left(\frac{m\delta^2}{1-\frac{1}{2m}} + \frac{4P^2n^2m\rho^2}{1-\frac{1}{2m}}\right)\right\} =$$

$$= \frac{M^2}{\alpha\delta^{2m+4}} Q(\rho, \delta) = M_1, \qquad \frac{A(\rho, \delta)}{Q(\rho, \delta)} \leqslant \frac{1-\frac{1}{2m}}{m\delta^2}.$$

Setting

$$E_1(h, \theta, \Delta(\Delta^{(1)})) = f_1(h, \theta, \Delta^{(1)}),$$

we see that the required estimates are indeed valid ; this completes the proof of the lemma.

Remarks. 1. If H is a real matrix, and $F(h, \varphi, \Delta)$ and $f(h, \varphi, \Delta)$ are real functions for real h, φ, Δ, then $u(h, \varphi, \Delta)$, $F_1(h, \varphi, \Delta)$ and $f_1(h, \varphi, \Delta)$ are also real for real h, φ, Δ.

2. If $F(h, \varphi, \Delta)$ and $f(h, \varphi, \Delta)$ are analytic functions of μ on the domain ϑ and if all the estimates of Lemma 2 hold uniformly on this domain, then $u(h, \varphi, \Delta)$, $F_1(h, \varphi, \Delta)$ and $f_1(h, \varphi, \Delta)$ are also analytic functions of μ on ϑ.

§ 4. Theorem on Quasi-Periodic Solutions

Lemma 2 proved in the preceding section enables us to construct a rapidly convergent iteration process for the transformation of the system (3.1) to the form (3.3) and, consequently, to solve the problem of the existence of quasi-periodic solutions of the original system. Proceeding to construct such a process, in the system

$$\frac{dh}{dt} = Hh + F(h, \varphi, \Delta),$$
$$\frac{d\varphi}{dt} = \omega + \Delta + f(h, \varphi, \Delta), \tag{4.1}$$

we make the change of variables

$$\varphi = \varphi^{(1)} + u^{(1)}(h, \varphi^{(1)}, \Delta^{(1)}),$$
$$\Delta = \Delta(\Delta^{(1)}),$$

and take $u^{(1)}$, $\Delta(\Delta^{(1)})$ as indicated in Lemma 2. For h and $\varphi^{(1)}$, we obtain the equations

$$\frac{dh}{dt} = Hh + F_1(h, \varphi^{(1)}, \Delta^{(1)}),$$

$$\frac{d\varphi^{(1)}}{dt} = \omega + \Delta^{(1)} + f_1(h, \varphi^{(1)}, \Delta^{(1)}).$$

Applying the transformation

$$\varphi^{(1)} = \varphi^{(2)} + u^{(2)}(h, \varphi^{(2)}, \Delta^{(2)}), \quad \Delta^{(1)} = \Delta^{(1)}(\Delta^{(2)}),$$

we arrive at the equations

$$\frac{dh}{dt} = Hh + F_2(h, \varphi^{(2)}, \Delta^{(2)}),$$

$$\frac{d\varphi^{(2)}}{dt} = \omega + \Delta^{(2)} + f_2(h, \varphi^{(2)}, \Delta^{(2)}).$$

At the s-th step of this process, by making the transformation

$$\varphi^{(s-1)} = \varphi^{(s)} + u^{(s)}(h, \varphi^{(s)}, \Delta^{(s)}),$$
$$\Delta^{(s-1)} = \Delta^{(s-1)}(\Delta^{(s)}),$$

we obtain the equations

$$\frac{dh}{dt} = Hh + F_s(h, \varphi^{(s)}, \Delta^{(s)}),$$

$$\frac{d\varphi^{(s)}}{dt} = \omega + \Delta^{(s)} + f_s(h, \varphi^{(s)}, \Delta^{(s)}).$$

It is evident that at each step it is necessary to ensure that the hypotheses of Lemma 2 remain valid. This can be accomplished in the following manner. Assume that the functions $F(h, \varphi, \Delta)$ and $f(h, \varphi, \Delta)$ are analytic in the domain (3.2) and satisfy the inequalities (3.4), while the frequencies $\omega = (\omega_1, \ldots, \omega_m)$ satisfy the conditions (2.5). Assume further that the constants N, M, L, ρ, σ are related by the inequalities

$$N \leqslant \frac{\alpha m}{2}, \qquad L \leqslant \frac{\alpha}{4}, \qquad M \frac{Q(\rho, \gamma)}{\alpha} \left(\frac{1}{\gamma^{2m+4}}\right)^2 \leqslant r_0 < 1,$$

where

$$\gamma = \frac{\rho}{\rho + 4}, \qquad M \leqslant \frac{\sigma}{4m}, \qquad \gamma^{2m+4} r_0 \leqslant \frac{1}{4m},$$

$$\frac{\gamma^{2m+4}}{2} \sum_{s=0}^{\infty} r_0^{2^s} < \ln \frac{3}{2},$$

$$\frac{\left(1 - \frac{1}{2m}\right)^{-1} P^2 n^2 m \rho^2}{Q(\rho, \gamma)} \sum_{s=0}^{\infty} r_0^{2^s} \gamma^{(s+1)(2m+1)} \gamma^{(2m+4)} \leqslant \frac{1}{8}.$$

At the s-th step consider the domain

$$\begin{aligned} &|\operatorname{Im} \varphi| \leqslant \rho_s = \rho - 2(\delta + \delta_1 + \ldots + \delta_{s-1}), \\ &\|h\| \leqslant \eta_s = \frac{\eta}{\rho} \rho_s, \qquad |\Delta^{(s)}| \leqslant M_{s-1}. \end{aligned} \tag{4.2}$$

To ensure that while contracting the domain $\rho \to \rho_1 \to \ldots \to \rho_s$ we keep the quantities ρ_s bounded below by a positive constant, as in § 1 (p. 10), we put

$$\delta = \gamma, \quad \delta_1 = \gamma^2, \ldots, \delta_s = \gamma^s, \ldots$$

and choose γ from the equation

$$2\gamma(1 + \gamma + \gamma^2 + \ldots) = \frac{2\gamma}{1 - \gamma} = \frac{\rho}{2}.$$

In this case, ρ_s will be greater than $\rho/2$ for all values of s.

We now show that for all $s = 1, 2, \ldots$, the inequalities

$$\begin{aligned} &\|F_s(h, \varphi, \Delta^{(s)})\| \leqslant N_s, \qquad n \left\| \frac{\partial F_s(h, \varphi, \Delta^{(s)})}{\partial h_q} \right\| \leqslant L_s, \\ &\|f_s(h, \varphi, \Delta^{(s)})\| \leqslant M_s, \qquad |\Delta^{(\nu)}(\Delta^{(\nu+1)}(\ldots \Delta^{(s)} \ldots) \ldots)| \leqslant M_{\nu-1} \end{aligned} \tag{4.3}$$

remain valid on the domain (4.2). Here

$$M_s = \frac{\alpha}{Q(\rho, \gamma)} r_0^{2^s} \gamma^{(s+2)(2m+4)}, \qquad N_s = \frac{\alpha}{2} \eta,$$

$$L_s = L + \frac{2P^2 n^2 m \rho^2 \alpha}{\left(1 - \frac{1}{2m}\right) Q(\rho, \gamma)} \sum_{p=0}^{s-1} r_0^{2^p} \gamma^{(p+1)(2m+1)+(2m+4)} \leqslant$$

$$\leqslant \frac{\alpha}{4} + \frac{\alpha}{4} = \frac{\alpha}{2}. \tag{4.4}$$

Assuming that all the estimates hold for $s \leqslant s_0$, we shall prove their validity for $s = s + 1$. We shall first verify that the hypotheses of Lemma 2 are satisfied as we pass from $s = s_0$ to $s = s_0 + 1$.

Replacing ρ, δ, σ in inequalities (3.4) and (3.5) respectively by

$$\rho_{s_0}, \quad \delta_{s_0} = \gamma^{s_0+1}, \quad M_{s_0-1},$$

we obtain the inequalities

$$\frac{N_{s_0}}{\alpha} \leqslant \eta_{s_0}\left(1 - \frac{\delta_{s_0}}{\rho_{s_0}}\right), \quad L_{s_0} \leqslant \frac{\alpha}{2}.$$

Furthermore, we get

$$\frac{M_{s_0}}{M_{s_0-1}} = r_0^{2^{s_0-1}} \gamma^{(2m+4)} \frac{Q(\rho_{s_0-1}, \gamma)}{Q(\rho_{s_0}, \gamma)} \leqslant \gamma^{(2m+4)} r_0 \leqslant \frac{1}{4m},$$

and

$$\frac{M_{s_0} Q(\rho_{s_0}, \delta_{s_0})}{\alpha \delta_{s_0}^{2m+4}} = \frac{M_{s_0} Q(\rho_{s_0}, \gamma^{(s_0+1)})}{\alpha \gamma^{(2m+4)(s_0+1)}} \leqslant \frac{M_{s_0} Q(\rho, \gamma)}{\alpha \gamma^{(2m+4)(s_0+1)}} =$$

$$= r_0^{2^{s_0}} \gamma^{(2m+4)} \leqslant r_0^2 \gamma^{(2m+4)} < \frac{1}{4m} < 1.$$

Thus as we pass from s_0 to $s_0 + 1$, we can make use of the estimates o Lemma 2.

On the D_s-domain given by

$$D_s \left\{ \| h \| \leqslant \eta \left(1 - \frac{\delta_s}{\rho_s}\right), \quad |\operatorname{Im} \varphi \leqslant \rho_s \right\},$$

the inequality

$$|f_{s_0+1}(h, \varphi, \Delta^{(s_0+1)})| \leqslant \frac{M_{s_0}^2 Q(\rho_{s_0}, \delta_{s_0})}{\alpha \delta_{s_0}^{2m+4}} \leqslant$$

$$\leqslant \frac{M_{s_0}^2 Q(\rho, \gamma)}{\alpha \gamma^{(2m+4)(s_0+1)}} =$$

$$= \frac{\alpha}{Q(\rho, \gamma)} \frac{r_0^{2(s_0+1)} \gamma^{2(s_0+2)(2m+4)}}{\gamma^{(2m+4)(s_0+1)}}$$

holds. Consequently, we may write

$$M_{s_0+1} = \frac{\alpha}{Q(\rho, \gamma)} r_0^{2^{(s_0+1)}} \gamma^{(s_0+3)(2m+4)}.$$

In addition, we have

$$\frac{M_{s_0+1}+M_{s_0}}{M_{s_0-1}}=\frac{M_{s_0+1}}{M_{s_0}}\,\frac{M_{s_0}}{M_{s_0-1}}+\frac{M_{s_0}}{M_{s_0-1}}\leqslant\left(\frac{1}{4m}\right)^2+\frac{1}{4m}<\frac{1}{2m},$$

and

$$M_{s_0+1}+M_{s_0}<\frac{1}{2m}\,M_{s_0-1}<M_{s_0-1}.$$

Thus, if

$$|\Delta^{(s_0+1)}|<M_{s_0},$$

then

$$|\Delta^{(s_0)}(\Delta^{(s_0+1)})|\leqslant\frac{1}{2m}\,M_{s_0-1}<M_{s_0-1},$$

and, consequently,

$$|\Delta^{(\nu)}(\Delta^{(\nu+1)}\ldots(\Delta^{(s_0+1)})\ldots)|<M_{\nu-1}$$

(formally we put $M_{-1}\equiv\sigma$).

It now remains to verify the validity of the formulas obtained for N_{s_0+1} and L_{s_0+1}.

By virtue of Lemma 2, on the domain D_{s_0+1}, we have

$$u^{(s_6+1)}(h,\varphi,\Delta^{(s_0+1)})|\leqslant\frac{\delta_{s_0+1}}{m}\leqslant\frac{\gamma^{(s_0+2)}}{m}.$$

However,

$$F_{s_0+1}(h,\varphi,\Delta^{(s_0+1)})=F_{s_0}(h,\varphi+u^{(s_0+1)},\Delta^{(s_0)}(\Delta^{(s_0+1)})).$$

In addition,

$$|\operatorname{Im}(\varphi+u^{(s_0+1)})|\leqslant\rho_{s_0+4}+\frac{\gamma^{(s_0+2)}}{m}=\rho_{s_0}-2\gamma^{(s_0+1)}+\frac{\gamma^{(s_0+2)}}{m}<\rho_{s_0}$$

on the same domain. Hence,

$$\|F_{s_0+1}(h,\varphi,\Delta^{(s_0+1)})\|\leqslant\frac{\alpha}{2}\,\eta.$$

Furthermore,

$$n\left\|\frac{\partial F_{s_0+1}(h,\varphi,\Delta^{(s_0+1)})}{\partial h_q}\right\|\leqslant L_{s_0}+\frac{2M_{s_0}P^2n^2m\rho^2}{\gamma^{3(s_0+1)}\left(1-\frac{1}{2m}\right)}=$$

$$=L_{s_0}+\frac{2\alpha}{Q(\rho,\gamma)}\,P^2n^2m\rho^2\left(1-\frac{1}{2m}\right)\times$$

$$\times\,\gamma^{(s_0+2)(2m+4)-3(s_0+1)}\,r_0^{2^{s_0}};$$

consequently, we may write

$$L_{s_0+1} = L_{s_0} + \frac{2\alpha}{Q(\rho, \gamma)} P^2 n^2 m \rho^2 \left(1 - \frac{1}{2m}\right)^{-1} \gamma^{(s_0+1)(2m+1)+(2m+4)} r_0^{2^{s_0}} =$$

$$= L + \frac{2Pn^2 m\rho^2 \alpha}{Q(\rho, \gamma)\left(1 - \frac{I}{2m}\right)} \sum_{p=0}^{s_0} r_0^{2^p} \gamma^{(p+1)(2m+1)+(2m+4)}.$$

Thus, Lemma 2 remains valid under the passage from $s = s_0$ to $s = s_0 + 1$ and, by induction, holds for all $s = 1, 2, \ldots$.

Making use of the estimates of Lemma 2, we now prove the convergence o the iteration process. For this, we express φ and Δ in terms of $\varphi^{(s)}$ and $\Delta^{(s)}$:

$$\varphi = \varphi^{(s)} + \Phi^{(s)}(h, \varphi^{(s)}, \Delta^{(s)}),$$

$$\Delta = D_1^{(s)}(\Delta^{(s)}).$$

We get the identities

$$D_1^{(s)}(\Delta^{(s)}) = D_1^{(s+1)}(\Delta^{(s+1)}),$$

$$\varphi^{(s)} + \Phi^{(s)}(h, \varphi^{(s)}, \Delta^{(s)}) = \varphi^{(s+1)} + \Phi^{(s+1)}(h, \varphi^{(s+1)}, \Delta^{(s+1)}).$$

From these, we obtain

$$D_1^{(s+1)}(\Delta^{(s+1)}) = D_1^{(s)}(\Delta^{(s)}(\Delta^{(s+1)})),$$

$$\Phi^{(s+1)}(h, \theta, \Delta^{(s+1)}) = u^{(s+1)}(h, \theta, \Delta^{(s+1)}) + \Phi^{(s)}(h, \theta + u^{(s+1)}$$

$$\times (h, \theta, \Delta^{(s+1)}), \Delta^{(s)}(\Delta^{(s+1)})). \quad (4.5)$$

We now proceed to find the estimates for the functions $D_1^{(s)}$, $\Phi^{(s)}$ and their derivatives. For $|\Delta^{(s+1)}| \leqslant M_s$ we have

$$\sum_{(q)} \left| \frac{\partial \Delta}{\partial \Delta_q^{(1)}} \right| \leqslant 1 + \frac{4Mm}{\sigma} \leqslant 2,$$

$$\sum_{(q)} \left| \frac{\partial \Delta_q^{(1)}}{\partial \Delta_q^{(2)}} \right| \leqslant 1 + \frac{4M_1 m}{M},$$

$$\vdots \qquad\qquad \vdots$$

$$\sum_{(q)} \left| \frac{\partial \Delta_q^{(s-1)}}{\partial \Delta_q^{(s)}} \right| \leqslant 1 + \frac{4M_s m}{M_{s-1}}.$$

From these, we obtain

$$\sum_{(q)} \left| \frac{\partial \Delta}{\partial \Delta_q^{(s)}} \right| \leqslant \left(1 + \frac{4Mm}{\sigma} \right) \left(1 + \frac{4M_1 m}{M} \right) \ldots \left(1 + \frac{4M_s m}{M_{s-1}} \right).$$

But

$$\frac{M_p}{M_{p-1}} \leqslant \gamma^{(2m+4)} r_0^{2^{p+1}}, \; \gamma^{(2m+4)} r_0 \leqslant \frac{1}{2m}.$$

Hence,

$$\left(1 + \frac{4Mm}{\sigma} \right) \left(1 + \frac{4M_1 m}{M} \right) \ldots \left(1 + \frac{4M_s m}{M_{s-1}} \right) \leqslant$$

$$\leqslant 4\,(1 + 2r_0)\,(1 + 2r_0^3) \ldots =$$

$$= 4 \prod_{(1 \leqslant p < \infty)} \left(1 + 2r_0^{2^{p-1}} \right) = C(r_0).$$

Consequently, for all s we get

$$\sum_q \left| \frac{\partial D_1^{(s)}(\Delta^{(s)})}{\partial \Delta_q^{(s)}} \right| \leqslant C(r_0).$$

Evidently, it is possible to set

$$\left| D_1^{(s+1)}(0) - D_1^{(s)}(0) \right| = \left| D_1^{(s)}(\Delta^{(s)}(0)) - D_1^{(s)}(0) \right| \leqslant$$

$$\leqslant C(r_0) \mid \Delta^{(s)}(0) \mid \leqslant$$

$$\leqslant C(r_0)\, Mm \leqslant C(r_0) \left(\frac{1}{4m} \right)^{m+1} \sigma,$$

and hence verify that the criterion for uniform convergence is satisfied:

$$\left| D_1^{(s+k)}(0) - D_1^{(s)}(0) \right| \leqslant C(r_0)\, \sigma \sum_{p=s+1} \left(\frac{1}{4m} \right)^{s+1} = \frac{C(r_0)\, \sigma}{1 - \dfrac{1}{4m}} \left(\frac{1}{4m} \right)^{s+1}.$$

Note that the inequality obtained is roughly majorised.

Further, let

$$\sum_{k=1}^{m} \left| \frac{\partial \Phi^{(s)}(h, \varphi, \Delta^{(s)})}{\partial \varphi_k} \right| \leqslant z_s$$

on the domain

$$\| h \| \leqslant \eta_s, \quad | \operatorname{Im} \varphi | \leqslant \rho_s, \quad | \Delta^{(s)} | \leqslant M_{s-1}.$$

Then from the expression (4.5) it follows that

$$\sum_{k=1}^{m} \left| \frac{\partial \Phi^{(s+1)}(h, \theta, \Delta^{(s+1)})}{\partial \theta_k} \right| \leqslant \sum_{k=1}^{m} \left| \frac{\partial u^{(s+1)}(h, \theta, \Delta^{(s+1)})}{\partial \theta_k} \right| + z_s$$
$$\times \left\{ 1 + \sum_{k=1}^{m} \left| \frac{\partial u^{(s+1)}(h, \theta, \Delta^{(s+1)})}{\partial \theta_k} \right| \right\},$$

and, making use of the estimates of Lemma 2, we obtain

$$\sum_{k=1}^{m} \left| \frac{\partial u^{(s+1)}}{\partial \theta_k} \right| \leqslant \frac{M_s Q(\rho_s, \delta_s)}{2\alpha \delta_s^{2m+4}} \leqslant \frac{M_s Q(\rho, \gamma)}{2\alpha \gamma^{(2m+4)(s+1)}} = \frac{1}{2} r_0^{2^s} \gamma^{(2m+4)}.$$

Consequently, we have the relations

$$z_{s+1} = \frac{1}{2} r_0^{2^s} \gamma^{(2m+4)} + z_s \left(1 + \frac{1}{2} r_0^{2^s} \gamma^{(2m+4)} \right),$$

$$(1 + z_{s+1}) = \left(1 + \frac{1}{2} r_0^{2^s} \gamma^{(2m+4)} \right) (1 + z_s),$$

$$z_s = \prod_{0 \leqslant p \leqslant s-1} \left(1 + \frac{1}{2} r_0^{2^p} \gamma^{(2m+4)} \right) - 1 <$$

$$< \exp \sum_{p=0}^{\infty} \frac{1}{2} r_0^{2^p} \gamma^{(2m+4)} - 1 = \frac{3}{2} - 1 = \frac{1}{2}.$$

Thus, on the domain

$$\| h \| \leqslant \eta_s, \quad | \operatorname{Im} \varphi | \leqslant \rho_s, \quad | \Delta^{(s)} | \leqslant M_{s-1},$$

we have the inequality

$$\sum_{k=1}^{m} \left| \frac{\partial \Phi^{(s)}(h, \varphi, \Delta^{(s)})}{\partial \varphi_k} \right| \leqslant \frac{1}{2}. \tag{4.6}$$

We now evaluate the sum

$$\sum_{(q)} \left| \frac{\partial \Phi^{(s)}(h, \varphi, \Delta^{(s)})}{\partial \Delta_q^{(s)}} \right| = Y_s.$$

By the estimate (3.10), we have

$$\sum_{k=1}^{m}\left|\frac{\partial u^{(s+1)}}{\partial \Delta^{(s+1)}}\right| \leqslant \frac{M_s 2m\left(1-\frac{1}{2m}\right)^{-1} A(\rho, \gamma)}{\alpha M_{s-1}\gamma^{(2m+1)\ (s+1)}} =$$

$$= r_0^{2^{s-1}} \gamma^{(2m+4)-(2m+1)(s+1)} \frac{2m\left(1-\frac{1}{2m}\right) A(\rho, \delta)}{\alpha},$$

$$s \geqslant 1,$$

$$\sum_{k=1}^{m}\left|\frac{\partial u^{(1)}}{\partial \Delta^{(1)}}\right| \leqslant \frac{M 2m\left(1-\frac{1}{2m}\right)^{-1} A(\rho, \gamma)}{\alpha\sigma\gamma^{(2m+1)}}.$$

Hence,

$$Y_{s+1} = \frac{3}{2} S_s + Y_s(1+\nu_s)\ ; \quad Y_0 \equiv 0, \tag{4.7}$$

with the notations

$$\nu_0 = 1 + \frac{4Mm}{\sigma}, \ldots, \nu_p = 1 + \frac{4M_p}{M_{p-1}} m, \quad p \geqslant 1,$$

$$S_0 = \frac{M}{\alpha\sigma} \frac{2m\left(1-\frac{1}{2m}\right)^{-1} A(\rho, \gamma)}{\gamma^{(2m+1)}},$$

$$S_s = r_0^{2^{s-1}} \gamma^{3-(2m+2)s}.$$

Set

$$Y_s = X_s \prod_{0 \leqslant p \leqslant s-1} (1+\nu_p).$$

From eqn. (4.7) we obtain for the constant X_s the equation

$$X_{s+1} = X_s + \frac{3}{2(1+\nu_s)} S_s.$$

Solving this, we arrive at the inequality

$$Y_s < \left\{\prod_{0 \leqslant p < \infty} (1+\nu_p)\right\} \frac{3}{2} \sum_{p=0}^{\infty} S_p \leqslant \frac{3}{2} C(r_0) \sum_{p=0}^{\infty} S_p =$$

$$= \frac{3C(r_0)\, m\left(1-\frac{1}{2m}\right)^{-1} A(\rho, \gamma)}{\alpha}$$

$$\times \left\{\frac{M}{\sigma\gamma^{2m+1}} + \sum_{p=0}^{\infty} r_0^{2^p} \gamma^{3-(2m+1)(p+1)}\right\} \equiv Y.$$

Using this inequality, we get the estimate

$$\begin{aligned}
&| \Phi^{(s+1)}(h, \theta, 0) - \Phi^{(s)}(h, \theta, 0) | \leqslant \\
&\quad \leqslant | u^{(s+1)}(h, \theta, 0) | + | \Phi^{(s)}(h, \theta + u^{(s+1)}, \Delta^{(s)}(0)) - \Phi^{(s)}(h, \theta, 0) \leqslant \\
&\quad \leqslant \frac{3}{2} \frac{M_s A(\rho, \gamma)}{\alpha \gamma^{(2m+1)(s+1)}} + Y M_{s-1} = \\
&\quad = \frac{3}{2} \frac{A(\rho, \gamma)}{Q(\rho, \gamma)} r_0^{2^s} \gamma^{(2m+4)+3(s+1)} + Y \frac{\alpha}{Q(\rho, \gamma)} r_0^{2^{s-1}} \gamma^{(s+1)(2m+4)},
\end{aligned}$$

as a consequence of which we obtain the inequality

$$\begin{aligned}
&| \Phi^{(s+k)}(h, \theta, 0) - \Phi^{(s)}(h, \theta, 0) | \\
&\quad \leqslant \frac{3}{2} \frac{A(\rho, \gamma)}{Q(\rho, \gamma)} \gamma^{2m+4} \sum_s^{\infty} r_0^{2^s} \gamma^{3(s+1)} + \\
&\qquad + \frac{3C(r_0)\, m \left(1 - \dfrac{1}{2m} \right) A(\rho, \gamma)}{Q(\rho, \gamma)} \sum_s^{\infty} r_0^{2^{s-1}} \gamma^{(s+1)(2m+4)} \leqslant \\
&\quad \leqslant \gamma^{2m+2} \left[\frac{3}{2m} \sum_s^{\infty} r_0^{2^s} \gamma^{3(s+1)} + 3C(r_0) \sum_s^{\infty} r_0^{2^{s-1}} \gamma^{(2m+4)s} \right],
\end{aligned}$$

which remains valid on the domain

$$\| h \| \leqslant \frac{\eta}{2}, \qquad | \operatorname{Im} \varphi | \leqslant \frac{\rho}{2}. \tag{4.8}$$

From this argument we see that, on the domain (4.8), the relations

$$\Phi^{(s)}(h, \theta, 0) \to \Phi^{(\infty)}(h, \theta, 0),$$
$$D_1^{(s)}(0) \to D_1^{(\infty)}(0)$$

hold uniformly and, in addition,

$$\begin{aligned}
&| \Phi^{(\infty)}(h, \theta, 0) - \Phi^{(s)}(h, \theta, 0) | \leqslant \\
&\quad \leqslant \gamma^{2m+2} \left[\frac{3}{2m} \sum_s^{\infty} r_0^{2^s} \gamma^{3(s+1)} + 3C(r_0) \sum_s^{\infty} r_0^{2^{s-1}} \gamma^{(2m+4)s} \right],
\end{aligned}$$

$$D_1^{(\infty)}(0) - D_1^{(s)}(0) | \leqslant \frac{C(r_0)\, \sigma}{1 - \dfrac{1}{4m}} \left(\frac{1}{4m} \right)^{s+1}. \tag{4.9}$$

The uniform convergence of the function $\Phi(h, \theta, 0)$ implies that the function

$\Phi^{(\infty)}(h, \theta, 0)$ is analytic on the domain (4.8) and the inequality (4.6) yields the estimate of its derivatives as

$$\sum_{k=1}^{m} \left| \frac{\partial \Phi^{(\infty)}(h, \theta, 0)}{\partial \theta_k} \right| \leqslant \frac{1}{2}. \tag{4.10}$$

Moreover, the expressions (4.5) and the estimates of Lemma 1 lead to the inequality

$$| \Phi^{(s+1)} | \leqslant \frac{\delta_s}{m} + | \Phi^{(s)} |,$$

and, consequently,

$$| \Phi^{(\infty)}(h, \theta, 0) | \leqslant \frac{1}{m} (\delta + \delta_1 + \ldots) = \frac{\rho}{4m} \tag{4.11}$$

for h, θ, varying in the domain (4.8).

Thus, by means of the substitution

$$\varphi = \theta + \Phi^{(\infty)}(h, \theta, 0),$$

for $\Delta = D_1^{(\infty)}$ the system of eqns. (4.1) reduces to the form

$$\frac{dh}{dt} = Hh + F'(h, \theta) = Hh + F(h, \theta + \Phi^{(\infty)}(h, \theta, 0), D_1^{(\infty)}), \tag{4.12}$$

$$\frac{d\theta}{dt} = \omega.$$

Since this system satisfies the conditions

$$\| F'(h, \theta) \| \leqslant \alpha \frac{\eta}{2}, \quad n \left\| \frac{\partial F'(h, \theta)}{\partial h_q} \right\| \leqslant L_\infty < \frac{\alpha}{2} \qquad (q = 1, \ldots, n)$$

for $\| h \| \leqslant \eta/2$, $| \operatorname{Im} \theta | \leqslant \rho/2$, it follows that the system obeys Theorem 1.

Thus,

$$h = S^{(\infty)}(\theta_0 + \omega t), \quad | S^{(\infty)} | \leqslant \frac{\eta}{2} \quad \text{for } | \operatorname{Im} \theta_0 | \leqslant \frac{\rho}{2},$$

the quasi-periodic solution with frequencies ω, exists for eqns. (4.12)

Reverting to the principal system (4.1), it is seen that it has quasi-periodic solutions

$$h = S^{(\infty)}(\theta_0 + \omega t),$$

$$\varphi = \theta_0 + \omega t + \Phi^{(\infty)}(S(\theta_0 + \omega t), \theta_0 + \omega t, 0)$$

for $\Delta = D_1^{(\infty)}$ with frequencies ω.

We shall show that this solution attracts solutions close to it. We note that as a result of the inequalities (4.10) and (4.11) the transformation

$$\varphi = \theta + \Phi^{(\infty)}(h, \theta, 0)$$

can be inverted on the domain

$$\| h \| \leqslant \frac{\eta}{2}, \quad | \operatorname{Im} \varphi | \leqslant \frac{\rho}{2}\left(1 - \frac{1}{2m}\right),$$

i.e. we can write

$$\theta = \varphi + \psi(h, \varphi).$$

Considering the solutions h_t, φ_t of the system (4.1), whose initial values h_0, φ_0 satisfy the conditions

$$\| h_0 \| \leqslant \frac{\eta}{2}, \quad | \operatorname{Im} \varphi_0 | \leqslant \frac{\rho}{2}\left(1 - \frac{1}{2m}\right),$$

and making use of Theorem 1, we obtain the estimates securing the attraction of nearby solutions :

$$\| h_t - S^{(\infty)}(\theta_0 + \omega t) \| \leqslant \eta e^{-\frac{\alpha}{2} t} \to 0,$$

$$\| \varphi_t - \theta_0 - \omega t - \Phi^{(\infty)}(S(\theta_0 + \omega t), \theta_0 + \omega t, 0) \| =$$

$$= | \Phi^{(\infty)}(h_t, \theta_0 + \omega t, 0) - \Phi^{(\infty)}(S(\theta_0 + \omega t), \theta_0 + \omega t, 0) | \to 0$$

as $t \to \infty$ and for $\theta_0 = \varphi_0 + \psi(h_0, \varphi_0)$.

Summing up the above arguments, we are led to formulate the following theorem.

Theorem 3. *If the functions $F(h, \varphi, \Delta)$ and $f(h, \varphi, \Delta)$ in eqns.* (4.1) *satisfy all the conditions on p.* 36, *then these equations, under a suitable choice of Δ, have quasi-periodic solutions, with frequencies ω, of the form*

$$h_\tau = S^{(\infty)}(\theta_0 + \omega t),$$

$$\varphi_\tau = \theta_0 + \omega t + \Phi^{(\infty)}(S(\theta_0 + \omega t), \theta_0 + \omega t, 0).$$

If h_t, φ_t are any solutions of system (4.1) *whose initial values h_0, φ_0 satisfy the conditions*

$$\| h_0 \| \leqslant \frac{\eta}{2}, \quad | \operatorname{Im} \varphi_0 | \leqslant \frac{\rho}{2}\left(1 - \frac{1}{2m}\right),$$

then h_t and φ_t asymptotically approach this quasi-periodic solution.

Remarks. **1.** If H is a real matrix, and $F(h, \theta, \Delta)$ and $f(h, \theta, \Delta)$ are real functions for real h, θ, Δ, then $F'(h, \theta)$, $\Phi^{(\infty)}(h, \theta, 0)$, and $\Delta = D_1^{(\infty)}(0)$ are also real for real h, θ.

2. If $F(h, \theta, \Delta)$ and $f(h, \theta, \Delta)$ are analytic functions of the parameter μ on the domain ϑ and if, on this domain, all the inequalities with fixed η, ρ, σ, r_0 hold, then $D_1^{(\infty)}(0)$, $\Phi^{(\infty)}(h, \theta, 0)$ are also analytic functions of μ on ϑ.

This is evident because according to thc expression (4.9), the convergence of $\Phi^{(s)}(h, \theta, 0)$ and $D_1^{(s)}(0)$ is defined by the quantities η, ρ, σ, r_0 and is, consequently, uniform with respect to μ on the domain ϑ.

5. Parametric Dependence of Quasi-Periodic Solutions. Asymptotic and Convergent Expansions

In the problems of nonlinear mechanics we encounter the systems depending on one or more parameters. While investigating the quasi-periodic solutions of such systems, it is essential to determine the dependence of these solutions on the parameters as well as to construct asymptotic expansions for them.

Consider the system of equations

$$\frac{dh}{dt} = Hh + F(h, \varphi, \Delta, \varepsilon),$$
$$\frac{d\varphi}{dt} = \omega + \Delta + f(h, \varphi, \Delta, \varepsilon), \tag{5.1}$$

where the functions $F(h, \varphi, \Delta, \varepsilon)$ and $f(h, \varphi, \Delta, \varepsilon)$ are analytic on the domain

$$| h | \leqslant \eta_0, \quad | \Delta | \leqslant \sigma_0, \quad | \varepsilon | \leqslant \varepsilon_0, \quad | \operatorname{Im} \varphi | \leqslant \rho_0. \tag{5.2}$$

The following theorem holds for the system (5.1)†.

Theorem 4. *Suppose that the following conditions are fulfilled for the system* (5.1) :

(i) *In the domain* (5.2), *the functions* $F(h, \varphi, \Delta, \varepsilon)$ *and* $f(h, \varphi, \Delta, \varepsilon)$ *satisfy the inequalities*

$$| F(h, \varphi, \Delta, \varepsilon) | \leqslant C_1 | h |^2 + C_2 | \Delta | + C_3 | \varepsilon |,$$
$$| f(h, \varphi, \Delta, \varepsilon) | \leqslant B_1 | h | + B_2 | \Delta | + B_3 | \varepsilon |, \tag{5.3}$$

where C_i, B_i $(i = 1, 2, 3)$ *are constants.*

(ii) *The functions* $F(h, \varphi, \Delta, \varepsilon)$ *and* $f(h, \varphi, \Delta, \varepsilon)$ *are real when the arguments are real.*

†See Appendix I.

(iii) *The real parts α of all the eigenvalues of the matrix H are smaller than* $-\alpha < 0$.

(iv) *The frequencies ω satisfy the inequalities*

$$|(k, \omega)| \geqslant K|k|^{-(m+1)} \quad (|k| \neq 0) \tag{5.4}$$

for all integral vectors $k = (k_1, \ldots, k_m)$.

Then it is possible to find positive constants $\varepsilon_1 < \varepsilon_0$, $\eta_1 < \eta_0$, $\rho_1 < \rho_0$ for which the following assertions remain valid:

(*a*) *On the domain*

$$|\varepsilon| \leqslant \varepsilon_1 \tag{5.5}$$

it is possible to define an analytic function $\Delta(\varepsilon)$, which is real for real ε, such that eqns. (5.1) *with $\Delta = \Delta(\varepsilon)$ have quasi-periodic solutions, with frequencies ω, of the form*

$$\begin{aligned} h &= S(\omega t + \psi, \varepsilon), \\ \varphi &= \omega t + \psi + \Phi(\omega t + \psi, \varepsilon), \end{aligned} \tag{5.6}$$

where $S(\theta, \varepsilon)$ and $\Phi(\theta, \varepsilon)$ are analytic functions of θ and ε on the domain

$$|\operatorname{Im} \theta| \leqslant \rho_1, \quad |\varepsilon| \leqslant \varepsilon_1, \tag{5.7}$$

and are real for real θ, ε.

(*b*) *Any solutions h_t, φ_t of the system* (5.1), *whose initial values h_0, φ_0 satisfy the conditions*

$$|h_0| \leqslant \eta, \quad |\operatorname{Im} \varphi_0| \leqslant \rho_1,$$

approach the solution (5.6) *asymptotically as $t \to \infty$.*

Proof. By the hypotheses of the theorem, the inequality

$$|e^{-Ht}| \leqslant Pe^{-\alpha t}, \quad t \geqslant 0 \tag{5.8}$$

holds. Furthermore, the relations

$$\begin{aligned} \|F(h, \varphi, \Delta, \varepsilon)\| &\leqslant N = PC_1\eta^2 + PC_2\sigma + PC_3|\varepsilon|, \\ |f(h, \varphi, \Delta, \varepsilon)| &\leqslant M = B_1\eta + B_2G + B_3|\varepsilon|, \end{aligned} \tag{5.9}$$

and

$$\varepsilon n \left\| \frac{\partial F(h, \varphi, \Delta, \varepsilon)}{\partial h_q} \right\| \leqslant L = \frac{n}{\eta}\{4PC_1\eta^2 + C_2\sigma + C_3|\varepsilon|\}, \tag{5.10}$$

obviously, remain valid on the domain

$$|h| \leqslant \eta \leqslant \frac{\eta_0}{2}, \qquad |\Delta| \leqslant \sigma_0, \qquad |\varepsilon| \leqslant \varepsilon_0.$$

We wish to find $r_0 < 1$ such that the inequalities

$$\gamma^{2m+4} r_0 \leqslant \frac{1}{4m}, \quad \frac{\gamma^{2m+4}}{2} \sum_{s=0}^{\infty} r_0^{2^s} < \ln \frac{3}{2}, \quad \gamma = \frac{\rho_0}{\rho_0 + 4},$$

$$\left(1 - \frac{1}{2m}\right)^{-1} \frac{P^2 n^2 m \rho_0^2}{Q(\rho_0, \gamma)} \sum_{s=0}^{\infty} r_0^{2^s} \gamma^{(s+1)(2m+1)+(2m+4)} \leqslant \frac{1}{8} \tag{5.11}$$

are satisfied.

Then, in order that all the hypotheses of Theorem 3 are satisfied, it will suffice to ensure that the conditions

$$N \leqslant \frac{\alpha\eta}{2}, \quad L \leqslant \frac{\alpha}{4}, \quad M \leqslant \frac{\sigma}{4m},$$
$$\frac{M}{\alpha} \frac{Q(\rho, \gamma)}{\gamma^{2m+8}} \leqslant r_0 \tag{5.12}$$

are satisfied. Obviously, these are to be satisfied for sufficiently small η, σ, $|\varepsilon|$.

Now consider the system of equations

$$\frac{dh}{dt} = \varepsilon A h + \varepsilon F(h, \varphi, \Delta, \varepsilon),$$
$$\frac{d\varphi}{dt} = \omega + \Delta + \varepsilon f(h, \varphi, \Delta, \varepsilon). \tag{5.13}$$

Suppose that the functions $F(h, \varphi, \Delta, \varepsilon)$ and $f(h, \varphi, \Delta, \varepsilon)$ satisfy all the hypotheses of Theorem 4 and that all the eigenvalues of the matrix A have negative real parts.

Taking advantage of Theorem 3 above, we write

$$H = \varepsilon A, \quad 0 < \varepsilon \leqslant \varepsilon_0. \tag{5.14}$$

Then

$$|e^{Ht}| \leqslant P e^{-\alpha t}, \tag{5.15}$$

where P is a constant and $\alpha = \varepsilon\beta > 0$.

Again, we find $r_0 < 1$ such that the inequalities (5.11) are satisfied.

Further, for the hypotheses of Theorem 3 to be met it will suffice that the conditions (5.12) are satisfied. Since $\alpha = \varepsilon\beta$, it follows that these conditions can be met only for small η, σ, ε_0, because N, L, M each contains a power in ε.

Summing up the above arguments, we arrive at the following theorem.

Theorem 5. *Suppose that the functions* $F(h, \varphi, \Delta, \varepsilon)$ *and* $f(h, \varphi, \Delta, \varepsilon)$ *appearing in the right-hand side of the system of eqns.* (5.13) *and the matrix* A *satisfy the hypotheses of Theorem* 4.

Then it is possible to find positive constants $\varepsilon_1 < \varepsilon_0$, $\eta_1 < \eta_0$, $\rho_1 < \rho_0$, such that:

(i) *The real function $\Delta = \Delta(\varepsilon)$ can be defined on the interval $0 < \varepsilon \leqslant \varepsilon_1$ so that eqns.* (5.13) *have, for $\Delta = \Delta(\varepsilon)$, the quasi-periodic solution*

$$\begin{aligned} h &= S(\omega t + \psi, \varepsilon), \\ \varphi &= \omega t + \psi + \Phi(\omega t + \psi, \varepsilon) \end{aligned} \tag{5.16}$$

with frequencies ω, where $S(\theta, \varepsilon)$ and $\Phi(\theta, \varepsilon)$ are analytic functions of the angular variable θ on the domain

$$|\operatorname{Im} \theta| < \rho_1, \quad 0 < \varepsilon \leqslant \varepsilon_0; \tag{5.17}$$

(ii) *The solutions h_t, φ_t whose initial values satisfy the conditions*

$$h_0 | \leqslant \eta_1, \quad |\operatorname{Im} \varphi_0| \leqslant \rho_1, \tag{5.18}$$

asymptotically approach this quasi-periodic solution as $t \to \infty$.

We now consider the problem of proving the analyticity of the functions $S(\theta, \varepsilon)$ and $\Phi(\theta, \varepsilon)$ which characterizes the quasi-periodic solution. We remark that in contrast to our earlier formulations, we consider here only positive ε lying in a sufficiently small interval. This is because of the fact that to use the theorems proved already it is necessary that the inequality

$$|e^{\varepsilon A t}| \leqslant P e^{-\beta |\varepsilon| t} \tag{5.19}$$

is satisfied.

If a are eigenvalues of the matrix A, then

$$\varepsilon a = |\varepsilon| a e^{i \arg \varepsilon} \tag{5.20}$$

will be the eigenvalues of the matrix εA and hence, whatever the value of a, it is possible, by a suitable choice of $\arg \varepsilon$, to make the eigenvalues of the matrix εA lie on the imaginary axis. This fact excludes the consideration of all complex ε lying in any arbitrarily small domain $|\varepsilon| \leqslant \varepsilon_0$. However, it is possible to establish that the functions, F, f, Δ are analytic in ε, taking $\arg \varepsilon$ sufficiently small and such that

$$\operatorname{Re} \varepsilon a < - \quad \varepsilon \ \beta < 0, \quad |\varepsilon| \leqslant \varepsilon_0. \tag{5.21}$$

It is easy to verify that all the estimates of Theorem 3 are satisfied on the domain (5.21) and, consequently, on this domain the functions $S(\theta, \varepsilon)$, $\Phi(\theta, \varepsilon)$ and $\Delta(\varepsilon)$ are analytic functions of ε. But the convergence of the power series expansions does not follow, as is known, from analyticity on a wedge-shaped domain. This may give rise to the question as to whether such a restrictive condition is not the result of insufficient rigour in our discussions. It is, however, not difficult to construct an elementary example of equations of the type

(5.13) for which the considered functions will not be analytic in any arbitrarily small circle.

Consider the system of equations

$$\frac{dh}{dt} = -\varepsilon\beta h + \varepsilon^2 \sum_{|k| \neq 0} \rho^{|k|} e^{i(k,\varphi)},$$
$$\frac{d\varphi}{dt} = \omega, \tag{5.22}$$

with a certain h, where $\varphi = (\varphi_1, \varphi_2)$, $0 < \rho < 1$, $\beta > 0$.

From the system of eqns. (5.22), it is easy to obtain

$$h = S(\varphi, \varepsilon) = \varepsilon^2 \sum_{|k| \neq 0} \frac{\rho^{|k|} e^{i(k,\varphi)}}{i(k, \omega) + \varepsilon\beta}. \tag{5.23}$$

Consider the denominator occurring in the expression (5.23),

$$i\,(k_1\omega_1 + k_2\omega_2) + \varepsilon\beta. \tag{5.24}$$

Setting

$$k_1 = p > 0,$$
$$k_2 = -q < 0 \tag{5.25}$$

in (5.24), we obtain

$$\omega_2 \left\{ i\left(p\,\frac{\omega_1}{\omega_2} - q \right) + \varepsilon\,\frac{\beta}{\omega_2} \right\}. \tag{5.26}$$

Since the number ω_1/ω_2 is irrational, it follows that there exists a sequence of integers p and q which tends to infinity, such that $p\,(\omega_1/\omega_2) - q \to 0$. Consequently, the expression (5.26) vanishes for an infinite sequence of ε lying on the imaginary axis :

$$\varepsilon = -\frac{i\omega_2}{\beta}\left(p\,\frac{\omega_1}{\omega_2} - q \right). \tag{5.27}$$

Hence $S(\varphi, \varepsilon)$, as a function of ε, has an infinite number of poles lying on the imaginary axis with the condensation point at zero. It is evident that such a function cannot be expanded as a convergent power series in ε.

It may, however, be remarked that a formal power series expansion of the function $S\,(\varphi, \varepsilon)$ in ε is asymptotically convergent. This is due to the following reasons. The inequality (5.4) in the case under consideration assumes the form

$$|(k, \omega)| \geqslant K\,|\,k\,|^{-3}. \tag{5.28}$$

Therefore, the expression (5.24) can vanish only under the condition

$$|\varepsilon\beta| \geqslant K|k|^{-3}, \tag{5.29}$$

that is, when

$$|k| \geqslant \left(\frac{K}{\beta|\varepsilon|}\right)^{\frac{1}{3}} = C|\varepsilon|^{-\frac{1}{3}}, \qquad C = \left(\frac{K}{\beta}\right)^{\frac{1}{3}}. \tag{5.30}$$

On the other hand, the corresponding numerator is of higher order than

$$\rho^{C|\varepsilon|^{-\frac{1}{3}}}. \tag{5.31}$$

Hence it tends to zero faster than any exponent of ε.

This property of the asymptotic convergence of formal power expansions is valid not only for the simple example given above but also for the general case of eqns. (5.13) under consideration.

To prove this, consider the formal expansions

$$h = S(\psi, \varepsilon) = \sum_{(k)} \varepsilon^k S_k(\psi),$$

$$\varphi = \psi + \Phi(\psi, \varepsilon) = \psi + \sum_{(k)} \varepsilon^k \Phi_k(\psi), \tag{5.32}$$

$$\Delta(\varepsilon) = \sum_{(k)} \varepsilon^k \Delta_k.$$

It is, in fact, possible to determine all the coefficients of respective functions by substituting the expressions (5.32) in the original eqns. (5.13) and, in the course of differentiation, taking into account that

$$\frac{d\psi}{dt} = \omega. \tag{5.33}$$

Consider the l-th approximation that can be obtained from the formal expansions (5.32) by disregarding in their right-hand sides terms of order ε^{l+1} (and higher) :

$$h^{(l)} = S^{(l)}(\psi, \varepsilon) = \sum_{(k)}^{l} \varepsilon^k S_k(\psi),$$

$$\varphi^{(l)} = \psi + \Phi^{(l)}(\psi, \varepsilon) = \sum_{(k)}^{l} \varepsilon^k \Phi_k(\psi) + \psi, \tag{5.34}$$

$$\Delta^{(l)}(\varepsilon) = \sum_{(k)}^{l} \varepsilon^k \Delta_k.$$

Now, carry out in eqns. (5.13) the change of variables

$$h = h^{(1)} + S^{(l)}(\psi, \varepsilon),$$
$$\varphi = \psi + \Phi^{(l)}(\psi, \varepsilon), \tag{5.35}$$
$$\Delta = D + \Delta^{(l)}(\varepsilon).$$

It is then evident that, for $h^{(1)} = 0$, $D = 0$, the expressions (5.35) satisfy the system of equations obtained to within the terms of $(l + 1)$-th order.

Thus, with the aid of the estimation technique developed above, it is easy to show that for the quasi-periodic solution considered, $h^{(1)}$ and D will be $(l+1)$-th order terms.

Apart from the asymptotic series (5.32), convergent expansions can be obtained by various methods.

Consider the equations of the form

$$\frac{dh}{dt} = \varepsilon Ah + \varepsilon F(h, \varphi, \Delta, \mu),$$
$$\frac{d\varphi}{dt} = \omega + \Delta + \varepsilon f(h, \varphi, \Delta, \mu), \tag{5.36}$$

which coincide with the original eqns. (5.13) for $\mu = \varepsilon$. Because of the conditions imposed on the functions $\varepsilon F(h, \varphi, \Delta, \mu)$ and $\varepsilon f(h, \varphi, \Delta, \mu)$ for sufficiently small $|\mu| \leqslant \varepsilon_0$ and $0 < \varepsilon \leqslant \varepsilon_0$, the conditions for the applicability of Theorem 3 are satisfied.

Therefore, $S(\varphi, \mu)$, $\Phi(\varphi, \mu)$, and Δ are analytic functions of μ on the domain

$$|\mu| \leqslant \varepsilon_0, \quad 0 < \varepsilon \leqslant \varepsilon_0, \tag{5.37}$$

and, consequently, can be expanded as uniformly convergent power series in μ.

Putting $\mu = \varepsilon$ in these series, we obtain convergent expressions for the functions $\Delta(\varepsilon)$, $S(\varphi, \varepsilon)$ and $\Phi(\varphi, \varepsilon)$, which characterize the quasi-periodic solution.

We will indicate another exceedingly simple method for practical calculations.

Write the initial system of equations in the form

$$\frac{dh}{dt} = \varepsilon Ah + \varepsilon zF(h, \varphi, \Delta, \varepsilon z),$$
$$\frac{d\varphi}{dt} = \omega + \Delta + \varepsilon zf(h, \varphi, \Delta, \varepsilon z). \tag{5.38}$$

Note that it coincides with the system (5.36) for $z = 1$, $\mu = \varepsilon$. It is possible to verify that the theorems proved for $|\varepsilon| \leqslant \varepsilon_0$, $|z| \leqslant \mathrm{const}/\sqrt[3]{\varepsilon}$ are applicable to this system (5.38). Therefore, for sufficiently small ε, the functions

$\Delta(\varepsilon z)$, $S(\varphi, \varepsilon z)$ and $\Phi(\varphi, \varepsilon z)$ can be expanded as a convergent power series in z. The desired point $z = 1$ lies in the circle of convergence. It is not difficult to see that in the expansions for h and φ, ε enters both as a cofactor to a definite power and also in the denominators of the type $[\varepsilon A + i(k, \omega)]^{-1}$.

§ 6. Quasi-Periodic Solutions in Second Order Systems

We shall apply the results obtained above to investigate an oscillatory system, defined by the system of n second order differential equations

$$\frac{d^2 q_k}{dt^2} + \lambda_k^2 q_k = \varepsilon f_k \left(q_1, \ldots, q_n, \quad \frac{dq_1}{dt}, \ldots, \frac{dq_n}{dt} \right), \tag{6.1}$$

$$(k = 1, 2, \ldots, n),$$

or, in the vector form

$$\frac{d^2 q}{dt^2} + \lambda^2 q = \varepsilon f\left(q, \frac{dq}{dt}\right), \tag{6.2}$$

$$q = (q_1, \ldots, q_n), \quad f = (f_1, \ldots, f_n), \quad \lambda = (\lambda_1, \ldots, \lambda_n).$$

We now carry out in the system of eqns. (6.2) the transformations customary in the nonlinear mechanics. We introduce by way of variables, the amplitude $a = (a_1, \ldots, a_n)$ and the phase $\theta = (\theta_1, \ldots, \theta_n)$ of the oscillations, by the formulas

$$q = a \sin\theta, \qquad \frac{dq}{dt} = a\lambda \cos\theta, \qquad \lambda > 0. \tag{6.3}$$

Then, in place of the system of eqns. (6.2), we obtain equations in the amplitude and phase, of the form

$$\begin{aligned} \frac{da}{dt} &= \varepsilon A(a, \theta, \lambda) = \frac{\varepsilon}{\lambda} f(a \sin\theta, a\lambda\cos\theta) \cos\theta, \\ \frac{d\theta}{dt} &= \lambda + \varepsilon B(a, \theta, \lambda) = \lambda - \frac{\varepsilon}{a\lambda} f(a\sin\theta, a\lambda\cos\theta) \sin\theta. \end{aligned} \tag{6.4}$$

As is known [12], the first approximation equations corresponding to the system (6.4) in the non-resonance case, assume the form

$$\begin{aligned} \frac{da}{dt} &= \varepsilon A^{(0)}(a, \lambda), \\ \frac{d\theta}{dt} &= \lambda + \varepsilon B^{(0)}(a, \lambda), \end{aligned} \tag{6.5}$$

where

$$A^{(0)}(a, \lambda) = \frac{1}{(2\pi)^n} \int_0^{2\pi} \cdots \int_0^{2\pi} A(a, \theta, \lambda)\, d\theta_1 \ldots d\theta_n,$$

$$B^{(0)}(a, \lambda) = \frac{1}{(2\pi)^n} \int_0^{2\pi} \cdots \int_0^{2\pi} B(a, \theta, \lambda)\, d\theta_1 \ldots d\theta_n. \tag{6.6}$$

Assume that the system of first approximation eqns. (6.5) has a stationary asymptotically stable solution with constant amplitude

$$a = a^{(0)}, \qquad A^{(0)}(a^{(0)}, \lambda) = 0, \tag{6.7}$$

and, consequently, that the real parts of all the roots of the characteristic equation

$$\text{Det}\left\{p - \varepsilon \left(\frac{\partial A^{(0)}(a, \lambda)}{\partial a} \right)_{a=a^{(0)}} \right\} = 0 \tag{6.8}$$

are negative.

Then, every solution of the first equation of the system (6.5), sufficiently close to $a^{(0)}$, will asymptotically approach the stationary solution as $t \to \infty$. Further assume that each of the coordinates $a_1^{(0)}, \ldots, a_n^{(0)}$ of the vector $a^{(0)}$ is different from zero.†

In the stationary solution considered, the phases $\theta = (\theta_1, \theta_2, \ldots, \theta_n)$ rotate with constant frequencies

$$\omega_k = \lambda_k + \varepsilon B_k^{(0)}(a_1^{(0)}, \ldots, a_n^{(0)}, \lambda_1, \ldots, \lambda_n) \quad (k = 1, 2, \ldots, n) \tag{6.9}$$

and, consequently, the functions $q_k = q_k(t)$ $(k = 1, 2, \ldots, n)$ are periodic with frequencies ω_k.

After making these assumptions about the solutions of the first approximation equations (6.5), we proceed to investigate the solutions of the original exact system of differential eqns. (6.1)††.

By drawing on Theorem 4 proved in the preceding section, it is possible to establish the following theorem.

Theorem 6. *Suppose that the following conditions are satisfied for the system of differential eqns.* (6.1) :

(i) *The functions $f(q, q')$ are analytic in the variables q, q' in the neighbourhood of the torus*

$$q = a^{(0)} \sin \theta, \qquad q' = a^{(0)} \omega \cos \theta, \tag{6.10}$$

$0 \leqslant \theta \leqslant 2\pi$.

†See Appendix 2.

††See Appendix 3.

(ii) *For all integral vectors* $k = (k_1, \ldots, k_n)$, *the real frequencies* ω *satisfy the inequalities*

$$|(k, \omega)| \geqslant K |k|^{-(n+1)} \quad (|k| \neq 0).$$

(iii) *The real parts of all the eigenvalues of the matrix*

$$\left\| \left(\frac{\partial A^{(0)}(a, \lambda)}{\partial a} \right)_{a=a^{(0)}} \right\|$$

are negative.

Then, it is possible to determine a positive ε_0 *such that for all* $0 < \varepsilon < \varepsilon_0$ *the following propositions remain valid*:

(*a*) *It is possible to define* ν *such that for*

$$\lambda = \omega + \nu$$

eqns. (6.1) *have the quasi-periodic solutions*

$$q = Q(\omega t + \psi, \varepsilon), \quad \nu = \nu(\varepsilon)$$

with frequencies ω, *and every solution sufficiently close to a quasi-periodic solution approaches it as* $t \to \infty$.

(*b*) *The formal expansions of the function* $Q(\omega t + \psi, \varepsilon)$ *in powers of* ε *are asymptotically convergent and can be transformed into uniformly convergent expansions with the aid of partial summations by introducing auxiliary parameters* z *or* μ.

Proof. As usual, we separate the systematic motion from the small oscillations. For this, we make in eqns. (6.4) the change of variables

$$\begin{aligned} a &= b + \varepsilon u(b, \varphi, \nu), \\ \theta &= \varphi + \varepsilon v(b, \varphi, \nu), \end{aligned} \tag{6.11}$$

where

$$\begin{aligned} u(b, \varphi, \nu) &= \sum_{|k| \neq 0} \frac{A^{(k)}(b, \nu)\, e^{i(k,\varphi)}}{i(k, \omega)}, \\ v(b, \varphi, \nu) &= \sum_{|k| \neq 0} \frac{B^{(k)}(b, \nu)\, e^{i(k,\varphi)}}{i(k, \omega)}, \end{aligned} \tag{6.12}$$

$A^{(k)}(b, \nu)$, $B^{(k)}(b, \nu)$ being the Fourier coefficients of functions $A(b, \theta, \omega + \nu)$ and $B(b, \theta, \omega + \nu)$, defined as

$$\begin{aligned} A(b, \theta, \omega + \nu) &= \sum_{(k)} A^{(k)}(b, \nu)\, e^{i(k,\theta)}, \\ B(b, \theta, \omega + \nu) &= \sum_{(k)} B^{(k)}(b, \nu)\, e^{i(k,\theta)}. \end{aligned} \tag{6.13}$$

We have the identities

$$\left(\omega, \frac{\partial u}{\partial \varphi}\right) = A\,(b, \varphi, \omega + \nu) - A^{(0)}\,(b, \omega + \nu),$$
$$\left(\omega, \frac{\partial v}{\varphi}\right) = B\,(b, \varphi, \omega + \nu) - B^{(0)}\,(b, \omega + \nu). \tag{6.14}$$

Substituting the expression (6.11) in eqns. (6.4), we obtain

$$\frac{db}{dt} + \varepsilon\,\frac{\partial u}{\partial b}\,\frac{db}{dt} + \varepsilon\,\frac{\partial u}{\partial \varphi}\,\frac{d\varphi}{dt} = \varepsilon A\,(b + \varepsilon u, \varphi + \varepsilon v, \omega + \nu),$$
$$\varepsilon\,\frac{\partial v}{\partial b}\,\frac{db}{dt} + \frac{d\varphi}{dt} + \varepsilon\,\frac{\partial v}{\partial \varphi}\,\frac{d\varphi}{dt} = \varepsilon B\,(b + \varepsilon u, \varphi + \varepsilon v, \omega + \nu) + \omega + \nu. \tag{6.15}$$

Thereupon, taking note of the identities (6.14), we obtain

$$\left(E + \varepsilon\,\frac{\partial u}{\partial b}\right)\frac{db}{dt} + \varepsilon\frac{\partial u}{\partial \varphi}\left(\frac{d\varphi}{dt} - \omega\right) =$$
$$= \varepsilon A^{(0)}\,(b, \omega + \nu) + \varepsilon\,[A\,(b + \varepsilon u, \varphi + \varepsilon v, \omega + \nu)$$
$$- A\,(b, \varphi, \omega + \nu)] =$$
$$= \varepsilon A^{(0)}\,(b, \omega + \nu) + \varepsilon^2 A^{(1)}\,(b, \varphi, \nu, \varepsilon), \tag{6.16}$$
$$\varepsilon\,\frac{\partial v}{\partial b}\,\frac{db}{dt} + \left(E + \varepsilon\,\frac{\partial v}{\partial \varphi}\right)\left(\frac{d\varphi}{dt} - \omega\right) =$$
$$= \nu + \varepsilon B^{(0)}\,(b, \omega + \nu) + \varepsilon\,[B\,(b + \varepsilon u, \varphi + \varepsilon v, \omega + \nu)$$
$$- B\,(b, \varphi, \omega + \nu)] =$$
$$= \nu + \varepsilon B^{(0)}\,(b, \omega + \nu) + \varepsilon^2 B^{(1)}\,(b, \varphi, \nu, \varepsilon).$$

We must solve the system (6.16) for db/dt and $d\varphi/dt$. For this, we consider the auxiliary system of equations

$$\left(E + \varepsilon\,\frac{\partial u}{\partial b}\right)x + \varepsilon\,\frac{\partial u}{\partial \varphi}\,y = \xi,$$
$$\varepsilon\,\frac{\partial v}{\partial b}\,x + \left(E + \varepsilon\,\frac{\partial v}{\partial \varphi}\right)y = \eta, \tag{6.17}$$

from which it is easy to deduce

$$x = [E + \varepsilon T^{(1,1)}\,(b, \varphi, \nu, \varepsilon)]\,\xi + \varepsilon T^{(1,2)}\,(b, \varphi, \nu, \varepsilon)\,\eta,$$
$$y = \varepsilon T^{(2,1)}\,(b, \varphi, \nu, \varepsilon)\,\xi + [E + \varepsilon T^{(2,2)}\,(b, \varphi, \nu, \varepsilon)]\,\eta, \tag{6.18}$$

where, for sufficiently small $b - a^{(0)}$, $|\,\mathrm{Im}\,\varphi\,|$, ν, and ε, $T^{(i,j)}\,(b, \varphi, \nu, \varepsilon)$

$(i, j = 1, 2)$ are analytic functions of $b, \varphi, \nu, \varepsilon$.

We now find

$$\begin{aligned}\frac{db}{dt} &= [E + \varepsilon T^{(1,1)}(b, \varphi, \nu, \varepsilon)][\varepsilon A^{(0)}(b, \omega + \nu) + \varepsilon^2 A^{(1)}(b, \varphi, \nu, \varepsilon)] + \\ &\quad + \varepsilon T^{(1,2)}(b, \varphi, \nu, \varepsilon)[\nu + \varepsilon B^{(0)}(b, \omega + \nu) + \varepsilon^2 B^{(1)}(b, \varphi, \nu, \varepsilon)], \\ \frac{d\varphi}{dt} &= \omega + [E + \varepsilon T^{(2,2)}(b, \varphi, \nu, \varepsilon)][\nu + \varepsilon B^{(0)}(b, \omega + \nu) + \\ &\quad + \varepsilon^2 B^{(1)}(b, \varphi, \nu, \varepsilon)] + \varepsilon T^{(2,1)}(b, \varphi, \nu, \varepsilon) \times \\ &\quad \times [\varepsilon A^{(0)}(b, \omega + \nu) + \varepsilon^2 A^{(1)}(b, \varphi, \nu, \varepsilon)],\end{aligned} \tag{6.19}$$

or,

$$\begin{aligned}\frac{db}{dt} &= \varepsilon A^{(0)}(b, \omega) + \varepsilon X(b, \varphi, \nu, \varepsilon), \\ \frac{d\varphi}{dt} &= \omega + \nu + \varepsilon B^{(0)}(b, \omega)\, \varepsilon Y(b, \varphi, \nu, \varepsilon).\end{aligned} \tag{6.20}$$

Here, as is easy to see, the functions $X(b, \varphi, \nu, \varepsilon)$ and $Y(b, \varphi, \nu, \varepsilon)$ are of order ν and ε.

We now introduce into equations (6.20) the new variable $h = (h_1, \ldots, h_n)$ by the formula

$$b = a^{(0)} + h, \tag{6.21}$$

and also adopt the notation

$$\left(\frac{\partial A^{(0)}(b, \omega)}{\partial b}\right)_{a=a^{(0)}} = A, \quad \nu + \varepsilon B^{(0)}(a^{(0)}, \omega) = \Delta. \tag{6.22}$$

Then, the system (6.20) takes the form

$$\begin{aligned}\frac{dh}{dt} &= \varepsilon A h + \varepsilon F(h, \varphi, \Delta, \varepsilon), \\ \frac{d\varphi}{dt} &= \omega + \Delta + \varepsilon f(h, \varphi, \Delta, \varepsilon),\end{aligned} \tag{6.23}$$

where $F(h, \varphi, \Delta, \varepsilon)$ is a function of order h^2, Δ, ε, while the function $f(h, \varphi, \Delta, \varepsilon)$ is of order h, Δ, ε.

Since the functions $f(q, q')$ are analytic in the neighbourhood of the torus

$$\begin{aligned}q &= a^{(0)} \sin \theta, \\ q' &= a^{(0)}\, \omega \cos t,\end{aligned} \tag{6.24}$$

it follows that the functions $F(h, \varphi, \Delta, \varepsilon)$ and $f(h, \varphi, \Delta, \varepsilon)$ are also analytic in

h, φ, Δ, ε on the domain

$$|h| \leqslant \eta_0, \quad |\operatorname{Im} \varphi| \leqslant \rho_0, \quad |\Delta| \leqslant \sigma_0, \quad |\varepsilon| \leqslant \mu_0 \tag{6.25}$$

(for sufficiently small η_0, ρ_0, σ_0, μ_0), and that the inequalities

$$\begin{aligned} |F(h, \varphi, \Delta, \varepsilon)| &\leqslant C_1 |h|^2 + C_2 |\Delta| + C_3 |\varepsilon|, \\ |f(h, \varphi, \Delta, \varepsilon)| &\leqslant B_1 |h| + B_2 |\Delta| + B_3 |\varepsilon|, \end{aligned} \tag{6.26}$$

are satisfied on this domain, where C_i, B_i $(i = 1, 2, 3)$ are constants not depending on h, Δ, ε.

The system (6.23) is a system of the type (5.36) and hence we can apply to it the results obtained in the previous section, which permit to choose Δ such that equations (6.23) possess a quasi-periodic solution with the properties stated. This also completes the proof of the theorem†.

†See Appendix 4.

Chapter 2

GENERAL SOLUTIONS OF NONLINEAR DIFFERENTIAL EQUATIONS IN THE NEIGHBOURHOOD OF QUASI-PERIODIC SOLUTIONS

§ 7. Statement of the Problem

In solving various problems in nonlinear mechanics, a system of differential equations containing a small parameter is encountered:

$$\frac{dx}{dt} = X(t, x, \varepsilon). \tag{7.1}$$

As already indicated, in a number of cases it is convenient to substitute for the variables $x = (x_1, \ldots, x_{n+m})$ the variables $h = (h_1, \ldots, h_n)$ and $\varphi = (\varphi_1, \ldots, \varphi_m)$ so that the system (7.1) reduces to the form

$$\begin{aligned} \frac{dh}{dt} &= Hh + F(h, \varphi), \\ \frac{d\varphi}{dt} &= \nu + f(h, \varphi). \end{aligned} \tag{7.2}$$

As remarked in the preceding chapter, it is possible, under specific restrictions on the right-hand side of the system (7.2), to find a family of quasi-periodic solutions

$$h = S^{(\infty)}(\theta + \omega t), \quad \varphi = \theta + \omega t + \Phi^{(\infty)}(S(\theta + \omega t), \ \theta + \omega t, 0), \tag{7.3}$$

in whose neighbourhood the system of eqns. (7.2) assumes the form

$$\begin{aligned} \frac{dh}{dt} &= Hh + F(h, \varphi), \\ \frac{d\varphi}{dt} &= \omega. \end{aligned} \tag{7.4}$$

Furthermore, it can be shown that for the system (7.4), subject to a number of restrictions, there exists an invariant manifold $S(\varphi)$ that attracts every solution of the system (7.4) whose initial values lie in some neighbourhood of $S(\varphi)$.

For ease of calculation, we start with the system of equations (7.4) assuming that H is a constant matrix, whose eigenvalues have: (i) negative real parts, $F(h, \varphi) = (F_1(h, \varphi), F_2(h, \varphi), \ldots, F_n(h, \varphi))$, which are analytic vector functions of complex arguments $h = (h_1, \ldots, h_n)$, $\varphi = (\varphi_1, \ldots, \varphi_m)$ in the sufficiently small domain

$$|h| \leqslant \eta, \qquad |\operatorname{Im} \varphi| \leqslant \rho, \tag{7.5}$$

and 2π-periodic in the angular variable φ, and (ii) natural frequencies $\omega = (\omega_1, \ldots, \omega_m)$ satisfying the condition

$$|(k, \omega)| \geqslant K |k|^{-(m+1)}, \tag{7.6}$$

where, just as in the preceding chapter, m is the dimension of the space ω, $(k, \omega) = k_1\omega_1 + k_2\omega_2 + \ldots + k_m\omega_m$, $|k| = |k_1| + |k_2| + \ldots + |k_m|$, $k_1, k_2, \ldots, k_m$ are any integers (positive or negative).

For further simplification, we consider the case in which the square $n \times n$ matrix H in the system of differential equations (7.4) degenerates to a vector $\beta = (\beta_1, \beta_2, \ldots, \beta_n)$ and assume that all the β's have negative real parts. Then (7.4) takes the form

$$\frac{dh}{dt} = \beta h + F(h, \varphi), \qquad \frac{d\varphi}{dt} = \omega. \tag{7.7}$$

In constructing a solution of the system (7.7) we may make use of the successive substitution method described in Chap. 1. In dealing with (7.7) we have to let β vary accordingly. In the limit this tends to a definite value, which we denote by $\alpha = (\alpha_1, \alpha_2, \ldots, \alpha_n)$, the 'faithful' coefficient of a linear system of differential equations obtained after the transformation of (7.7). Exactly as in §1, it is obviously expedient to take α with respect to F and not β and, assuming this to be given, to determine some correction $\xi = \beta - \alpha$, $(\xi = (\xi_1, \xi_2, \ldots \ldots, \xi_n))$ as a function of α:

$$\xi = \xi(\alpha). \tag{7.8}$$

Thus, on introducing the correction ξ into (7.7) we see that this system assumes the form

$$\frac{dh}{dt} = (\alpha + \xi) h + F(h, \varphi, \xi), \qquad \frac{d\varphi}{dt} = \omega, \tag{7.9}$$

where $F(h, \varphi, \xi)$ is an analytic function of the complex variables h, φ, ξ in the domain

$$|h| \leqslant \eta, \quad |\operatorname{Im} \varphi| \leqslant \rho, \quad |\xi| \leqslant \sigma. \tag{7.10}$$

The problem posed is : to determine the transformation

$$h = g + U^{(\infty)}(g, \varphi), \tag{7.11}$$

analytic in g and φ and such that $\xi = \Xi^{(\infty)}$, for which (7.9) reduces to a system of linear differential equations with constant coefficients

$$\frac{dg}{dt} = \alpha g, \qquad \frac{d\varphi}{dt} = \omega. \tag{7.12}$$

Then, integrating the system (7.12), a general solution of (7.9) is given by

$$h = Ce^{\alpha t} + U^{(\infty)}(Ce^{\alpha t}, \omega t + \theta_0), \qquad \varphi = \omega t + \theta_0, \tag{7.13}$$

which contains $n + m$ arbitrary constants

$$C = (C_1, C_2, \ldots, C_n), \quad \theta_0 = (\theta_{01}, \theta_{02}, \ldots, \theta_{0m}).$$

This solution, in conjunction with the quasi-periodic solutions (7.3), yields a general solution of system (7.2) in the neighbourhood of a quasi-periodic solution.

It is noteworthy that by imposing a number of restrictions on α and using a well-known theorem of Poincaré we can always find, for a system of equations of type (7.9), but only for a single φ, with $F(h, \varphi, \xi)$ analytic and sufficiently small, an integral of the type

$$h = f(\omega t + \theta_0, Ce^{\alpha t}), \tag{7.14}$$

where the functions $f(\omega t + \theta_0, h)$ are analytic and regular for sufficiently small values of h, ω and θ_0 being scalars. However, even in this special case (in determining the solutions in the neighbourhood of nonperiodic solutions), the construction of the integral (7.14), as a rule, reduces to developing a power series in small parameters and is by no means always effective because of their slow rate of convergence.

An appeal to the accelerated convergence method is, therefore, made here to permit the construction of the general solutions of the system (7.9) or (7.1) in the neighbourhood of quasi-periodic solutions [$\varphi = (\varphi_1, \varphi_2, \ldots, \varphi_m)$]†

†It may be remarked that the possibility of solving a similar problem is indicated in Kolmogorov's paper [28] and some results bordering on those set forth in this paper are given in [6].

§ 8. Some Auxiliary Statements

For obtaining the estimates needed in the sequel, we consider the system of equations

$$\left(\frac{\partial u}{\partial \varphi}, \omega\right) + \frac{\partial u}{\partial h}(\alpha h + F(h, \varphi, \xi)) = \alpha u + F^{(1)}(h, \varphi, \xi), \tag{8.1}$$

using the notation

$$F^{(1)} = (F_1^{(1)}, F_2^{(1)}, \ldots, F_n^{(1)}),$$

$$F_{k_0}^{(1)}(h, \varphi, \xi) = F_{k_0}(h, \varphi, \xi) - \sum_{s=1}^{n} \overline{\frac{\partial F_{k_0}(h, \varphi, \xi)}{dh_s}}\Bigg|_{h=h_0} h_s \quad (k_0 = 1, 2, \ldots, n) \tag{8.2}$$

Evidently, $F^{(1)}(h, \varphi, \xi)$ does not contain first order terms in $h = (h_1, \ldots, h_n)$ with constant coefficients.

Concerning the solutions of (8.1), we have the following theorem.

Theorem 7. *Suppose that $F(h, \varphi, \xi)$ is an analytic function of complex variables h, φ, ξ on the domain*

$$h\,| \leqslant \eta, \quad |\operatorname{Im} \varphi| \leqslant \rho, \quad |\xi| < \sigma, \tag{8.3}$$

and satisfies, on this domain, the inequalities

$$|F(h, \varphi, \xi)| \leqslant N, \quad n\left|\frac{\partial F(h, \varphi, \xi)}{\partial h_q}\right| \leqslant L \quad (q = 1, \ldots, n), \tag{8.4}$$

where η, ρ, N, L are constants satisfying the conditions

$$L \leqslant \frac{\bar{\alpha}}{2}, \quad N < \bar{\alpha}(\eta - \varkappa), \quad 0 \leqslant 2\varkappa < \eta, \quad \bar{\alpha} = \min_{1 \leqslant k_0 \leqslant n} |\operatorname{Re} \alpha_{k_0}|, \quad \varkappa < 1. \tag{8.5}$$

In addition, suppose that α_{k_0} $(k_0 = 1, 2, \ldots, n)$ each has a negative and a distinct real part $|\operatorname{Re} \alpha_{k_0}| = \bar{\alpha}_{k_0}$, satisfying the conditions

$$\sum_{k_0=1}^{n} l_{k_0} \bar{\alpha}_{k_0} - \bar{\alpha}_q \geqslant \gamma > 0 \tag{8.6}$$

for any $q = 1, 2, \ldots, n$ and all nonnegative integers $l_1, \ldots, l_n$ for which $l_1 +, \ldots, + l_n \geqslant 2$†.

†We observe that contrary to the Liouville condition (7.6), the conditions (8.6) have no arithemetical character. The contrast between the two is well marked in that (8.6) are continuous. i.e. values close to α_{k_0} also satisfy the conditions (8.6).

Then the system of equations (8.1) *has a solution*

$$u(h, \varphi, \xi) = \sum_k \frac{F_k^{(1)}(\xi)\, e^{i(k,\varphi)}}{i(k, \omega) - \alpha} + \int_{-\infty}^{0} e^{-\alpha\tau} \times$$

$$\times [F^{(1)}(h_\tau, \varphi_\tau, \xi) - F^{(1)}(S(\varphi_\tau), \varphi_\tau, \xi)]\,(d\tau) \qquad (8.7)$$

2π-*periodic in* φ, *where* $\varphi_\tau = \varphi + \omega\tau$, h_τ *is a solution of the equation*

$$\frac{dh}{d\tau} = \alpha h + F(h, \varphi_\tau, \xi) \qquad (8.8)$$

with initial condition $h_\tau|_{\tau=0} = h$, *and* $S(\varphi)$ *is an invariant manifold of the system of eqns.* (7.7) :

$$F_k^{(1)}(\xi) = \frac{1}{(2\pi)^m} \int_0^{2\pi} \dots \int_0^{2\pi} F^{(1)}(S(\varphi), \varphi, \xi) e^{-i(k,\varphi)}\, d\varphi_1 \dots d\varphi_m. \qquad (8.9)$$

This solution is analytic on the domain

$$|h| \leqslant \eta - 2\varkappa, \quad |\operatorname{Im} \varphi| \leqslant \rho - 2\varkappa, \quad |\xi| \leqslant \sigma, \qquad (8.10)$$

and satisfies the inequalities

$$|u(h, \varphi, \xi)| \leqslant N \left\{ \frac{8\eta n}{3\alpha\varkappa} + \frac{(\varkappa + \eta)(1 + e)^m}{\alpha\varkappa^{m+1}} \right\},$$

$$\sum_{q=1}^{n} \left| \frac{\partial u(h, \varphi, \xi)}{\partial h_q} \right| \leqslant \frac{8Nn^2\eta}{3\alpha\varkappa^2}. \qquad (8.11)$$

Proof. To show first that (8.7) satisfies eqns. (8.1), we express the function $F^{(1)}(S(\varphi), \varphi, \xi)$ in the form of the series

$$F^{(1)}(S(\varphi), \varphi, \xi) = \sum_{(k)} F_k^{(1)}(\xi)\, e^{i(k,\varphi)}. \qquad (8.12)$$

Introduce the notation

$$\tilde{F}^{(1)}(S(\varphi), \varphi, \xi) = \sum_{(k)} \frac{F_k^{(1)}(\xi)\, e^{i(k,\varphi)}}{i(k, \omega) - \alpha}. \qquad (8.13)$$

Then,

$$\frac{d\tilde{F}^{(1)}(S(\varphi), \varphi, \xi)}{dt} - \alpha\tilde{F}^{(1)}(S(\varphi), \varphi, \xi) = F^{(1)}(S(\varphi), \varphi, \xi). \qquad (8.14)$$

The expression (8.7) is representable as the sum

$$u(h, \varphi, \xi) = u_1(\varphi, \xi) + u_2(h, \varphi, \xi), \tag{8.15}$$

which admits, by (8.7) and (8.13), the notations

$$u_1(\varphi, \xi) = F^{(1)}(S(\varphi), \varphi, \xi), \tag{8.16}$$

$$u_2(h, \varphi, \xi) = \int_{-\infty}^{0} e^{-\alpha\tau}\,[F^{(1)}(h_\tau, \varphi_\tau, \xi) - F^{(1)}(S(\varphi_\tau), \varphi_\tau, \xi)]\,d\tau. \tag{8.17}$$

Differentiating (8.16), we obtain

$$\frac{\partial u_1}{\partial h} = 0, \quad \left(\frac{\partial u_1}{\partial \varphi}, \omega\right) = \frac{du_1}{dt} = \alpha u_1 + F^{(1)}(S(\varphi), \varphi, \xi). \tag{8.18}$$

Hence, it remains to show that the function $u_2(h, \varphi, \xi)$, defined by (8.17), identically satisfies the equation

$$\left(\frac{\partial u_2}{\partial \varphi}, \omega\right) + \frac{\partial u_2}{\partial h}(\alpha h + F(h, \varphi, \xi)) = \alpha u_2 + F^{(1)}(h, \varphi, \xi) - F^{(1)}(S(\varphi), \varphi, \xi). \tag{8.19}$$

For this let $\varphi_t = \varphi + \omega t$ and h_t satisfy

$$\frac{dh_t}{dt} = \alpha h_t + F(h_t, \varphi_t, \xi)$$

with the initial condition that $h_t = h$ when $t = 0$. The substitution $h = h_t$, $\varphi = \varphi_t$, reduces (8.19) to

$$\frac{du_2(h_t, \varphi_t, \xi)}{dt} = \alpha u_2(h_t, \varphi_t, \xi) + F^{(1)}(h_t, \varphi_t, \xi) - F^{(1)}(S(\varphi_t), \varphi_t, \xi). \tag{8.20}$$

On the other hand, the replacement of h by h_t and φ by φ_t in (8.17), leads to

$$u_2(h_t, \varphi_t, \xi) = \int_{-\infty}^{0} e^{-\alpha t}\,[F^{(1)}(h_{t+\tau}, \varphi_{t+\tau}, \xi) - F^{(1)}(S(\varphi_{t+\tau}), \varphi_{t+\tau}, \xi)]\,d\tau. \tag{8.21}$$

Now, set $t + \tau = z$ in the integrand in (8.21), to obtain

$$u_2(h_t, \varphi_t, \xi) = \int_{-\infty}^{0} e^{\alpha(t-z)}\,[F^{(1)}(h_z, \varphi_z, \xi) - F^{(1)}(S(\varphi_z), \varphi_z, \xi)]\,dz. \tag{8.22}$$

Evidently, (8.22) satisfies eqn. (8.20) and, consequently, (8.17) satisfies eqn. (8.19).

Thus it is clear that the expression (8.15), where $u_1(h, \varphi, \xi)$ is defined by (8.16) and $u_2(h, \varphi, \xi)$ by (8.17), satisfies eqn. (8.1).

The stage is now set to establish the estimate (8.11). Consider the domain

$$|h| \leqslant \eta - \varkappa, \quad |\operatorname{Im} \varphi| \leqslant \rho, \quad |\xi| \leqslant \sigma. \tag{8.23}$$

where, as a result of the inequality (8.5),

$$|h_t| \leqslant \eta - \varkappa, \quad |S(\varphi_t)| \leqslant \eta - \varkappa. \tag{8.24}$$

Recall that if $f(h)$ is an analytic function of h on the domain $\Sigma_\eta(|h| \leqslant \eta)$, where

$$|f(h)| \leqslant N \quad \text{for} \; |h| \leqslant \eta,$$

then, considering the Cauchy integral

$$f(h) = \frac{1}{2\pi i} \oint \frac{f(\xi)\, d\xi}{\xi - h},$$

we deduce that

$$\left| \frac{\partial f(h)}{\partial h_q} \right| \leqslant \frac{N}{\eta - \eta_1} \qquad \text{for} \; |h| \leqslant \eta_1 < \eta. \tag{8.25}$$

Hence, as a result of the inequalities (8.4) and (8.25) we deduce for the function $F^{(1)}(h, \varphi, \xi)$ on the domain (8.23), the inequality

$$\begin{gathered} \left| \frac{\partial F^{(1)}(h, \varphi, \xi)}{\partial h_q} \right| \leqslant \left| \frac{\partial F(h, \varphi, \xi)}{\partial h_q} \right| + \left| \overline{\frac{\partial F(h, \varphi, \xi)}{\partial h_q}} \bigg|_{h=0} \right| \leqslant \\ \leqslant \frac{N}{\varkappa} + \frac{N}{\varkappa} = \frac{2N}{\varkappa}, \end{gathered} \tag{8.26}$$

which yields further the inequality

$$|F^{(1)}(h_\tau, \varphi_\tau, \xi) - F^{(1)}(S(\varphi_\tau), \varphi_\tau, \xi)| \leqslant \frac{2Nn}{\varkappa} |h_\tau - S(\varphi_\tau)|, \tag{8.27}$$

or, taking account of the inequality (1.92),

$$|F^{(1)}(h_\tau, \varphi_\tau, \xi) - F^{(1)}(S(\varphi_\tau), \varphi_\tau, \xi)| \leqslant \frac{4Nn\eta}{\varkappa} e^{-\frac{\bar{\alpha}}{2} t}, \qquad t \geqslant 0. \tag{8.28}$$

This yields for $u_2(h, \varphi, \xi)$ on the domain (8.23) the estimate

$$|u_2(h, \varphi, \xi)| \leqslant \frac{4Nn\eta}{\varkappa} \int\limits_{-\infty}^{0} e^{-\alpha\tau} e^{-\frac{\bar{\alpha}}{2}\tau} d\tau \leqslant \frac{8Nn\eta}{3\bar{\alpha}\varkappa}, \tag{8.29}$$

or,

$$\left| \frac{\partial u_2 (h, \varphi, \xi)}{\partial h_q} \right| \leqslant \frac{8 N n \eta}{3 \bar{\alpha} \varkappa^2}, \tag{8.30}$$

on the smaller domain (8.10).

An estimate is now sought for $u_1 (\varphi, \xi)$. With the notation of (8.16) and (8.13) we have on the domain $| \operatorname{Im} \varphi | \leqslant \rho - 2\varkappa$,

$$| u_1 (\varphi, \xi) | = \sum_{(k)} \left| \frac{F_k^{(1)} (\xi) \, e^{i(k, \varphi)}}{i (k, \omega) - \alpha} \right| \leqslant \frac{1}{\bar{\alpha}} \sum_{(k)} | F_k^{(1)} (\xi) | \, e^{(\rho - 2\varkappa) | k |}, \tag{8.31}$$

where

$$F_k^{(1)} (\xi) = \frac{1}{(2\pi)^m} \int_0^{2\pi} \dots \int_0^{2\pi} F^{(1)} (S (\theta), \theta, \xi) \, e^{-i(k, \theta)} d\theta_1 \dots d\theta_m =$$

$$= \frac{1}{(2\pi)^m} \int_0^{2\pi} \dots \int_0^{2\pi} F^{(1)} (\theta + i\varphi) \, e^{-i(k, \theta)} \, e^{(k, \varphi)} \, d\theta_1 \dots d\theta_m. \tag{8.32}$$

This admits the notation

$$F^{(1)} (\theta + i\varphi) = F^{(1)} (S (\theta + i\varphi), \theta + i\varphi, \xi).$$

Set $\varphi_k = -\rho \operatorname{sign} k$ in (8.32) ; then $(k, \varphi) = -\rho | k |$ and we can deduce from (8.32) that

$$| F_k^{(1)} (\xi) | \leqslant N \left(1 + \frac{\eta}{\varkappa} \right) e^{-\rho | k |}. \tag{8.33}$$

Furthermore,

$$\sum_{(k)} e^{-\rho | k |} = \left(1 + 2 \sum_{1 \leqslant q < \infty} e^{-q\rho} \right)^m = \frac{(1 + e^{-\rho})^m}{(1 - e^{-\rho})^m};$$

however, $1 - e^{-\rho} > \rho e^{-\rho}$ and, consequently,

$$\frac{1 + e^{-\rho}}{1 - e^{-\rho}} \leqslant \frac{1}{\rho} (1 + e^{\rho}) < \frac{1 + e}{\rho}.$$

Thus, finally, the estimate for $u_1 (\varphi, \xi)$ on the domain (8.10) is given by

$$| u_1 (\varphi, \xi) | \leqslant \frac{N \left(1 + \frac{\eta}{\varkappa} \right) (1 + e)^m}{\bar{\alpha} \varkappa^m} = \frac{N (\varkappa + \eta) (1 + e)^m}{\bar{\alpha} \varkappa^{m+1}}. \tag{8.34}$$

This estimate combined with (8.29) and (8.30) also gives the desired estimate (8.11), which completes the proof.

§ 9. Inductive Theorem

The auxiliary theorem that follows is of basic importance in constructing the transformation which reduces the system of equations (7.9) to the integral system (7.12).

Theorem 8. *For the system of eqns.* (7.9),

$$\frac{dh}{dt} = (\alpha + \xi)\, h + F(h, \varphi, \xi),$$
$$\frac{d\varphi}{dt} = \omega, \tag{9.1}$$

where, as before, h, α, ξ, are n-dimensional vectors, suppose that $F(h, \varphi, \xi)$ is an n-dimensional analytic vector function of complex variables, h, φ, ξ, which is sufficiently small for sufficiently small h, Im *φ, ξ and satisfies on the domain*

$$|h| \leqslant \eta, \quad |\operatorname{Im} \varphi| \leqslant \rho, \quad |\xi| \leqslant \sigma \tag{9.2}$$

the inequality

$$|F(h, \varphi, \xi)| \leqslant N, \quad n \left| \frac{\partial F(h, \varphi, \xi)}{\partial n_q} \right| \leqslant L; \tag{9.3}$$

here the constants N, L, η, σ are related by the inequalities

$$\frac{N}{\varkappa} \leqslant \frac{\sigma}{4n}, \quad \frac{nN}{\varkappa} \leqslant L \leqslant \frac{\bar{\alpha}}{2}, \quad \frac{N}{\bar{\alpha}} \leqslant \eta - \varkappa, \tag{9.4}$$

$$\frac{NQ(\eta, \varkappa)}{\bar{\alpha}\varkappa^{2n+4}} \leqslant 1, \quad 0 \leqslant 2\varkappa \leqslant \eta, \quad \varkappa < 1; \tag{9.5}$$

$$Q(\eta, \varkappa) = \frac{16n^2\eta\varkappa^{2n+1}}{3}(\eta + \varkappa) + \left[\frac{8\eta n}{3} + \frac{(\eta + \varkappa)(1 + e)^m}{\varkappa^m}\right] \times$$
$$\times \left(4 + \frac{n}{\left(1 - \frac{1}{2n}\right)}\right) \varkappa^{2n+2}. \tag{9.6}$$

In addition, assume that α has a negative real part and satisfies the inequality (8.6), *and that $\bar{\alpha}$ is a positive constant satisfying the condition*

$$|e^{\alpha t}| \leqslant e^{-\bar{\alpha} t}.$$

Then, we get the analytic transformation

$$h = h^{(1)} + u\,(h^{(1)}, \varphi, \xi^{(1)}),$$
$$\xi = \xi(\xi^{(1)}), \tag{9.7}$$

which reduces the system of eqns. (9.1) *to the form*

$$\frac{dh^{(1)}}{dt} = (\alpha + \xi^{(1)})\, h^{(1)} + F_1\,(h^{(1)}, \varphi, \xi^{(1)}),$$
$$\frac{d\varphi}{dt} = \omega\,; \tag{9.8}$$

here $\xi(\xi^{(1)})$ *and* $F_1\,(h^{(1)}, \varphi, \xi^{(1)})$ *are analytic functions of* $h^{(1)}, \varphi, \xi^{(1)}$ *satisfying, on the domain*

$$|\,h\,| \leqslant \eta - 2\varkappa, \quad |\,\mathrm{Im}\,\varphi\,| \leqslant \rho, \quad |\,\xi^{(1)}\,| \leqslant \frac{N}{\varkappa}, \tag{9.9}$$

the inequalities

$$|\,\xi(\xi^{(1)}) - \xi^{(1)}\,| \leqslant \frac{N}{\varkappa}, \quad \sum_{q=1}^{n} \left| \frac{\partial \xi(\xi^{(1)})}{\partial \xi_q^{(1)}} \right| \leqslant 1 + \frac{4N}{\varkappa \sigma_2} \leqslant 2,$$

$$|\,u\,(h^{(1)}, \varphi, \xi^{(1)})\,| \leqslant \frac{Na\,(\eta, \varkappa)}{\bar{\alpha} \varkappa^{2n+1}} \leqslant \frac{\varkappa}{n},$$

$$\sum_{q=1}^{n} \left| \frac{\partial u(h^{(1)}, \varphi, \xi^{(1)})}{\partial h_q^{(1)}} \right| \leqslant \frac{NQ\,(\eta, \varkappa)}{2\bar{\alpha} \varkappa^{2n+1}} \leqslant \frac{1}{2}, \tag{9.10}$$

$$\sum_{q=1}^{n} \left| \frac{\partial u\,(h^{(1)}, \varphi, \xi^{(1)})}{\partial \xi_q^{(1)}} \right| \leqslant \frac{Nna\,(\eta, \varkappa)}{\bar{\alpha} \sigma_2 \varkappa^{2n+1} \left(1 - \frac{1}{2n} \right)},$$

$$|\,F_1\,(h^{(1)}, \varphi, \xi^{(1)})\,| \leqslant N_1, \quad n \left| \frac{\partial F_1\,(h^{(1)}, \varphi, \xi^{(1)})}{\partial h_q^{(1)}} \right| \leqslant L_1,$$

where

$$N_1 = \frac{N^2 Q\,(\eta, \varkappa)}{\bar{\alpha} \varkappa^{2n+4}}, \qquad L_1 = \frac{nNQ\,(\eta, \varkappa)}{\bar{\alpha} \varkappa^{2n+4} \varkappa} \leqslant \frac{\bar{\alpha}}{2}, \tag{9.11}$$

$$a\,(\eta, \varkappa) = \frac{8\eta n}{3}\, \varkappa^{2n} + \frac{(\eta + \varkappa)\,(1 + e)^m}{\varkappa^m}\, \varkappa^{2n}. \tag{9.12}$$

Proof. The objective is first to reduce the system (9.1) to a form where the vector function $F\,(h, \varphi, \xi)$ is of a small magnitude of second order. For this, choose the transformation $h \to h^{(1)}$ in the form

$$h = h^{(1)} + u\,(h^{(1)}, \varphi, \xi), \tag{9.13}$$

where $u\,(h^{(1)}, \varphi, \xi)$ is the function defined by (8.7). Substitute (9.13) into (9.1), to obtain

$$\left(\frac{\partial u}{\partial \varphi}, \omega\right) + \left(E + \frac{\partial u}{\partial h}\right)\frac{dh^{(1)}}{dt} = (\alpha + \xi)\,(h^{(1)} + u) + F(h^{(1)} + u, \varphi, \xi). \tag{9.14}$$

Using the fact that $u\,(h^{(1)}, \varphi, \xi)$ is a solution of the system (9.1), equation (9.14) implies that

$$\left(E + \frac{\partial u}{\partial h^{(1)}}\right)\left(\frac{dh^{(1)}}{dt} - \alpha h^{(1)}\right) = \xi^{(1)}h^{(1)} + \xi u + F\,(h^{(1)} + u, \varphi, \xi)$$
$$- F\,(h^{(1)}, \varphi, \xi) + \frac{\partial u}{\partial h^{(1)}}\,F\,(h^{(1)}, \varphi, \xi). \tag{9.15}$$

Set

$$\xi^{(1)} = \{\xi^{(1)}_{k_0}\} = \left\{\xi_{k_0} + \sum_{s=1}^{n} \frac{\partial F_{k_0}\,(h, \varphi, \xi)}{\partial h_s}\bigg|_{h=0}\right\} \quad (k_0 = 1, 2, \ldots, n). \tag{9.16}$$

Solving the system (9.15) in $(dh^{(1)}/dt) - \alpha h^{(1)}$, we obtain

$$\frac{dh^{(1)}}{dt} = (\alpha + \xi^{(1)})\,h^{(1)} + F_1\,(h^{(1)}, \varphi, \xi^{(1)}), \tag{9.17}$$

where

$$F_1\,(h^{(1)}, \varphi, \xi^{(1)}) = \left[\left(E + \frac{\partial u}{\partial h^{(1)}}\right)^{-1} - E\right]\xi^{(1)}h^{(1)} + \left(E + \frac{\partial u}{\partial h^{(1)}}\right)^{-1} \times$$
$$\times \left[\xi u + F\,(h^{(1)} + u, \varphi, \xi) - F\,(h^{(1)}, \varphi, \xi) + \frac{\partial u}{\partial h^{(1)}}\,F\,(h^{(1)}, \varphi, \xi)\right]. \tag{9.18}$$

Before evaluating (9.18), we consider the dependence of ξ on $\xi^{(1)}$, defined by (9.16). We introduce the notation

$$b\,(\xi) = \{b_{k_0}\,(\xi)\} = \left\{\sum_{s=1}^{n} \frac{\partial F_{k_0}\,(h, \varphi, \xi)}{\partial h_s}\bigg|_{h=0}\right\} \quad (k_0 = 1, \ldots, n).$$

By hypothesis $b\,(\xi)$ is an analytic function of ξ for $|\,\xi\,| \leqslant \sigma$, and satisfies the inequality

$$|\,b\,(\xi)\,| \leqslant \frac{N}{\varkappa}, \tag{9.19}$$

since, because of (8.4) and (8.25), $|\partial F/\partial h_q| \leqslant N/\varkappa$ for $|h| \leqslant \eta - \varkappa$. Hence, the inequality

$$\sum_{q=1}^{n} \left| \frac{\partial b(\xi)}{\partial \xi_q} \right| \leqslant \frac{2Nn}{\sigma\varkappa} \leqslant \frac{1}{2} \quad \text{for } |\xi| \leqslant \frac{\sigma}{2}$$

holds for $\dfrac{\partial b(\xi)}{\partial \xi_q}$ and, consequently, the equation

$$\xi + b(\xi) = \xi^{(1)} \tag{9.20}$$

is solvable in ξ and $\xi(\xi^{(1)})$ is an analytic function of $\xi^{(1)}$ for

$$|\xi^{(1)}| < \frac{N}{\varkappa}. \tag{9.20'}$$

Furthermore, in view of the relations (9.19) and (9.20), the inequalities

$$|\xi(\xi^{(1)})| \leqslant \frac{2N}{\varkappa} \leqslant \frac{\sigma}{2n} \leqslant \frac{\sigma}{2}, \qquad |\xi(\xi^{(1)}) - \xi^{(1)}| \leqslant \frac{N}{\varkappa},$$

$$\sum_{q=1}^{n} \left| \frac{\partial \xi(\xi^{(1)})}{\partial \xi_q^{(1)}} \right| \leqslant 1 + \frac{4Nn}{\sigma\varkappa} \leqslant 2; \tag{9.21}$$

$$|\xi(\xi^{(1)}) - \xi^{(1)}| \leqslant \frac{N}{\varkappa} \tag{9.22}$$

remain valid for (9.20′).

It is now intended to evaluate the function $u(h^{(1)}, \varphi, \xi^{(1)}) = u(h^{(1)}, \varphi, \xi(\xi^{(1)}))$ appearing in the change-of-variable formula (9.13).

By Theorem 7 and the notations defined by (9.12) and (9.6) we have, on the domain

$$|h^{(1)}| \leqslant \eta - 2\varkappa, \quad |\operatorname{Im} \varphi| \leqslant \rho - 2\varkappa, \quad |\xi| \leqslant \sigma, \tag{9.23}$$

the inequalities

$$|u(h^{(1)}, \varphi, \xi^{(1)})| \leqslant \frac{Na(\eta, \varkappa)}{\bar{\alpha}\varkappa^{2n+1}} \leqslant \frac{\varkappa}{n} \cdot \frac{NQ(\eta, \varkappa)}{\bar{\alpha}\varkappa^{2(n+2)}} \leqslant \frac{\varkappa}{n};$$

$$\sum_{(\alpha, q)} \left| \frac{\partial u_\alpha(h^{(1)}, \varphi, \xi^{(1)})}{\partial h_q^{(1)}} \right| \leqslant \frac{8Nn^2\eta}{3\bar{\alpha}\varkappa^2} < \frac{NQ(\eta, \varkappa)}{2\bar{\alpha}\varkappa^{2n+4}} \leqslant \frac{1}{2}, \tag{9.24}$$

and also

$$\left| \frac{\partial u(h^{(1)}, \varphi, \xi)}{\partial \xi_q} \right| \leqslant \frac{Na(\eta, \varkappa)}{\bar{\alpha}\varkappa^{2n+1}\sigma_2\left(1 - \dfrac{1}{2n}\right)} \qquad (q = 1, 2, \ldots, n) \tag{9.25}$$

for

$$|h^{(1)}| \leqslant n - 2\varkappa, \quad |\operatorname{Im} \varphi| \leqslant \rho - 2\varkappa, \quad |\xi| \leqslant \frac{\sigma}{2n}. \tag{9.26}$$

Now, consider the function $F_1(h^{(1)}, \varphi, \xi^{(1)})$. This is defined by (9.18) and is analytic on the domain (9.9).

To evaluate this function, we take account of the inequalities (9.23), to obtain

$$\left|\left(E + \frac{\partial u}{\partial h^{(1)}}\right)^{-1} - E\right| \leqslant 2\sum_q \left|\frac{\partial u_2}{\partial h_q^{(1)}}\right| \leqslant \frac{16Nn^2\eta}{3\bar{\alpha}\varkappa^2},$$

$$\left|\left(E + \frac{\partial u}{\partial h^{(1)}}\right)^{-1}\right| \leqslant 2. \tag{9.27}$$

For $h^{(1)}, \varphi, \xi^{(1)}$ in the domain (9.24), we have the inequalities

$$|F(h^{(1)} + u, \varphi, \xi^{(1)}) - F(h^{(1)}, \varphi, \xi^{(1)}) \leqslant \sum_{q=1}^{n} \left|\frac{\partial F(\tilde{h}, \varphi, \xi)}{\partial h_q}\right|,$$

$$|u_q| \leqslant \sum_{q=1}^{n} \left|\frac{\partial F(\tilde{h}, \varphi, \xi)}{\partial h_q}\right| \frac{Na(\eta, \varkappa)}{\bar{\alpha}\varkappa^{2n+1}},$$

where $\tilde{h}$ is some value of h in the interval $(h^{(1)}, h^{(1)} + u)$.

In the domain (9.24), we have

$$|\tilde{h}| \leqslant |h^{(1)}| + |u| \leqslant \eta - 2\varkappa + \frac{\varkappa}{n},$$

so that

$$\left|\frac{\partial F(\tilde{h}, \varphi, \xi)}{\partial h_q}\right| \leqslant \frac{N}{2\varkappa\left(1 - \frac{1}{2n}\right)} = \frac{Nn}{\varkappa(2n-1)},$$

and, consequently, we finally get

$$|F(h^{(1)} + u, \varphi, \xi^{(1)}) - F(h^{(1)}, \varphi, \xi^{(1)})| \leqslant \frac{N^2n^2a(\eta, \varkappa)}{\varkappa(2n-1)\bar{\alpha}\varkappa^{2n+1}}. \tag{9.28}$$

The inequalities (9.27) and (9.28) yield the required estimate

$$|F^{(1)}(h^{(1)}, \varphi, \xi^{(1)})| \leqslant \frac{16Nn^2\eta}{3\bar{\alpha}\varkappa^2} \cdot \frac{N}{\varkappa}\eta + 2\left[\frac{2N}{\varkappa} \cdot \frac{Na(\eta, \varkappa)}{\bar{\alpha}\varkappa^{2n+1}} + \right.$$

$$\left. + \frac{N^2n^2a(\eta, \varkappa)}{\varkappa(2n-1)\bar{\alpha}\varkappa^{2n+1}} + \frac{8Nn^2\varkappa}{3\bar{\alpha}\varkappa^2} N\right] =$$

$$= \frac{N^2Q(\eta, \varkappa)}{\bar{\alpha}\varkappa^{2n+4}}.$$

This completes the proof of Theorem 8.

§ 10. Iteration Process and its Convergence

The inductive Theorem 8 proved in the preceding section is applied here to construct an iteration process with accelerated convergence.

With this aim, in eqns. (9.1), we set

$$\begin{aligned} h &= h^{(1)} + u^{(1)}(h^{(1)}, \varphi, \xi^{(1)}), \\ \xi &= \xi(\xi^{(1)}), \end{aligned} \tag{10.1}$$

where $u^{(1)}(h^{(1)}, \varphi, \xi^{(1)})$, $\xi(\xi^{(1)})$ are defined by equations (8.7) and (9.20).

Then, eqns. (9.1) take the form

$$\begin{aligned} \frac{dh^{(1)}}{dt} &= (\alpha + \xi^{(1)})\, h^{(1)} + F^{(1)}(h^{(1)}, \varphi, \xi^{(1)}), \\ \frac{d\varphi}{dt} &= \omega. \end{aligned} \tag{10.2}$$

To these equations we apply again a transformation of the form (10.1), this time starting from the right-hand side of the system (10.2) :

$$\begin{aligned} h^{(1)} &= h^{(2)} + u^{(2)}(h^{(2)}, \varphi, \xi^{(2)}), \\ \xi^{(1)} &= \xi^{(1)}(\xi^{(2)}). \end{aligned} \tag{10.3}$$

This leads to the system

$$\begin{aligned} \frac{dh^{(2)}}{dt} &= (\alpha + \xi^{(2)})\, h^{(2)} + F^{(2)}(h^{(2)}, \varphi, \xi^{(2)}), \\ \frac{d\varphi}{dt} &= \omega, \end{aligned} \tag{10.4}$$

etc.

Continuing this process, at the s-th step the transformation yields

$$\begin{aligned} h^{(s-1)} &= h^{(s)} + u^{(s)}\, h^{(s)}, \varphi, \xi^{(s)}, \\ \xi^{(s-1)} &= \xi^{(s-1)}(\xi^{(s)}), \end{aligned} \tag{10.5}$$

which implies the equations

$$\begin{aligned} \frac{dh^{(s)}}{dt} &= (a + \xi^{(s)})\, h^{(s)} + F^{(s)}(h^{(s)}, \varphi, \xi^{(s)}), \\ \frac{d\varphi}{dt} &= \omega. \end{aligned} \tag{10.6}$$

At every step of the iteration process, we have to verify that the equations satisfy the hypotheses of the inductive Theorem 8. For this, we assume that the system (9.1) does so and, in addition, that the following inequalities are satisfied :

$$\frac{N}{\bar{\alpha}} \leqslant \frac{\eta}{2}, \quad \frac{nN}{\varkappa} \leqslant \frac{\bar{\alpha}}{2}, \quad \frac{NQ(\eta, \varkappa)}{\bar{\alpha}} \left(\frac{1}{\gamma^{2n+4}}\right)^2 \leqslant r_0 < 1, \tag{10.7}$$

with

$$\gamma = \frac{\eta}{\eta + 4}, \quad \frac{N}{\gamma} \leqslant \frac{\sigma}{4n}, \quad \gamma^{2n+3} r_0 \leqslant \frac{1}{4n}, \quad \frac{\gamma^{2n+4}}{2} \sum_{s=0}^{\infty} r_0^{2^s} < \ln \frac{3}{2}.$$

To construct the iteration process at the s-th step, consider the domain

$$|h| \leqslant \eta_s = \eta - 2(\varkappa_0 + \varkappa_1 + \varkappa_2 + \ldots + \varkappa_{s-1}),$$

$$|\operatorname{Im} \varphi| \leqslant \rho_s = \rho - 2(\varkappa_0 + \varkappa_1 + \varkappa_2 + \ldots + \varkappa_{s-1}), \quad |\xi^{(s)}| \leqslant \frac{N_{s-1}}{\varkappa_{s-1}}; \tag{10.8}$$

here $\varkappa_s = \gamma^{s+1}$ and γ is defined by

$$2(\gamma + \gamma^2 + \ldots + \gamma^s + \ldots) = \frac{2\gamma}{1-\gamma} = \frac{\eta}{2}.$$

It is required to show that, on the domain (10.8), the following inequalities hold :

$$|F^{(s)}(h^{(s)}, \varphi, \xi^{(s)})| \leqslant N_s, \quad |\xi^{(v)}(\xi^{(v-1)}(\ldots \xi^{(s)} \ldots))| \leqslant \frac{N_{v-1}}{\varkappa_{v-1}},$$

$$n \left| \frac{\partial F^{(s)}(h^{(s)}, \varphi, \xi^{(s)})}{\partial h_q} \right| \leqslant L_s, \tag{10.9}$$

with

$$N_s = \frac{\bar{\alpha}\, r_0^{2^s} \gamma^{(s+2)(2n+4)}}{Q(\eta, \varkappa)}, \quad L_s = \frac{nN_{s-1}}{\varkappa_{s-1}}. \tag{10.10}$$

In fact, for $s = 1$ these inequalities are satisfied by the inductive theorem and the conditions (10.7).

Assuming this to be satisfied for $1 \leqslant s \leqslant s_0$, we show that it is true for $s = s_0 + 1$.

The substitution of $\eta_{s_0}, \varkappa_{s_0}, N_{s_0}$ for $\eta, \varkappa, N$ in the inequalities (9.4) and (9.5), yields the relations

$$\frac{N_{s_0}}{\varkappa_{s_0}} \leqslant \frac{1}{4n} \frac{N_{s_0-1}}{\varkappa_{s_0-1}}, \quad L_{s_0} = \frac{nN_{s_0-1}}{\varkappa_{s_0-1}} \leqslant \frac{\bar{\alpha}}{2}, \quad \frac{N_{s_0}}{\bar{\alpha}} \leqslant \eta_{s_0} - \varkappa_{s_0},$$

$$\frac{N_{s_0} Q(\eta_{s_0}, \alpha_{s_0})}{\bar{\alpha} \varkappa_{s_0}^{2n+4}} \leqslant 1, \quad 0 \leqslant 2\alpha_{s_0} < \eta_{s_0}, \quad \varkappa_{s_0} < 1. \tag{10.11}$$

We show that these inequalities are indeed satisfied under the condition (10.7) with the notation (10.10).

Obviously,

$$\frac{\dfrac{N_{s_0}}{\varkappa_{s_0}}}{\dfrac{N_{s_0-1}}{\varkappa_{s_0-1}}} = \frac{N_{s_0}}{N_{s_0-1}\,\gamma} = r_0^{2}\gamma^{2n+3} \leqslant \frac{1}{4n}. \tag{10.12}$$

Furthermore, if $\dfrac{nN}{\bar{\alpha}} \leqslant \dfrac{\bar{\alpha}}{2}$, then it also follows that

$$L_{s_0} = \frac{nN_{s_{0-1}}}{\varkappa_{s_{0-1}}} \leqslant \frac{\bar{\alpha}}{2}, \tag{10.13}$$

where $N_{s_{0-1}}$ is defined by (10.10).

Since the inequalities (10.7) and the formula (10.10) imply for $s = 0$ $(N = N_0)$ that

$$\frac{N}{\bar{\alpha}} = \frac{r_0\gamma^{2(2n+4)}}{Q(\eta, \gamma)} \leqslant \frac{\eta}{2} \qquad (\varkappa = \gamma),$$

we have

$$\frac{N_{s_0}}{\bar{\alpha}} = \frac{r_0^{2^{s_0}}\gamma^{(s_0+2)(2n+4)}}{Q(\eta, \gamma)} < \frac{\eta}{2} < \eta_{s_0} - \varkappa_{s_0}, \tag{10.14}$$

because by (10.8) $\eta_{s_0} > \dfrac{\eta}{2} + \varkappa_{s_0}$.

Furthermore,

$$\frac{N_{s_0}\,Q(\eta_{s_0}, \varkappa_{s_0})}{\bar{\alpha}\varkappa^{2n+4}} = \frac{N_{s_0}\,Q(\eta_{s_0}, \gamma^{s_0+1})}{\bar{\alpha}\gamma^{(s_0+1)(2n+4)}} \leqslant \frac{N_{s_0}\,Q(\eta, \gamma)}{\bar{\alpha}\gamma^{(s_0+1)(2n+4)}} =$$

$$= r_0^{2^{s_0}}\gamma^{(2n+4)} < \frac{1}{4n} < 1, \tag{10.15}$$

and, consequently, on the domain

$$|h| \leqslant \eta_{s_0}, \quad |\operatorname{Im}\varphi| \leqslant \rho, \quad |\xi^{(s_0)}| \leqslant \frac{N_{s_0-1}}{\varkappa_{s_0-1}} \tag{10.16}$$

all the inequalities necessary for the inductive theorem hold.

This permits us to use all the estimates of the inductive theorem in the passage from s_0 to $s_0 + 1$. By these estimates, for the step following $s_0 + 1$, on the smaller domain

$$|h| \leqslant \eta_{s_0} - 2\varkappa_{s_0}, \quad |\operatorname{Im}\varphi| \leqslant \rho, \quad |\xi^{(s_0+1)}| \leqslant \frac{N_{s_0}}{\varkappa_{s_0}}, \tag{10.17}$$

the inequalities

$$| F^{(s_0+1)} (h^{(s_0+1)}, \varphi, \xi^{(s_0+1)} | \leqslant \frac{N_{s_0}^2 Q(\eta_{s_0}, \varkappa_{s_0})}{\bar{\alpha} \varkappa_{s_0}^{2n+4}} \leqslant \frac{N_{s_0}^2 Q(\eta, \gamma)}{\bar{\alpha} \gamma^{(s_0+1)(2n+4)}} <$$

$$< \frac{\bar{\alpha}}{Q(\eta, \gamma)} r_0^{2^{s_0+1}} \gamma^{(s_0+3)(2n+4)} = N_{s_0+1}, \qquad (10.18)$$

and

$$n \left| \frac{\partial F^{(s_0)} (h, \varphi, \xi^{(s_0+1)})}{\partial h_q} \right| \leqslant \frac{n N_{s_0}}{\varkappa_{s_0}} = L_{s_0+1}, \qquad (10.19)$$

remain valid.

Also, since $\xi^{(s_0)} (\xi^{(s_0+1)})$ is a solution of the equation

$$\xi^{(s_0)} + b^{(s_0)} (\xi^{(s_0)}) = \xi^{(s_0+1)}, \qquad (10.20)$$

where

$$b^{(s_0)}(\xi) = \{b_{k_0}^{(s_0)} (\xi)\} = \left\{ \sum_{s=1}^{n} \overline{\frac{\partial F_{k_0}^{(s_0)} (h, \varphi, \xi)}{\partial h_s}} \bigg|_{h=0} \right\} \quad (k_0 = 1, 2, \ldots, n),$$

it follows that

$$| \xi^{(s_0)} (\xi^{(s_0+1)}) | \leqslant | \xi^{(s_0+1)} | + | b^{(s_0)} (\xi^{(s_0)}) |.$$

We remark that

$$| b_q^{(s_0)} (\xi^{(s_0)}) | = \left| \sum_{s=1}^{n} \overline{\frac{\partial F_q^{(s_0)} (h, \varphi, \xi^{(s_0)})}{\partial h_s}} \bigg|_{h=0} \right| \leqslant \frac{N_{s_0}}{\varkappa_{s_0}} \quad (q = 1, 2, \ldots, n) \quad (10.21)$$

on the domain (10.16), from which it follows that

$$| \xi^{(s_0)} (\xi^{(s_0+1)}) | \leqslant 2 \frac{N_{s_0}}{\varkappa_{s_0}}.$$

Thereupon, taking (10.12) into account we obtain, finally,

$$| \xi^{(s_0)} (\xi^{(s_0+1)}) | \leqslant \frac{1}{2n} \frac{N_{s_0-1}}{\varkappa_{s_0-1}} < \frac{N_{s_0-1}}{\varkappa_{s_0-1}} \qquad (10.22)$$

for

$$| \xi^{(s_0+1)} | \leqslant \frac{N_{s_0}}{\varkappa_{s_9}}.$$

Consequently, for every ν, we have

$$|\xi^{(\nu)}(\xi^{(\nu+1)}\dots(\xi^{(s_0+1)}))| < \frac{N_{\nu-1}}{\varkappa_{\nu-1}}, \tag{10.23}$$

where $N_{-1}=\sigma$, $\varkappa_{-1}=1$.

This proves the validity of the inductive theorem for the passage from $s=s_0$ to the $s=(s_0+1)$-th step of iteration. Since at every step of the iteration process, all the estimates of Theorem 8 hold, these can be employed in proving the convergence of the derived iteration process. For this, just as in the previous chapter, we express h and ξ in the terms of $h^{(s)}$ and $\xi^{(s)}$ by the change-of-variable formula (10.5),

$$\begin{gathered} h = h^{(s)} + U^{(s)}(h^{(s)}, \varphi, \xi^{(s)}), \\ \xi = \xi(\xi^{(1)}(\xi^{(2)}(\dots\xi^{(s)}))) = \Xi^{(s)}(\xi^{(s)}). \end{gathered} \tag{10.24}$$

Obviously,

$$\begin{gathered} \Xi^{(s)}(\xi^{(s)}(\xi^{(s+1)})) = \Xi^{(s+1)}(\xi^{(s+1)}), \\ U^{(s+1)}(h, \varphi, \xi^{(s+1)}) = u^{(s+1)}(h, \varphi, \xi^{(s+1)}) + U^{(s)}(h + u^{(s+1)} \times \\ \times (h, \varphi, \xi^{(s+1)}), \xi^{(s)}(\xi^{(s+1)})). \end{gathered} \tag{10.25}$$

To evaluate the expression (10.25) and its derivatives we consider equation (10.20) and take account of the relations (10.9) and (10.10), to obtain

$$\begin{aligned} \sum_{q=1}^{n}\left|\frac{\partial\xi}{\partial\xi_q^{(s)}}\right| &= \sum_{q_1=1}^{n}\left|\frac{\partial\xi}{\partial\xi_{q_1}^{(1)}}\right|\sum_{q_2=1}^{n}\left|\frac{\partial\xi^{(1)}}{\partial\xi_{q_2}^{(2)}}\right|\dots\sum_{q_s=1}^{n}\left|\frac{\partial\xi^{(s-1)}}{\partial\xi_{q_s}^{(s)}}\right| \leqslant \\ &\leqslant \left(1+\frac{4Nn}{\sigma\gamma}\right)\left(1+\frac{4N_1n}{N\gamma}\right)\dots\left(1+\frac{4N_{s-1}n}{N_{s-2}\gamma}\right) \leqslant \\ &\leqslant 4\prod_{1\leqslant p<\infty}\left(1+r_0^{2^{p-1}}\right) = c(r_0) \end{aligned} \tag{10.26}$$

for

$$|\xi^{(s)}| \leqslant \frac{N_{s-1}}{2\varkappa_{s-1}}.$$

Consequently, for every s, the inequality

$$\sum_{q=1}^{n}\left|\frac{\partial\Xi^{(s)}(\xi^{(s)})}{\partial\xi_q^{(s)}}\right| \leqslant c(r_0)$$

holds, implying

$$
\begin{aligned}
&|\Xi^{(s+k)}(0) - \Xi^{(s)}(0)| \leqslant \\
&\leqslant \sum_{p=1}^{k} |\Xi^{(s+p)}(0) - \Xi^{(s+p-1)}(0)| = \\
&= \sum_{p=1}^{k} |\Xi^{(s+p-1)}(\xi^{(s+p-1)}(0)) - \Xi^{(s+p-1)}(0)| \leqslant \\
&\leqslant c(r_0) \sum_{p=1}^{k} |\Xi^{(s+p-1)}(0)| \leqslant \\
&\leqslant c(r_0) \sum_{p=1}^{k} \frac{N_{s+p-2}}{\varkappa_{s+p-2}} \leqslant c(r_0) \frac{\bar{\alpha}\sigma}{Q(\eta, \gamma)} \sum_{p=s+1}^{\infty} \left(\frac{1}{4n}\right)^{p} = \\
&= \frac{c(r_0)\,\sigma\bar{\alpha}}{\left(1 - \frac{1}{4n}\right) Q(\eta, \gamma)} \left(\frac{1}{4n}\right)^{s+1}.
\end{aligned}
\tag{10.27}
$$

The inequality (10.27) implies

$$\Xi^{(s)}(0) \to \Xi^{(\infty)}(0) \quad \text{as } s \to \infty, \tag{10.28}$$

and also

$$|\Xi^{(\infty)}(0) - \Xi^{(s)}(0)| \leqslant \frac{c(r_0)\,\sigma\alpha}{Q(\eta, \gamma)\left(1 - \frac{1}{4n}\right)} \left(\frac{1}{4n}\right)^{s+1}. \tag{10.29}$$

We next prove the uniform convergence of the function $(U^{(s)}(h, \varphi, \xi^{(s)})$. Since by the estimates of the inductive Theorem 8 we have

$$\sum_{q=1}^{n} \left| \frac{\partial u^{(s+1)}}{\partial h_q} \right| \leqslant \frac{N_s Q(\eta_s, \varkappa_s)}{2\bar{\alpha}\varkappa_s^{2n+1}} \leqslant \frac{1}{2} r_0^{2^s} \gamma^{(2n+4)}, \tag{10.30}$$

the identity (10.25) implies the inequality

$$
\begin{aligned}
\sum_{q=1}^{n} \left| \frac{\partial U(h, \varphi, \xi^{(s+1)})}{\partial h_q} \right| &\leqslant \frac{1}{2} r_0^{2^s} \gamma^{(2n+4)} + \sum_{q=1}^{n} \left| \frac{\partial U^{(s)}(h + u^{(s+1)}, \xi^{(s)})}{\partial h_q} \right| \times \\
&\times \left(1 + \frac{1}{2} r_0^{2^s} \gamma^{(2n+4)}\right).
\end{aligned}
\tag{10.31}
$$

Solving the inequality (10.31) in the domain

$$|h| \leqslant \eta_s, \quad |\operatorname{Im} \varphi| \leqslant \rho, \quad |\xi^{(s)}| \leqslant \frac{N_{s-1}}{\varkappa_{s-1}},$$

we get

$$\sum_{q=1}^{n} \left| \frac{\partial U^{(s)}(h, \varphi, \xi^{(s)})}{\partial h_q} \right| \leqslant \exp \sum_{p=0}^{\infty} \frac{1}{2} r_0^{2^p} \gamma^{(2n+4)} - 1 \leqslant \frac{1}{2}. \tag{10.32}$$

To evaluate the sum

$$\sum_{q=1}^{n} \left| \frac{\partial U^{(s)}(h, \varphi, \xi^{(s)})}{\partial \xi_q^{(s)}} \right| = y_s, \tag{10.33}$$

we differentiate the identity (10.25), to obtain

$$\sum_{q=1}^{n} \left| \frac{\partial U^{(s+1)}(h, \varphi, \xi^{(s+1)})}{\partial \xi_q^{(s+1)}} \right| \leqslant \sum_{q=1}^{n} \left| \frac{\partial u^{(s+1)}(h, \varphi, \xi^{(s+1)})}{\partial \xi_{q_1}^{(s+1)}} \right| +$$
$$+ \sum_{q=1}^{n} \left| \frac{\partial U^{(s)}(h, \varphi, \xi^{(s+1)})}{\partial \xi_q^{(s)}} \right| \sum_{q_1=1}^{n} \left| \frac{\partial \xi^{(s)}(\xi^{(s+1)})}{\partial \xi_{q_1}^{(s+1)}} \right| +$$
$$+ \sum_{q=1}^{n} \left| \frac{\partial U^{(s)}(h, \varphi, \xi^{(s+1)})}{\partial h_q} \right| \sum_{q_1=1}^{n} \left| \frac{\partial u^{(s+1)}(h, \varphi, \xi^{(s+1)})}{\partial \xi_{q_1}^{(s+1)}} \right|. \tag{10.34}$$

By the inequalities (9.10), (10.7) and with the notation (10.10),

$$\sum_{q=1}^{n} \left| \frac{\partial u^{(s+1)}(h, \varphi, \xi^{(s+1)})}{\partial \xi_q^{(s+1)}} \right| \leqslant r_0^{2^{s-1}} \gamma^{2n+3-2n(s+1)} \cdot \frac{na(\eta, \gamma)}{\bar{\alpha}\left(1 - \frac{1}{2n}\right)}, \quad s \geqslant 1, \tag{10.35}$$

and

$$\sum_{q=1}^{n} \left| \frac{\partial u^{(1)}(h, \varphi, \xi^{(1)})}{\partial \xi_q^{(1)}} \right| \leqslant \frac{Nna(\eta, \gamma)}{\bar{\alpha}\sigma\gamma^{2n+1}\left(1 - \frac{1}{2n}\right)}, \quad s = 0. \tag{10.36}$$

Furthermore, by (10.32) we have

$$y_{s+1} \leqslant \frac{3}{2} S_s + y_s(1 + v_s), \qquad y_0 \equiv 0; \tag{10.37}$$

where we are using the notation of (10.33), and have taken into account the estimates of Theorem 8 and the inequalities (10.26), and have introduced the notations

$$\nu_0 = \frac{4Nn}{\sigma\gamma}, \quad \nu_1 = \frac{4N_1 n}{N\gamma}, \ldots, \nu_p = \frac{4N_p n}{N_{p-1}\gamma}, \quad p \geqslant 1, \tag{10.38}$$

$$S_0 = \frac{Nna(\eta, \gamma)}{\bar{\alpha}\sigma\gamma^{2n+1}\left(1 - \frac{1}{4n}\right)}, \ldots, S_p = r_0^{2^{p-1}}\gamma^{3-(2n+1)p}\frac{na(\eta, \gamma)}{\bar{\alpha}\left(1 - \frac{1}{2n}\right)}.$$

Solving the inequality (10.37), we arrive at the estimate

$$\sum_{q=1}^{n}\left|\frac{\partial U^{(s)}(h, \varphi, \xi^{(s)})}{\partial \xi_q^{(s)}}\right| < \frac{3}{2}\prod_{0 \leqslant p < \infty}(1 + \nu_p)\sum_{p=0}^{\infty} S_p \leqslant$$

$$\leqslant \frac{3}{2}c(r_0)\sum_{p=0}^{\infty} S_p =$$

$$= \frac{3c(r_0)\,na(\eta, \gamma)}{2\bar{\alpha}\left(1 - \frac{1}{2n}\right)}\left[\frac{N}{\sigma\gamma^{2n+1}} + \sum_{p=1}^{\infty} r_0^{2^{p-1}}\gamma^{3-(2n+1)s}\right] \equiv Y. \tag{10.39}$$

Hence, by the expressions (10.25), (10.32), (10.33), (10.39) and the estimates of Theorem 8, on the domain

$$|h| \leqslant \frac{\eta}{2}, \quad |\operatorname{Im}\varphi| \leqslant \frac{\rho}{2}, \tag{10.40}$$

we obtain the inequality

$$|U^{(s+1)}(h, \varphi, 0) - U^{(s)}(h, \varphi, 0)| \leqslant$$

$$\leqslant |u^{(s+1)}(h, \varphi, 0)| + |U^{(s)}(h + u^{(s+1)}(h, \varphi, 0), \quad \xi^{(s)}(0))$$

$$- U^{(s)}(h, \varphi, 0)| \leqslant$$

$$\leqslant \frac{3}{2}\frac{N_s a(\eta, \gamma)}{\bar{\alpha}\gamma^{(2n+1)(s+1)}} + Y\frac{N_{s-1}}{\gamma^s} =$$

$$= \frac{3}{2}\frac{a(\eta, \gamma)}{Q(\eta, \gamma)}r_0^{2^s}\gamma^{(2n+4)+3(s+1)} + Y\frac{\bar{\alpha}\, r_0^{2^{s-1}}}{Q(\eta, \gamma)}\gamma^{(2n+4)(s+1)-s}. \tag{10.41}$$

Thus, starting from the inequalities (10.41) and (10.39), on the domain

(10.40), we have

$$U^{(s+k)}(h, \varphi, 0) - U^{(s)}(h, \varphi, 0) | \leqslant \frac{3a(\eta, \gamma)}{2Q(\eta, \gamma)} \gamma^{(2n+4)} \sum_{p=m}^{\infty} r_0^{2^p} \gamma^{3(p+1)} +$$

$$+ \frac{3c(r_0) na(\eta, \gamma)}{2Q(\eta, \gamma)\left(1 - \frac{1}{2n}\right)} \left[\frac{N}{\sigma\gamma^{2n+1}} + \sum_{p=1}^{\infty} r_0^{2^{p-1}} \gamma^{3-(2n+1)p}\right] \sum_{p=m}^{\infty} r_0^{2^{p-1}} \gamma^{(2n+4)(p+1)-p},$$

or,

$$U^{(s+k)}(h, \varphi, 0) - U^{(s)}(h, \varphi, 0) | \leqslant \frac{3\left(1 - \frac{1}{2}\right)n}{2(n^2 + 4n - 2)} \gamma^{2n+2} \sum_{p=m}^{\infty} r_0^{2^p} \gamma^{3(p+1)} +$$

$$+ \frac{3c(r_0) n^2}{2(n^2 + 4n - 2)} \left[\frac{1}{4n\gamma^{2n+2)}} + \sum_{p=1}^{\infty} r_0^{2^{p-1}} \gamma^{1-(2n+1)p}\right] \gamma \sum_{p=m}^{\infty} r_0^{2^{p-1}} \gamma^{(2n+3)(p+1)}. \tag{10.42}$$

Here we have used the relation

$$\frac{a(\eta, \gamma)}{Q(\eta, \gamma)} < \frac{\left(1 - \frac{1}{2n}\right)n}{\gamma^2 (n^2 + 4n - 2)},$$

and the fact that $N/\sigma \leqslant \gamma/4n$.

The inequality (10.42) implies that the convergence of the functions

$$U^{(s)}(h, \varphi, 0) \to U^{(\infty)}(h, \varphi, 0) \qquad \text{as } s \to \infty \tag{10.43}$$

is uniform on the domain (10.40) ; the rate of convergence is characterized by the inequality (10.42), where it is necessary to substitute $U^{(s+k)}(h, \varphi, 0)$ by $U^{(\infty)}(h, \varphi, 0)$.

§ 11. Theorem on Reducibility of Nonlinear Equations

The results achieved above are summed up in this section. In the system

$$\begin{aligned} \frac{dh}{dt} &= (\alpha + \xi) h + F(h, \varphi, \xi), \\ \frac{d\varphi}{dt} &= \omega, \end{aligned} \tag{11.1}$$

set $\xi = \Xi^{(\infty)}(0)$. Then, by means of the analytic change of variables

$$h = g + U^{(\infty)}(g, \varphi, 0) \tag{11.2}$$

in the domain

$$|h| \leqslant \frac{\eta}{2}, \qquad |\operatorname{Im} \varphi| \leqslant \frac{\rho}{2}, \tag{11.3}$$

the system of equations (11.1) reduces to the form

$$\frac{dg}{dt} = \alpha g,$$
$$\frac{d\varphi}{dt} = \omega. \tag{11.4}$$

Integrating the system (11.4), we obtain

$$g = Ce^{\alpha t}, \qquad \varphi = \omega t + \varphi_0, \tag{11.5}$$

where $C = (C_1, C_2, \ldots, C_n)$, $\varphi_0 = (\varphi_{01}, \varphi_{02}, \ldots, \varphi_{0m})$ are constants of integration.

Substituting the values of g and φ in the right-hand side of (11.2), we obtain the general solutions

$$h = Ce^{\alpha t} + U^{(\infty)}(Ce^{\alpha t}, \omega t + \varphi_0, 0),$$
$$\varphi = \omega t + \varphi_0, \tag{11.6}$$

of the original system of equations (11.2) which depend on $n + m$ arbitrary constants.

These results are summed up in the following theorem†.

Theorem 9. *For the system of equations* (11.1), *suppose that the function* $F(h, \varphi, \xi)$ *and constants* α *satisfy all the hypotheses of Theorem* 8 *proved in* § 9. *In addition, suppose that the inequalities* (10.7) *are satisfied and that* $\xi = \Xi^{(\infty)}(0)$. *Then, equations* (11.1) *have general solutions of the form* (11.6), *where the function* $U^{(\infty)}(h, \varphi, 0)$ *is analytic in* h, φ *on the domain* (11.3).

Recalling now the system of equations

$$\frac{dh}{dt} = (\alpha + \xi) h + F(h, \varphi, \Delta, \xi),$$
$$\frac{d\varphi}{dt} = \omega + \Delta + f(h, \varphi, \Delta, \xi), \tag{11.7}$$

we assume this to satisfy all the hypotheses of Theorems 3 and 8. Then, by the former theorem, the transformation (analytic in h and θ)

$$\varphi = \theta + \Phi^{(\infty)}(h, \theta, 0) \tag{11.8}$$

†See Appendix 5.

reduces the system (11.7) for $\Delta = D_1^{(\infty)}(0)$ to the form

$$\frac{dh}{dt} = (\alpha + \xi)\, h + F(h, \theta, \xi),$$
$$\frac{d\theta}{dt} = \omega, \tag{11.9}$$

where $F(h, \theta, \xi) = F(h, \theta + \Phi^{(\infty)}(h, \theta, 0), D_1^{(\infty)}(0), \xi)$ is an analytic function of h, θ, ξ on the domain

$$|h| \leqslant \frac{\eta_1}{2}, \qquad |\operatorname{Im} \theta| \leqslant \frac{\rho_1}{2}, \qquad |\xi| \leqslant \sigma. \tag{11.10}$$

Set $\eta = \frac{\eta_1}{2}$, $\rho = \frac{\rho_1}{2}$, then, for $\xi = \Xi^{(\infty)}(0)$, the change of variable

$$h = g + U^{(\infty)}(g, \theta, 0) \tag{11.11}$$

reduces the system of equations (11.9) to the form

$$\frac{dg}{dt} = \alpha g,$$
$$\frac{d\theta}{dt} = \omega. \tag{11.12}$$

Combining these substitutions, it is seen that the system (11.7) with $\Delta = D_1^{(\infty)}(0)$ and $\xi = \Xi^{(\infty)}(0)$ is transformed into the system (11.12) by means of the change of variables

$$h = g + U^{(\infty)}(g, \theta, 0),$$
$$\varphi = \theta + \Phi^{(\infty)}(g + U^{(\infty)}(g, \theta, 0), \theta, 0), \tag{11.13}$$

this substitution being analytic on the domain (11.3).

Integrate the system (11.12) and substitute the solution so obtained in the expression (11.13), to get the general solutions

$$h_t = Ce^{\alpha t} + U^{(\infty)}(Ce^{\alpha t}, \omega t + \theta_0, 0),$$
$$\varphi_t = \omega t + \theta_0 + \Phi^{(\infty)}(Ce^{\alpha t} + U^{(\infty)}(Ce^{\alpha t}, \omega t + \theta_0, 0), \omega t + \theta_0, 0), \tag{11.14}$$

of the system (11.7); here C, θ_0 are arbitrary constants, which must satisfy

$$|C| \leqslant \frac{\eta}{2}, \qquad |\operatorname{Im} \theta_0| \leqslant \frac{\rho}{2}\left(1 - \frac{1}{2m}\right). \tag{11.15}$$

Set $C = 0$ in (11.14), to obtain a particular quasi-periodic solution with the frequency bases

$$\omega = (\omega_1, \omega_2, \ldots, \omega_m):$$

$$h(\omega t) = U^{(\infty)}(0, \omega t + \theta_0, 0), \tag{11.16}$$

$$\varphi(\omega t) = \omega t + \theta_0 + \Phi^{(\infty)}(U^{(\infty)}(0, \omega t + \theta_0, 0), \omega t + \theta_0, 0),$$

whose existence has been established in the preceding chapter.

We now evaluate the differences $|h_t - h(\omega t)|$, $|\varphi_t - \varphi(\omega t)|$. The relations (10.31) and (10.43) on the domain (10.40) imply

$$\sum_{q=1}^{n} \left| \frac{\partial U^{(\infty)}}{\partial h_q} \right| \leqslant \frac{1}{2}, \tag{11.17}$$

and, consequently, it is easy to obtain

$$|h_t - h(\omega t)| \leqslant \frac{3}{2} |C| e^{-\bar{\alpha} t} \tag{11.18}$$

for $|C| \leqslant \frac{\eta}{2}$, $|\operatorname{Im} \theta_0| \leqslant \frac{\rho}{2}$.

Furthermore, by the estimates (2.27) and (4.4), we have

$$\sum_{q=1}^{n} \left| \frac{\partial v^{(s+1)}}{\partial h_q} \right| \leqslant \frac{M_s \cdot 4\rho_1^2 n}{\bar{\alpha} \eta_1 \delta_s^2}, \tag{11.19}$$

where

$$M_s \leqslant \frac{\bar{\alpha} r_0^{2^s} \gamma^{(s+2)(2m+4)}}{Q_1(\rho, \gamma)}, \qquad \delta_s = \gamma^{s+1}.$$

Hence, from the identity (4.5) and the estimates used in the proof of Theorem 3, we can write

$$\left| \frac{\partial \Phi^{(s+1)}}{\partial h_q} \right| \leqslant \left| \frac{\partial v^{(s+1)}}{\partial h_q} \right| + \left| \frac{\partial \Phi^{(s)}}{\partial h_q} \right| + \left| \frac{\partial \Phi^{(s)}}{\partial \theta} \right| \cdot \left| \frac{\partial v^{(s+1)}}{\partial h_q} \right| \leqslant$$

$$\leqslant \left| \frac{\partial \Phi^{(s)}}{\partial h_q} \right| + \frac{6\rho_1^2 m r_0^{2^s}}{\eta_1 Q_1(\rho_1, \gamma)} \gamma^{(s+2)(2m+2)+2}. \tag{11.20}$$

Introduce the notation

$$\sum_{q=1}^{n} \left| \frac{\partial \Phi^{(s)}}{\partial h_q} \right| = z_s. \tag{11.21}$$

Then, we obtain

$$z_{s+1} \leqslant z_s + \frac{6\rho_1^2 m^2 r_0^{2^s}}{\eta_1 Q_1(\rho_1, \gamma)} \gamma^{(s+2)(2m+2)+2}, \tag{11.22}$$

whence

$$z_s \leqslant \frac{6\rho_1^2 m^2}{\eta_1 Q_1(\rho_1, \gamma)} \gamma^2 \sum_{s=0}^{\infty} r_0^{2^s} \gamma^{(s+2)(2m+2)} = c_1(r_0), \tag{11.23}$$

and, consequently,

$$\sum_{q=1}^{n} \left| \frac{\partial \Phi^{(s)}}{\partial h_q} \right| \leqslant c_1(r_0) \tag{11.24}$$

for any s. Hence, the difference $|\varphi_t - \varphi(\omega t)|$ obeys the estimate

$$|\varphi_t - \varphi(\omega t)| = |\Phi^{(\infty)}(Ce^{\alpha t} + U^{(\infty)}(Ce^{\alpha t}, \omega t + \theta_0, 0), \omega t + \theta_0, 0)$$

$$- \Phi^{(\infty)}(U^{(\infty)}(0, \omega t + \theta_0, 0), \omega t + \theta_0, 0)| \leqslant$$

$$\leqslant \frac{3}{2} c_1(r_0) |C| e^{-\bar{\alpha} t} \tag{11.25}$$

for

$$|C| \leqslant \frac{\eta}{2}, \quad |\operatorname{Im} \theta_0| \leqslant \frac{\rho}{2}\left(1 - \frac{1}{2m}\right).$$

The above considerations lead to the following fundamental theorem.

Theorem 10. *Suppose that all the hypotheses of Theorems* 3 *and* 9 *are satisfied for the system of equations*

$$\frac{dh}{dt} = (\alpha + \xi) h + F(h, \varphi, \Delta, \xi),$$
$$\frac{d\varphi}{dt} = \omega + \Delta + f(h, \varphi, \Delta, \xi). \tag{11.26}$$

Then by a convenient choice of $\Delta = D_1^{(\infty)}(0)$ *and* $\xi = \Xi^{(\infty)}(0)$ *the change of variables*

$$h = g + U^{(\infty)}(g, \theta, 0),$$
$$\varphi = \theta + \Phi^{(\infty)}(g + U^{(\infty)}(g, \theta, 0), \theta, 0) \tag{11.27}$$

reduces the system (11.26) *to a linear system with constant coefficients*

$$\frac{dg}{dt} = \alpha g,$$
$$\frac{d\theta}{dt} = \omega. \tag{11.28}$$

Integrating this system we obtain the general solutions of the original system (11.26) *in the form*

$$h_t = Ce^{\alpha t} + U^{(\infty)}(Ce^{\alpha t}, \omega t + \theta_0, 0),$$
$$\varphi_t = \omega t + \theta_0 + \Phi^{(\infty)}(Ce^{\alpha t} + U^{(\infty)}(Ce^{\alpha t}, \omega t + \theta_0, 0), \omega t + \theta_0, 0), \tag{11.29}$$

where the $n + m$ *arbitrary constants* C, θ_0 *satisfy*

$$| C | \leqslant \frac{\eta}{2}, \qquad | \operatorname{Im} \theta_0 | \leqslant \frac{\rho}{2}. \tag{11.30}$$

With the flow of time the solution (11.29) *approximates to the stationary quasi-periodic solution*

$$h(\omega t) = U^{(\infty)}(0, \omega t + \theta_0, 0),$$
$$\varphi(\omega t) = \omega t + \theta_0 + \Phi^{(\infty)}(U^{(\infty)}(0, \omega t + \theta_0, 0), \omega t + \theta_0, 0), \tag{11.31}$$

where $\operatorname{Im} \theta_0 | \leqslant \frac{\rho}{2}\left(1 - \frac{1}{2n}\right)$, *the rate of convergence being given by*

$$| h_t - h(\omega t) | \leqslant \frac{3}{2} | C | e^{-\bar{\alpha} t},$$
$$| \varphi_t - \varphi(\omega t) | \leqslant \frac{3}{2} c_1(r_0) | C | e^{-\bar{\alpha} t}. \tag{11.32}$$

If, in addition to the restrictions indicated, the functions $F(h, \varphi, \Delta, \xi)$ *and* $f(h, \varphi, \Delta, \xi)$ *are analytic in some parameter* ε *on the domain* $| \varepsilon | \leqslant \varepsilon_0$ *and the inequalities* (10.29) *and* (10.42) *are satisfied on this domain with* η, γ, r_0 *and* σ *fixed, then* $\Xi^{(\infty)}(0)$ *and* $U^{(\infty)}(h, \varphi, 0)$ *are analytic functions of* ε *on the domain* $| \varepsilon | \leqslant \varepsilon_0$.

Since the convergence of the expressions (10.29) *and* (10.42) *is defined only for* η, γ, r_0 *and* σ, *it is uniform with respect to* ε *in the domain* $| \varepsilon | \leqslant \varepsilon_0$. *Hence, by Theorem* 3, $D_1^{(\infty)}(0)$, $\Xi^{(\infty)}(0)$, $U^{(\infty)}(h, \varphi, 0)$ *and* $\Phi^{(\infty)}(h, \varphi, 0)$ *are analytic functions of* ε *for* $| \varepsilon | \leqslant \varepsilon_0$.

The proof of the theorem on the reducibility of the system of equations (11.26) and, consequently, also of the system (7.1), to equations with constant coefficients (11.28) and on the construction of general solutions in a neighbourhood of the stable stationary quasi-periodic solution (11.32), gives us the possibility of investigating the behaviour of solutions in the neighbourhood of quasi-periodic solutions and opens up the prospect of exploring in depth different forms of equations containing a small parameter. Thus, the technique developed and the results obtained in the present chapter will be useful in the sequel for investigating in greater detail systems of linear differential equations with quasi-periodic coefficients, the behaviour of the solution of a system of differential equations in the neighbourhood of an invariant manifold, and allied fields.

Chapter 3

A SMOOTHING TECHNIQUE

§ 12. Loss of Derivatives

In the preceding chapter, it was essential for the application of the iteration process with accelerated convergence that the right-hand side of the equation under consideration should be analytic on some complex domain of the type

$$|\operatorname{Im} \varphi| < \rho.$$

As is known, in most equations encountered in nonlinear mechanics, the right-hand side is not analytic. For the methods employed in nonlinear mechanics it is, therefore, impossible to establish the analyticity of the invariant tori. One can, however, prove, subject to suitable restrictions, that the functions defining a torus have a sufficient number of derivatives. The tediousness that arises in this connection can be circumvented by introducing the smoothing of the defining functions.

Let D be a domain of an n-dimensional Euclidean space E_n. We denote by $C^r(D)$ a space of r-times continuously differentiable functions defined on D and having the norm

$$|\cdot|_r = \max_{0 \leqslant d \leqslant r} \sup_D |D^d \cdot|, \tag{12.1}$$

where D^d is any derivative of order d.

Let A be an operator, which carries the elements of the space $C^r(D)$ into $C^{r_1}(D)$, where $r_1 \leqslant r$. Then, evidently, it is possible to seek a solution of the equation

$$Au = 0 \tag{12.2}$$

by an appeal to the iteration

$$u_{s+1} - u_s = Au_s \qquad (s = 0, 1, 2, \ldots, u_0 \in C^r(D)), \tag{12.3}$$

i.e. by using Picard's method of successive approximations. However, among equations of the type (12.2) prototypes are frequently encountered which are intractable by Picard's method. In such equations, as a rule, the operator affects the smoothness of functions on which it acts, i.e. it 'sheds' derivatives, carrying the space $C^r(D)$ into $C^{r-s_0}(D)$, where $s_0 > 0$.

For operators involving the loss of derivatives the process of constructing the iteration (12.3) breaks down when functions are not infinitely differentiable. In fact, if

$$u_0 \in C^{r+(k_0-1)\,s_0}(D) \diagdown C^{r+k_0 s_0}(D),$$

then

$$u_{k_0} - u_{k_0-1} = Au_{k_0-1} \in C^{r-s_0}(D) \diagdown C^r(D),$$

and, consequently, the construction of $(k_0 + 1)$-th term with respect to u_0 is impossible, because the number of derivatives of u_0 required for this purpose is exhausted. To evade this difficulty connected with the loss of derivatives, it is necessary to modify the iteration process itself by devising a convenient smoothing procedure described in the next section.

§ 13. Examples of Smoothing Operators

An operator T_θ depending on a parameter θ is called a smoothing operator if it acts from the space $C^r(D)$ into the space $C^{(\infty)}(D_1)$ $(D_1 \leqslant D)$ and converges in the norm (12.1) to the identity operator as $\theta \to \theta_0$.

Thus, the operator T_θ acting on an r-times continuously differentiable function $f(x)$, carries this into an infinitely differentiable function $T_\theta f(x)$ such that

$$| f(x) - T_\theta f(x) |_r \to 0$$

as $\theta \to \theta_0$.

Take S_N as an example of a smoothing operator, which for a function $f(\varphi) \in C^r(E_m)$, 2π-periodic in $\varphi = (\varphi_1, \ldots, \varphi_m)$, is definable as the truncated Fourier series

$$S_N f(\varphi) = \sum_{|k| \leqslant N} f_k e^{i(k,\varphi)}. \tag{13.1}$$

It is evident that $S_N f(\varphi) \in C^{(\infty)}(E_m)$.

Another form of a smoothing operator can be seen by considering a convolution of functions with a finite kernel :

$$T_\theta f = (K_\theta * f) = \int_{E_n} K_\theta(x-y) f(y)\, dy. \tag{13.2}$$

Further examples of such operators are provided by the operators, with the following kernels :

(*a*) The kernel [70]

$$K_\theta(x) = \begin{cases} \dfrac{1}{\theta^n I}\, e^{\frac{r^2}{r^2-\theta^2}} & \text{for } r \leqslant \theta, \\ 0 & \text{for } r \geqslant \theta, \end{cases} \tag{13.3}$$

where

$$r^2 = \sum_{\nu=1}^{n} x_\nu^2, \qquad I = \int_{E_n} K_1(x)\,dx;$$

(*b*) The Nach kernel [55]

$$K_\theta(x) = \theta^n K_1(\theta x), \tag{13.4}$$

where $K_1(x)$ is a function, whose Fourier transform $\hat{K}_1(\xi_1, \ldots, \xi_n)$ is defined by

$$\hat{K_1}(\xi_1, \ldots, \xi_n) = \Psi(\xi),$$

with

$$\xi = (\xi_1^2 + \ldots + \xi_n^2)^{1/2}, \quad \Psi(\xi) \in C^\infty(E_1),$$

$$\Psi((\xi) = \begin{cases} 1 & \text{for } \xi \leqslant 1, \\ \text{monotone decreasing} & \text{for } 1 \leqslant \xi \leqslant 2, \\ 0 & \text{for } \xi \geqslant 2; \end{cases}$$

(*c*) The Moser kernel [52]

$$K_{\theta_1\theta_2}(x_1, \ldots, x_d, \ldots, x_n) = \theta_1^d \chi(\theta_1 x_1) \ldots \chi(\theta_1 x_d)\, \theta_2^{n-d} \chi(\theta_2 x_{d+1}) \ldots \chi(\theta_2 x_n), \tag{13.5}$$

where

$$\chi(x) \in C^\infty(E_1), \quad \chi(x) \equiv 0 \quad \text{for } |x| > 1,$$

$$\int_{-\infty}^{\infty} x^{k_0} \chi(x)\,dx = \begin{cases} 1 & \text{for } k_0 = 0, \\ 0 & \text{for } 0 < k_0 < l, \end{cases} \tag{13.6}$$

l being a fixed number.

The kernel (13.5) depends upon a pair of parameters θ_1 and θ_2, in consequence of which the order of approximation relative to each of the arguments

$$x' = (x_1, \ldots, x_d) \quad \text{and} \quad x'' = (x_{d+1}, \ldots, x_n)$$

can be regulated independently.

Among the smoothing operators which are not convolutions, the operators T_θ^0 and T_θ^1 may be mentioned.

The operator T_θ^0 acts on the function $f(x', x'')$, and is such that

$$f(x', 0) \equiv 0, \tag{13.7}$$

and yields the relation [60]

$$T^0_\theta f = \iint\limits_{E_n} [K_0 (x' - y', x'' - y'') - K_0 (x' - y', -y'')] f (y', y'') \, dy' \, dy'', \tag{13.8}$$

where $K_\theta (x', x'')$ is the Moser kernel.

In the process of smoothing, the property (13.7) is preserved :

$$T^0_\theta f|_{x''=0} \equiv 0.$$

If the operation of T^1_θ extends to the function $f (x', x'')$, satisfying

$$f (x', 0) = \left. \frac{\partial f (x', x'')}{\partial x''} \right|_{x''=0} \equiv 0, \tag{13.9}$$

and is defined by the relation

$$T^1_\theta f = \iint\limits_{E_n} \Bigg[K_\theta (x' - y', x'' - y'') - K_\theta (x' - y', -y'') +$$

$$+ \left(\frac{\partial K_\theta (x' - y', -y'')}{\partial y''}, x'' \right) \Bigg] f (y', y'') \, dy' \, dy'', \tag{13.10}$$

then the property (13.9) is also preserved:

$$T^1_\theta f \Big|_{x'=0} = \left. \frac{\partial T^1_\theta f}{\partial x''} \right|_{x''=0} \equiv 0.$$

A more general smoothing process consists in smoothing the functions in $C^r (D)$ by means of operators acting from the space $C^r (D)$ into the space $C^{r+s_0} (D_1)$ with a finite s_0.

An example of such a process extensively used in the theory of ordinary differential equations and nonlinear mechanics is provided by smoothing the functions $f (x) \in C^0 (D)$ into their convolution with the kernel $\Delta_a (x) \in C^1 (E_n)$ defined by [10]

$$\Delta_a (x) = \begin{cases} A_a \left(1 - \dfrac{|x|^2}{a^2} \right)^2 & \text{for} \quad |x| \leqslant a, \\ 0 & \text{for} \quad |x| > a, \end{cases} \tag{13.11}$$

where $|x|^2 = x_1^2 + x_2^2 + \ldots + x_n^2$, and the positive constant A_a is determined from the equation

$$\int\limits_{E_n} \Delta_a (x) \, dx = 1. \tag{13.12}$$

§ 14. The Basic Properties of a Smoothing Operator

We now examine the operators T_θ, T_θ^0, T_θ^1 and S_N.

The set of functions on which these operators act is bounded ; for the operator S_N the set consists of nonperiodic functions $f(\varphi_1, \ldots, \varphi_d)$, representable by the Fourier series

$$f(\varphi) = \sum_{|k| \neq 0} f_k e^{i(k, \varphi)} ; \tag{14.1}$$

the operator $\mathcal{T}_\theta = \{T_\theta, \overset{0}{T_\theta}, \overset{1}{T_\theta}\}$ acts on functions $f(\varphi_1, \ldots, \varphi_d, x_1, \ldots, x_m)$, defined and continuous on the domain D :

$$-a \leqslant x_i \leqslant a \qquad (i = 1, \ldots, m). \tag{14.2}$$

The smoothing of functions $f(\varphi, x)$ is carried out by means of parameters N and M, if in the kernel (13.5) we put

$$\theta_1 = N, \quad \theta_2 = M, \quad x' = (\varphi_1, \ldots, \varphi_d), \quad x'' = (x_1, \ldots, x_m).$$

By the definition of the operators $\mathcal{T}_{NM}$ it follows that the smoothed functions $\mathcal{T}_{NM} f(\varphi, x)$ are defined in the domain D_1 :

$$-a + M^{-1} < x_i < a - M^{-1}, \quad M^{-1} < a ; \qquad i = 1, 2, \ldots, n.$$

The foregoing preliminary remarks are adequate to bring out the motivation for the following lemmas concerning the properties of these operators.

Lemma 3. *For any non-negative integers $\rho = (\rho_1, \ldots, \rho_d)$, $r = (r_1, \ldots, r_m)$ and every $x \in D_1$*

$$\left| \frac{\partial^{|\rho| + |r|}}{\partial \varphi_1^{\rho_1} \ldots \partial \varphi_d^{\rho_d} \partial x_1^{r_1} \ldots \partial x_m^{r_m}} \mathcal{T}_{NM} f \right| \leqslant C N^{|\rho|} M^{|r| + \delta} |f(\varphi, x)|_0, \tag{14.3}$$

where the constant C does not depend on N, M and f, and

$$\delta = \begin{cases} 0 & \text{for} \quad \mathcal{T}_{NM} = \left\{T_{NM}, T_{NM}^0\right\}, \\ 1 & \text{for} \quad \mathcal{T}_{NM} = T_{NM}^1. \end{cases}$$

Proof. The validity of inequality (14.3) for $\mathcal{T} = T$ follows immediately from (13.2) and (13.5). In fact,

$$\left| \frac{\partial^{|\rho|+|r|}}{\partial\varphi_1^{\rho_1}\ldots\partial\varphi_d^{\rho_d}\partial x_1^{r_1}\ldots\partial x_m^{r_m}} T_{NM} f(\varphi, x) \right| \leqslant$$

$$\leqslant |h|_0 \iint\limits_{E_n} \left| \frac{\partial^{|\rho|+|r|}}{\partial\varphi_1^{\rho_1}\ldots\partial x_m^{r_m}} K_{NM}(\varphi-\varphi', x-x') \right| d\varphi' dx' \leqslant$$

$$\leqslant |h|_0 N^{|\rho|} M^{|r|} \iint\limits_{E_n} \frac{\partial^{|\rho|+|r|}}{\partial\varphi_1^{\rho_1}\ldots\partial x_m^{r_m}} K_{11}(\varphi, x)\, d\varphi dx \leqslant$$

$$\leqslant CN^{|\rho|} M^{|r|} |h(\varphi, x)|_0. \tag{14.4}$$

For $\mathcal{T} = T^0$ the estimate (14.3) follows from (14.4) by the representation

$$T^0_{NM} f(\varphi, x) = T_{NM} f(\varphi, x) - T_{NM} f(\varphi, x)\Big|_{x=0}, \tag{14.5}$$

and for $\mathcal{T} = T^1$ (14.3) is implied by (14.4) and the representation

$$T^1_{NM} f(\varphi, x) = T_{NM} f(\varphi, x) - T_{NM} f(\varphi, x)\Big|_{x=0} - \left(\frac{\partial T_{NM} f(\varphi, x)}{\partial x}\Big|_{x=0}, x \right). \tag{14.6}$$

Lemma 4. *Suppose* $f(\varphi, x) \in C^l(D)$; *then*

$$|f(\varphi, x) - \mathcal{T}_{NM} f(\varphi, x)| \leqslant C \sup_{|\rho|+|r|=l} N^{-|\rho|} M^{-|r|+\delta} \left| \frac{\partial^{|\rho|+|r|} f(\varphi, x)}{\partial\varphi_1^{\rho_1}\ldots\partial x_m^{r_m}} \right|_0, \tag{14.7}$$

where, as before, $f(\varphi, 0) = 0$ *if* $\mathcal{T}_{NM} = T^0_{NM}$, *and*

$$f(\varphi, 0) = \frac{\partial f(\varphi, x)}{dx}\Big|_{x=0} = 0, \quad \text{if} \quad \mathcal{T}_{NM} = T^1_{NM}.$$

Proof. By (13.2) and (13.5), for $\mathcal{T}_{NM} = T_{NM}$, we have

$$T_{NM} f = (K_{NM} * f) = (f * K_{NM}) \iint\limits_{E_{d+m}} K_{NM}(\varphi', x') f(\varphi - \varphi', x - x')\, d\varphi' dx =$$

$$= \iint\limits_{\substack{|N\varphi'|<1 \\ |Mx'|<1}} N^d M^m K_{11}(N\varphi', Mx') f(\varphi - \varphi', x - x')\, d\varphi' dx' =$$

$$= \iint\limits_{\substack{|\xi|<1 \\ |\eta|<1}} K_{11}(\xi, \eta)\, f\left(\varphi - \frac{\xi}{N}, \; x - \frac{\eta}{M}\right) d\xi d\eta.$$

Hence the difference $f - T_{NM} f$ is representable as

$$f - T_{NM} f = \iint\limits_{\substack{|\xi|<1 \\ |\eta|<1}} K_{11}(\xi, \eta) \left[f\left(\varphi - \frac{\xi}{N}, \quad x - \frac{\eta}{M} \right) - f(\varphi, x) \right] d\xi d\eta. \tag{14.8}$$

Expanding

$$f\left(\varphi - \frac{\xi}{N}, \quad x - \frac{\eta}{M} \right) - f(\varphi, x)$$

at the point $\xi = \eta = 0$ in Taylor's series and taking the remainder term $r_l(\varphi, x, \xi, \eta)$, defined by the l-th order derivative, it is found that

$$|f - T_{NM} f| \leqslant \iint\limits_{\substack{|\xi|<1 \\ |\eta|<1}} |K_{11}(\xi, \eta)| \, |r_l(\varphi, x, \xi, \eta)| \, d\xi d\eta. \tag{14.9}$$

But

$$r_l(\varphi, x, \xi, \eta) = \frac{1}{l!} d^l_{\xi,\eta} f\left(\varphi - \frac{\xi}{N} \theta, \quad x - \frac{\eta}{M} \theta \right) \quad (0 < \theta < 1),$$

hence

$$|r_l(\varphi, x, \xi, \eta)| < C \sup_{\substack{|\rho|+|r|=l \\ \varphi, x \in D}} N^{-|\rho|} M^{-|r|} \left| \frac{\partial^{|\rho|+|r|} f(\varphi, x)}{\partial \varphi_1^{\rho_1} \ldots \partial x_m^{r_m}} \right|, \tag{14.10}$$

for $\xi| < 1$, $|\eta| < 1$, C a constant not depending on N, M and f.

The inequalities (14.9) and (14.10) imply the estimate (14.7), by setting

$$C = C_1 \iint\limits_{E_n} |K_{11}(\xi, \eta)| \, d\xi d\eta.$$

Let $f(\varphi, 0) = 0$. Then

$$f - T^0_{NM} f = f - T_{NM} f - [f - T_{NM} f]_{x=0},$$

and, consequently,

$$|f - T^0_{NM} f| \leqslant 2 |f - T_{NM} f| \leqslant 2C \sup_{\substack{|\rho|+|r|=l \\ \varphi, x \in D}} N^{-|\rho|} M^{-|r|} \left| \frac{\partial^{|\rho|+|r|} f}{\partial \varphi_1^{\rho_1} \ldots \partial x_m^{r_m}} \right|.$$

Further, let

$$f(\varphi, 0) = \left. \frac{\partial f(\varphi, 0)}{\partial x} \right|_{x=0} = 0.$$

Taking note of (14.6), we may set

$$f - T_{NM}^1 = f - T_{NM} - [f - T_{NM}]_{x=0} - \left(\left[\frac{\partial f}{\partial x} - \frac{\partial T_{NM} f}{\partial x}\right]_{x=0}, x\right). \quad (14.11)$$

However,

$$\frac{\partial}{\partial x} T_{NM} = T_{NM} \frac{\partial}{\partial x};$$

hence, because of the inequalities (14.7), (14.11) implies for $|f - T_{NM}f|$ that

$$f - T_{NM}^1 f \leqslant 2C \sup_{|\rho|+|r|=l} N^{-|\rho|} M^{-|r|} \left| \frac{\partial^{|\rho|+|r|} f(\varphi, x)}{\partial \varphi_1^{\rho_1} \ldots \partial x_m^{r_m}} \right|_0 +$$

$$+ C_2 \sup_i \left| T_{NM} \frac{\partial f}{\partial x_i} - \frac{\partial f}{\partial x_i} \right| \leqslant$$

$$\leqslant 2C \sup_{|\rho|+|r|=l} N^{-|\rho|} M^{-|r|} \left| \frac{\partial^{|\rho|+|r|} f(\varphi, x)}{\partial \varphi_1^{\rho_1} \ldots \partial x_m^{r_m}} \right|_0 +$$

$$+ C_2 C \sup_{|\rho|+|r|=l-1} N^{-|\rho|} M^{-|r|} \sup_i \left| \frac{\partial^{|\rho|+|r|} \partial f(\varphi, x)}{\partial \varphi_1^{\rho_1} \ldots \partial x_m^{r_m} \partial x_i} \right|_0 \leqslant$$

$$\leqslant C_3 \sup_{|\rho|+|r|=l} N^{-|\rho|} M^{-|r|+1} \left| \frac{\partial^{|\rho|+|r|} f(\varphi, x)}{\partial \varphi_1^{\rho_1} \ldots \partial x_m^{r_m}} \right|_0, \quad (14.12)$$

proving the lemma.

Lemma 5. *For every integer* $\lambda \geqslant 0$,

$$|S_N f(\varphi)|_\lambda \leqslant C N^{\lambda+\delta} |f(\varphi)|_0, \quad (14.13)$$

and for $0 \leqslant \lambda \leqslant l - d - 1$ *and* $f(\varphi) \in C^l(E_d)$,

$$|f(\varphi) - S_N f(\varphi)|_\lambda \leqslant C N^{-l+\lambda+\delta} |f(\varphi)|_l, \quad (14.14)$$

where $d < \delta \leqslant d + 1$.

Proof. For $\lambda = 0$, we have

$$|S_N f| \leqslant \sum_{|k| \leqslant N} |f_k| \leqslant |f|_0 \sum_{|k| \leqslant N} 1. \quad (14.15)$$

However, the sum $\sum_{|k| \leqslant N} 1$ is equal to the number of vectors in E_d with integral components for which $|k| \leqslant N$ and hence does not exceed the value $2^d N^d$. The expression (14.15), in view of what has been established, implies (14.13) for $\lambda = 0$ and $C = 2^d$, $\delta = d$.

For $\lambda = |\rho| > 0$, the differentiation of $S_N f(\varphi)$ gives

$$\left| \frac{\partial^{|\rho|} S_N f(\varphi)}{\partial\varphi_1^{\rho_1} \ldots \partial\varphi_d^{\rho_d}} \right| \leqslant \sum_{|k| \leqslant N} |f_k| \, |k_1|^{\rho_1} \ldots |k_d|^{\rho_d} \leqslant |f|_0 \sum_{|k| \leqslant N} |k_1|^{\rho_1} \ldots |k_d|^{\rho_d}.$$

Hence, making use of the well-known inequality [25]

$$|k_1|^{\rho_1} \ldots |k_d|^{\rho_d} \leqslant \left(\frac{\rho_1 \, |k_1| + \ldots + \rho_d \, |k_d|}{\rho_1 + \ldots + \rho_d} \right)^{\rho_1 + \ldots + \rho_d} \leqslant |k|^{|\rho|},$$

we see that the inequality (14.13) yields

$$\left| \frac{\partial^{|\rho|} S_N f(\varphi)}{\partial\varphi_1^{\rho_1} \ldots \partial\varphi_d^{\rho_d}} \right| \leqslant |f|_0 \sum_{|k| \leqslant N} |k|^{|\rho|} \leqslant |f|_0 N^{|\rho|} \sum_{|k| \leqslant N} 1 \leqslant$$

$$\leqslant 2^d N^{|\rho| + d} |f|_0 = C N^{\lambda + \delta} \, |f(\varphi)|_0.$$

Further, let $f(\varphi) \in C^l(E_d)$. Then

$$|f_k| \leqslant \min_{0 \leqslant \rho_1 + \ldots + \rho_d \leqslant l} |k_1|^{-\rho_1} \ldots |k_d|^{-\rho_d} \left| \frac{\partial^{|\rho|} f(\varphi)}{\partial\varphi_1^{\rho_1} \ldots \partial\varphi_d^{\rho_d}} \right|_0 \leqslant$$

$$\leqslant \left\{ \max_{i = 1, \ldots, d} |k_i|^{-l} \right\} |f(\varphi)|_l. \tag{14.16}$$

Using (14.16), we obtain the estimate

$$|f(\varphi) - S_N f(\varphi)|_\lambda \leqslant \sum_{|k| > N} |f_k| \, |k|^\lambda | \leqslant |f|_l \sum_{|k| > N} \frac{|k|^\lambda}{\left\{ \max_i |k_i| \right\}^l} \leqslant$$

$$\leqslant |f|_l d^l \sum_{|k| > N} |k|^{-l-\lambda} = d^l |f|_l \sum_{\nu > N} \nu^{-l+\lambda} \sum_{|k| = \nu} 1 \leqslant$$

$$\leqslant 2 d^l |f|_l \sum_{\nu > N} \nu^{-l+\lambda+d-1}. \tag{14.17}$$

However, for every ε with $0 < \varepsilon \leqslant 1$, we have

$$\sum_{\nu > N} \nu^{-l+\lambda+d+1} < N^{-l+\lambda+d+\varepsilon} \sum_{\nu > N} \nu^{-1-\varepsilon} \left(\frac{N}{\nu} \right)^{l-\lambda-d-\epsilon} <$$

$$< N^{-l+\lambda+d+\varepsilon} \sum_{\nu > N} \nu^{-1-\varepsilon} = C(\varepsilon) \, N^{-l+\lambda+d+\varepsilon}, \tag{14.18}$$

where

$$C(\varepsilon) = \sum_{\nu > N} \nu^{-1-\varepsilon} < \sum_{\nu=1}^{\infty} \nu^{-1-\varepsilon}.$$

The inequalities (14.17) and (14.18) imply the estimate (14.14), with $C = 2d^l C(\varepsilon)$, $\delta = d + \varepsilon$, so that the lemma is established.

The operator S_N is applicable not only to differentiable but also to analytic functions. Hence, we can derive estimates for analytic functions, analogous to (14.13) and (14.14).

Lemma 6 [4]. *Suppose that $f(\varphi)$ is an analytic function for*

$$|\operatorname{Im} \varphi| = \max_{\alpha} |\operatorname{Im} \varphi_\alpha| \leqslant \rho.$$

Then

$$|S_N f(\varphi)| \leqslant 2^d \frac{|f|_0}{\delta^d}, \quad |f(\varphi) - S_N f(\varphi)| \leqslant \left(\frac{2d}{l}\right)^d \frac{e^{-N\delta}|f|_0}{\delta^{d+1}}, \quad (14.19)$$

for

$$|\operatorname{Im} \varphi| = \max_{\alpha} |\operatorname{Im} \varphi_\alpha| \leqslant \rho - 2\delta, \qquad 0 < 2\delta \leqslant \rho < 1.$$

§ 15. Iteration Process with Smoothing

Recalling (12.2), we assume that the operator A involves the loss of derivatives. The construction of the iteration process by (12.3) then turns out to be impossible. In order to obtain the terms, a modification of the iteration process, therefore, becomes necessary. It is only natural that the modified procedure should rest heavily on the active use of smoothing operators. In fact, let $\mathcal{T}_N : C^{r+s_0}(D_N) \to C^\infty(D)$ be a smoothing operator, and let $\mathcal{T}_N \to I$, $D_N \to D$ as $N \to \infty$, where I is the identity operator.

We now choose an increasing sequence of numbers N_s ($N_{s+1} > N_s$, $N_s \to \infty$ as $s \to \infty$) and set

$$u_{s+1} - u_s = \mathcal{T}_{N_{s+1}} A u_s \quad (s = 0, 1, \ldots, u_0 \in C^r(D)). \qquad (15.1)$$

It is not difficult to see that this iteration process does not break down and also that if N_1 is sufficiently large, all the terms u_s belong to the space $C^r(D_0)$, where D_0 is some domain contained in D.

The modified iteration process represents a natural generalization of Picard's iteration process (12.3). The problem of the convergence of this process was investigated by Moser [51]. To obtain a convergent approximation, he drew upon the premise of accelerated convergence and a suitable smoothing process.

He demonstrated that if the operator $\mathcal{T}_{N_{s+1}}$ satisfies the inequality

$$| \mathcal{T}_{N_{s+1}} Au_s |_\rho \leqslant C \left\{ N_s^{2\sigma} \left| u_s - u_{s-1} \right|_\rho^2 + N_s^{-\lambda} N_{s-1}^{\lambda+t} \right\} \qquad (s = 1, 2, \ldots) \ (15.2)$$

for every λ belonging to some sufficiently large interval $0 \leqslant \lambda \leqslant l$, σ fixed, t and the constant C not depending on N and u_s, then the sequence u_s converges in the norm of the space $C^r(D_0)$, when we set

$$N_{s+1} = N_s^{\varkappa}, \quad 1 < \varkappa < 2 \qquad (s = 0, 1, \ldots) \qquad (15.3)$$

and λ and N_0 are taken sufficiently large, and $| u_1 - u_0 |_\rho = | \mathcal{T}_{N_1} Au_0 |_\rho$—sufficiently small.

It must be remarked that the verification of condition (15.2) is very difficult and demands, as a rule, that the original equation be suitably transformed. An important step in such a transformation is connected with solving the corresponding 'linearized equation'. Namely, suppose that the limit

$$\lim_{t \to 0} \frac{A(u + tv) - Au}{t} = A'(u)v$$

exists for suitable u, v and let us denote by $A' = A'(u)$ the operator occurring in the limit.

Assume that the operator A' has an inverse, i.e. that the equation

$$vA' = h \qquad (h \in C^{r-s_0}(D))$$

can be solved for v:

$$v = L(u)h.$$

Set $A_1 = L(u)A$ and consider the equation

$$A_1 u = 0. \qquad (15.4)$$

Identifying this with equation (12.2) we see that the inequality (15.2) is satisfied for A_1, essentially because the iteration process (15.1) for A_1 is an analogue of Newton's iteration process, which assures accelerated convergence.

Making use of the convergence process (15.1), Moser [52] obtained a series of important results in the field of qualitative study of differential equations and, in particular, solved Kolmogorov's theorem on the conservation of a conditionally periodic motion under small perturbations of the Hamiltonian functions for the case of a differentiable Hamiltonian.

The iteration process modified by smoothing holds out a promise for extending the application of this method to a range substantially wider than that surmised in the previous chapter, and constitutes a powerful tool for the investigation of the behaviour, on and in the neighbourhood of a smooth invariant manifold, of trajectories of differential equations encountered in nonlinear mechanics.

Chapter 4

TRAJECTORIES ON A TORUS

§ 16. Behaviour of Trajectories on a Two-Dimensional Torus

Consider in the three-dimensional Euclidean space E_3 the usual torus $\mathcal{T}_2$, i.e. a surface generated by the rotation of a circle S_1 about an axis in its plane but not cutting the circle. Introducing on $\mathcal{T}_2$ the natural system of coordinates $\varphi = (\varphi_1, \varphi_2)$, we define on this torus the system of dynamical equations

$$\begin{aligned} \frac{d\varphi_1}{dt} &= f_1(\varphi_1, \varphi_2), \\ \frac{d\varphi_2}{dt} &= f_2(\varphi_1, \varphi_2), \end{aligned} \tag{16.1}$$

where $f_1(\varphi_1, \varphi_2)$ and $f_2(\varphi_1, \varphi_2)$ are single-valued functions on $\mathcal{T}_2$, 2π-periodic in φ_1 and φ_2. Eliminating the time variable t from (16.1), we obtain

$$\frac{d\varphi_2}{d\varphi_1} = \frac{f_2(\varphi_1, \varphi_2)}{f_1(\varphi_1, \varphi_2)} = F(\varphi_1, \varphi_2), \tag{16.2}$$

which defines a trajectory on the torus.

Assume that $F(\varphi_1, \varphi_2)$ is a bounded twice continuously differentiable function. Let

$$\varphi_2 = \omega(\varphi_1, \theta) \tag{16.3}$$

be a solution of equation (16.2), satisfying $\omega(0, \theta) = \theta$. It is not difficult to see that, regarded as a function of θ, $\omega(\varphi_1, \theta)$ is monotone increasing, with the property :

$$\omega(\varphi_1, \theta + 2\pi) = \omega(\varphi, \theta) + 2\pi.$$

Set

$$f(\theta) = \omega(2\pi, \theta) - \theta,$$

and consider the transformation $T_1 : \theta \to T_1\theta$ defined by

$$T_1 : \theta \to T_1\theta = \theta + f(\theta). \tag{16.4}$$

By the properties deduced above, T_1 can be considered as a one-to-one and continuous mapping of the circle S_1 into itself. The properties of this transformation define the general character of the behaviour of trajectories of the dynamical system (16.1), for which Poincaré and Denjoy [57, 17] have demonstrated the following:

(*a*) The limit

$$2\pi\mu = \lim_{s\to\infty} \frac{T_1^s\theta_0}{s} \tag{16.5}$$

exists and is independent of θ_0;

(*b*) If μ, called the rotation number of the transformation T_1, is irrational then the motion on $\mathcal{T}_2$ is conditionally periodic: every trajectory 'sweeps out' the whole of the torus, in other words is everywhere dense on the torus.

(*c*) If μ is rational, $\mu = r/p$ (r, p integers, relatively prime), then the motions on the torus $\mathcal{T}_2$ are $2\pi p$-periodic, the remaining (nonperiodic) motions are attracted to it as $t \to +\infty$.

The original results of Poincaré and Denjoy gave a strong motivation and stimulus for later researches on trajectories on a torus. A relatively large part of these investigations, initiated by Bolyai [14] goes into clarifying the question, whether the system of equations (16.1) can be reduced, by a suitable transformation of the torus to the canonical form

$$\begin{aligned} \frac{d\psi_1}{dt} &= \lambda_1, \\ \frac{d\psi_2}{dt} &= \lambda_2, \end{aligned} \tag{16.6}$$

in which λ_1, λ_2 are constants.

In order to make the character of the results derived in this section more explicit, it is intended to consider a different form of the above problem. Suppose, as before, that T_1 is a continuous one-to-one mapping of the circle S_1 into itself:

$$\theta \to T_1\theta = \theta + f(\theta), \quad f(\theta + 2\pi) = f(\theta),$$

and that μ is the rotation number of

$$T_1 : \mu = \lim_{s=\infty} \frac{T_1^s\theta}{s}.$$

Denote by T_0 a rotation of S_1 through a constant angle $2\pi\mu$:

$$\theta \to T_0\,\theta = \theta + 2\pi\mu.$$

Then the problem of the transformation of the torus is equivalent to asking whether an l times continuously differentiable mapping T_1 is representable in the form

$$T_1 = S^{-1} T_0 S \tag{16.7}$$

by means of an s_0 $[l = l(s_0)]$ times continuously differentiable one-to-one mapping S:

$$\theta \to S\theta = \theta + g(\theta), \qquad g(\theta + 2\pi) = g(\theta).$$

The solution of this problem is trivial when μ is rational. The representation (16.7) is then possible for a cyclic T_1, that is, if for some power of T_1, all the points of the circle remain fixed; however, this is impossible in the opposite case [20], in which the mapping S for a cyclic T_1 is not one-to-one.

For an irrational μ the representation (16.7) is possible for all T_1 having a bounded rotation of first derivative, though only by means of a continuous mapping S [17]. Apart from taking l sufficiently large instead of small, it is necessary for enhancing the smoothness of S, as shown by a simple example, to restrict the mapping T_1 to exclude cases in which μ is represented by a slowly convergent continued fraction. In this connection, it is customary to assume that μ satisfies the inequality

$$\left| \mu - \frac{r}{p} \right| \geqslant Kp^{-3} \tag{16.8}$$

for all $r, p = 1, 2, \ldots$ and some $K > 0$.

It has been asserted by Finzi [20] that the representation (16.7) is possible for $l = 5/2$ and $s_0 = 1$.†

Kolmogorov [27] has proved the validity of relation (16.7) for analytic mappings T_1 and S, when the origin of system (16.1) is an analytic invariant integral. This has also been shown by Arnold [2] assuming T_1 to be distinct from the rotation T_0,

Subject to the same assumption, the representation (16.7) is proved in [48, 60] for an arbitrary

$$l = l(s_0) \leqslant l_0(s_0) = 1 + 24(s_0 + 2)^2. \tag{16.9}$$

Moser [54] has reduced the relation (16.9) to the simpler form

$$l = l(s_0) \leqslant l_0(s_0) = 1 + s_0.$$

Having sketched briefly the main results of Poincaré and Denjoy, connected with the study of the behaviour of trajectories on a two-dimensional torus, the character of motion on an m-dimensional torus will be examined in the next section.

† The validity of this statement remains an open question because the derivations in [20] are yet to be proved conclusively.

§ 17. Behaviour of Trajectories on an m-Dimensional Torus

It is customary to represent the torus as a direct product of two surfaces: $\mathcal{T}_2 = S_1 \times S_2$. A direct product of m surfaces is called an m-dimensional torus: $\mathcal{T}_m = S_1 \times S_2 \times \ldots \times S_m$. Introducing the angular coordinates $\varphi = (\varphi_1, \ldots, \varphi_m)$ on the torus, we define on $\mathcal{T}_m$ the dynamical system of equations

$$\frac{d\varphi}{dt} = f(\varphi), \tag{17.1}$$

where $f(\varphi) = (f_1(\varphi), \ldots, f_m(\varphi))$ is a function, 2π periodic in φ.

In contrast to a two-dimensional torus, the general character of motion on an m-dimensional torus has not so far been investigated in sufficient detail. Certain basic results due to Arnold [2] are set forth below.

Theorem 11 [2]. *Suppose that* $\mu = (\mu_1, \ldots, \mu_m)$ *is a vector with incommensurable components such that*

$$(\mu, k) | > K | k |^{-m} \quad (| k | \neq 0)$$

for every integral vector $k = (k_1, \ldots, k_m)$.

Then there exists $\varepsilon(R, K, m) > 0$ *such that for every analytic vector field* $f(\varphi) = (f_1, \ldots, f_m)$ *on the torus* (*that is, such that* $f(\varphi + 2\pi k) = f(\varphi)$) *with* $|f(\varphi)| < \varepsilon$ *for* $|\operatorname{Im} \varphi| < R$, *there is a vector* $a = (a_1, \ldots, a_m)$ *for which the system of differential equations*

$$\frac{d\varphi}{dt} = f(\varphi) + a$$

can be reduced by an analytic change of variables to the form

$$\frac{d\theta}{dt} = 2\pi\mu.$$

The extension of Arnold's results to the case of a smooth system was realized by the present authors [48, 60] and by Moser [54] independently.

We now formulate the basic theorem of the present chapter on the reducibility of a system of differential equations with smooth right-hand sides, defined on an m-dimensional torus with a proper rotation. This theorem is of great significance in solving many problems in nonlinear mechanics.

As is known, in nonlinear mechanics the process of proving existence theorems for tori entails the squaring of infinite intervals containing φ_t. Furthermore, because Im φ_t may approach infinity as $t \to \infty$, it may move out of the domain on which the integrand is bounded. This difficulty is resolved by resorting to the smoothing method which requires only differentiability and not analyticity, thereby making the complex domain a convenient one to handle.

Theorem 12. *Suppose that $f(\varphi) = (f_1, \ldots, f_m)$ is a function which is 2π-periodic in φ, and that $\mu = (\mu_1, \ldots, \mu_m)$ is a vector with incommensurable components such that*

$$|(\mu, k)| \geqslant K|k|^{-m} \qquad (|k| \neq 0) \tag{17.2}$$

for every vector $k = (k_1, \ldots, k_m)$.

Then, for arbitrary positive constants ε. c_0, and integer $s_0 \geqslant 1$ there exist $\delta_0(c_0, \varepsilon, s)$ and an integer $l = l(s_0)$ such that if $f(\varphi)$ has continuous derivatives upto order l inclusive and satisfies the inequalities

$$|f(\varphi)|_0 < \delta_0, \qquad |f(\varphi)|_l < c_0,$$

then there exist a constant vector $\Delta = (\Delta_1, \ldots, \Delta_n)$ satisfying the inequality $|\Delta_0| < \varepsilon$ and an s_0-times continuously differentiable function $\Phi(\theta) = (\Phi_1, \ldots, \Phi_m)$, 2π-periodic in $\theta = (\theta_1, \ldots, \theta_m)$, satisfying the inequality $\Phi(\theta)|_{s_0} < \varepsilon$, so that the system of equations

$$\frac{d\varphi}{dt} = \mu + \Delta + f(\varphi) \tag{17.3}$$

can be reduced by the change of variable $\varphi = \theta + \Phi(\theta)$ to the form

$$\frac{d\theta}{dt} = \mu. \tag{17.4}$$

The proof of this theorem is the chief concern of this chapter. However, before we take up this project, some comments are in order concerning the relationship of l and s_0, connecting the order of the differentiability of the functions $f(\varphi)$ and $\Phi(\theta)$. This relation is given [48, 60] by the estimate†

$$l(s_0) \leqslant l_0 = 1 + 8(s_0 + 2)^2(m + 1). \tag{17.5}$$

§ 18. Inductive Theorem

The proof of Theorem 12 is constructive. Given μ and $f(\varphi)$, we construct the function $\Phi(\varphi)$ via transformations, and also the vector Δ.

The following theorem plays an important role in such construction.

Theorem 13. *Suppose that the right-hand sides of the system of equations*

$$\frac{d\varphi}{dt} = \mu + \Delta + f(\varphi, \Delta) \tag{18.1}$$

†This estimate is too generous and was later improved upon by Moser to the form

$$l(s_0) \leqslant l_0(s_0) = 1 + s_0 + m.$$

satisfy the conditions :

(i) *The function* $f(\varphi, \Delta)$, 2π-*periodic in* $\varphi = (\varphi_1, \ldots, \varphi_m)$, *is* $l = l(s_0)$-*times continuously differentiable in* φ, Δ *for*

$$|\Delta| = \max_i |\Delta_i| < M_0^{-1}, \tag{18.2}$$

and satisfies the inequalities

$$|f(\varphi, \Delta)| = \max |f_i(\varphi, \Delta)| < \delta_0,$$

$$\left|\frac{\partial^{\rho_1 + \ldots + r_m} f(\varphi, \Delta)}{\partial \varphi_1^{\rho_1} \ldots \partial \Delta_m^{r_m}}\right| = |D_\varphi^\rho D_\Delta^r f(\varphi, \Delta)| <$$

$$< N_0^{|\rho|} M_0^{|r|} \quad \text{for } |\rho| + |r| = l. \tag{18.3}$$

(ii) *The frequency* μ *satisfies the inequality* (17.2).

Then, it is possible to find δ^0 *not depending on* N_0, M_0 *and* $f(\varphi, \Delta)$ *such that for every* $\delta_0 \leqslant \delta^0$ *there exist the transformations*

$$\varphi = \theta + u(\theta, \Delta^{(1)}),$$
$$\Delta = \Delta^{(1)} + v(\Delta^{(1)}), \tag{18.4}$$

which reduce the system (18.1) *to the form*

$$\frac{d\theta}{dt} = \mu + \Delta^{(1)} + f^{(1)}(\theta, \Delta^{(1)}). \tag{18.5}$$

In addition, $u(\theta, \Delta^{(1)})$, $v(\Delta^{(1)})$ *and* $f^{(1)}(\theta, \Delta^{(1)})$, *defined on the domain* $|\Delta^{(1)}| < M^{-1}$ and 2π-*periodic in* θ, *are* l-*times continuously differentiable and satisfy the inequalities*

$$|v(\Delta^{(1)})| < M^{-1}, \quad |v(\Delta^{(1)})|_{s_0+1} < N^{-1}, \quad |u(\theta, \Delta^{(1)})|_{s_0+1} < N^{-1}, \tag{18.6}$$

$$|f^{(1)}(\theta, \Delta^{(1)})| < \delta, \quad \left|D_\theta^\rho D_{\Delta^{(1)}}^r f^{(1)}(\theta, \Delta^{(1)})\right| \leqslant N^{|\rho|} M^{|r|}$$
$$for\ |\rho| + |r| = l, \tag{18.7}$$

where

$$N = N_0^\alpha, \quad M = N^\nu, \qquad \delta = M^{-\beta}, \ \alpha = 1 + \frac{1}{2(s_0+1)},$$

$$\beta = s_0 + 2, \quad \nu = 4(m+1) + \frac{1}{4(s_0+2)^2},$$

$$l(s_0) = 1 + 8(s_0+2)(m+1). \tag{18.8}$$

Proof. Let $F(\varphi)$ be a 2π-periodic function of φ, representable in the form of the Fourier series

$$F(\varphi) = \sum_k F_k e^{i(k,\varphi)},$$

where

$$F_k = \frac{1}{(2\pi)^m} \int_0^{2\pi} \ldots \int_0^{2\pi} F(\varphi)\, e^{-i(k,\varphi)}\, d\varphi_1 \ldots d\varphi_m.$$

Introduce, as usual, the notations

$$\overline{F}(\varphi) = \frac{1}{(2\pi)^m} \int_0^{2\pi} \ldots \int_0^{2\pi} F(\varphi)\, d\varphi_1 \ldots d\varphi_m,$$

$$\tag{18.9}$$

$$\widetilde{F}(\varphi) = \sum_{|k| \neq 0} \frac{F_k}{i(k,\mu)} e^{i(k,\varphi)}.$$

It is not difficult to verify the following lemma.

Lemma 7. *Let $F(\varphi)$ be a $\tau = (2m + \sigma + 1)$-times continuously differentiable unction. Then $\widetilde{F}(\varphi)$ is a σ-times continuously differentiable solution of the equation*

$$\left(\frac{\partial w}{\partial \varphi}, \mu\right) = F(\varphi) - \overline{F}(\varphi), \tag{18.10}$$

satisfying the inequality

$$\left| \widetilde{F}(\varphi) \right|_\sigma \leqslant c_1 \max_\alpha \left| D^\tau_{\varphi_\alpha} F(\varphi) \right|_0, \tag{18.11}$$

where the constant c_1 does not depend on $F(\varphi)$.

Proof. In fact, since $F(\varphi)$ is a τ-times continuously differentiable function, we have

$$\left| F_k \right| \leqslant \left(\max_\alpha |k_\alpha| \right)^{-\tau} \max_\alpha \left| D^\tau_{\varphi_\alpha} F(\varphi) \right|_0, \tag{18.12}$$

for $k\ | \neq 0$. Making use of the inequalities (17.2) and (18.12), we obtain the estimate

$$\left| D_\varphi^\rho \tilde{F}(\varphi) \right| = \left| \sum_{|k| \neq 0} \frac{(ik_1)^{\rho_1} \dots (ik_m)^{\rho_m}}{i(\mu, k)} F_k e^{i(k, \varphi)} \right| \leqslant K^{-1} \sum_{|k| \neq 0} \max_\alpha k_\alpha^{|\rho|} \times$$

$$\times |k|^m |F_k| \leqslant K^{-1} \max_\alpha \left| D_{\varphi_\alpha}^\tau F(\varphi) \right|_0 \sum_{|k| \neq 0} \left(\max_\alpha |k_\alpha| \right)^{-\tau + |\rho|} \times$$

$$\times |k|^m \leqslant K^{-1} \max_\alpha \left| D_{\varphi_\alpha}^\tau F(\varphi) \right|_0 \sum_{l=1}^{\infty} m^\tau \sum_{|k|=l} |k|^{-\tau+m+|\rho|} \leqslant$$

$$\leqslant K^{-1} m^\tau \max_\alpha \left| D_{\varphi_\alpha}^\tau F(\varphi) \right|_0 \sum_{l=1}^{\infty} l^{-\tau+m+|\rho|} \sum_{|k|=l} 1 \leqslant$$

$$\leqslant K^{-1} m^\tau \max_\alpha \left| D_{\varphi_\alpha}^\tau F(\varphi) \right|_0 \sum_{l=1}^{\infty} l^{-\tau+2m+|\rho|-1} \leqslant$$

$$\leqslant c_1 \max_\alpha \left| D_{\varphi_\alpha}^\tau F(\varphi) \right|_0,$$

which is valid on the domain $0 \leqslant \varphi \leqslant \sigma$. These inequalities prove the convergence of the series (18.9), its σ-times differentiability and the continuity of (18.11).

Denote by T the smoothing operator T_{NM} considered in § 13, and in the system of equations (18.1) put

$$\varphi = \theta + \tilde{T}f(\theta, \Delta), \tag{18.13}$$

$$\Delta^{(1)} = \Delta + \overline{Tf}(\theta, \Delta), \tag{18.14}$$

where

$$\overline{Tf}(\theta, \Delta) = \frac{1}{(2\pi)^m} \int_0^{2\pi} \dots \int_0^{2\pi} Tf(\theta, \Delta)\, d\theta_1 \dots d\theta_m.$$

Since

$$\left(\frac{\partial \tilde{T} f}{\partial \theta}, \mu \right) = Tf(\theta, \Delta) - \overline{Tf}(\theta, \Delta),$$

with the new variable θ the system (18.1) takes the form

$$\frac{d\theta}{dt} = \mu + \Delta^{(1)} + f_1^{(1)}(\theta, \Delta, \Delta^{(1)}), \tag{18.15}$$

with the notation

$$f_1^{(1)}(\theta, \Delta, \Delta^{(1)}) = \left[\left(E + \frac{\partial \tilde{T}f}{\partial \theta}\right)^{-1} - E\right]\Delta^{(1)} + \left(E + \frac{\partial \tilde{T}f}{\partial \theta}\right)^{-1} \times$$

$$\times\, [f(\theta + \tilde{T}f, \Delta) - Tf(\theta, \Delta)]. \tag{18.16}$$

Solving equations (18.14) for Δ we obtain $\Delta = \Delta(\Delta^{(1)})$, and substituting $\Delta = \Delta(\Delta^{(1)})$ in the change-of-variable formula (18.13), we arrive at the system of equations (18.5), where

$$f^{(1)}(\theta, \Delta^{(1)}) = f_1^{(1)}(\theta, \Delta(\Delta^{(1)}), \Delta^{(1)}), \tag{18.17}$$

Thus, to prove the inductive theorem it is necessary to set

$$u = \tilde{T}f(\theta, \Delta(\Delta^{(1)})), \quad v = \Delta(\Delta^{(1)}) - \Delta^{(1)} \tag{18.18}$$

and determine the estimates (18.6) and (18.7).

The problem of the solvability of (18.14) in Δ may be tackled first. By the properties of the smoothing operator T_{NM} the function $a(\Delta) = \overline{Tf(\theta, \Delta)}$ is infinitely differentiable in the domain $|\Delta| < M_0^{-1} - M^{-1}$ and satisfies the inequalities

$$|a(\Delta)|_{s_1} \leqslant |Tf(\theta, \Delta)|_{s_1} \leqslant M^{s_1}|f|_0 \leqslant M^{s_1}\delta_0 = N^{v\left(s_1 - \frac{\beta}{\alpha}\right)} < N^{s_1 - s_0 - 2} \leqslant \frac{1}{2m} \tag{18.19}$$

for $s_1 = 1, 2, \ldots, s_0 + 1$. Consequently, for large N, the equation

$$v = -a(\Delta^{(1)} + v) \tag{18.20}$$

is solvable for v in the domain

$$|\Delta^{(1)}| < M^{-1}, \quad |v| < M^{-1} \quad (2M^{-1} < M_0^{-1}), \tag{18.21}$$

and its solution $v = v(\Delta^{(1)})$ belongs to the space $C^\infty(|\Delta^{(1)}| < M^{-1})$. To obtain the estimate (18.6), we differentiate the identity

$$v(\Delta^{(1)}) + a(\Delta^{(1)} + v(\Delta^{(1)})) = 0.$$

We have

$$\left(E + \frac{\partial a}{\partial \Delta}\right)\frac{\partial v}{\partial \Delta_\alpha^{(1)}} + \frac{\partial a}{\partial \Delta_\alpha} = 0 \quad (\alpha = 1, 2, \ldots, n),$$

or, solving for the vector $\partial v/\partial \Delta_\alpha^{(1)}$,

$$\frac{\partial v}{\partial \Delta_\alpha^{(1)}} = -\left(E + \frac{\partial a}{\partial \Delta}\right)^{-1} \frac{\partial a}{\partial \Delta_\alpha}. \tag{18.22}$$

Taking into account the estimate (18.19), we find that

$$\left|\frac{\partial v}{\partial \Delta_\alpha^{(1)}}\right| \leqslant \sum_{k_0=0}^{\infty} \left(\frac{1}{2}\right)^{k_0} a\,|_1 \leqslant 2M\delta_0 \leqslant N^{-s_0-1}. \tag{18.23}$$

The inequalities (18.21) and (18.23) imply the estimate $|v|_1 \leqslant N^{-s_0-1}$. Furthermore, if we assume that

$$|v|_{s_1} < N^{-s_0-1+s_0} \tag{18.24}$$

for $0 \leqslant s_1 \leqslant s_0 + 1$, we can show that

$$|v|_{s_1+1} < N^{-s_0+s_1}. \tag{18.25}$$

In fact, by differentiating the identity (18.22), it is a straightforward matter to verify that every $s_1 + 1$-th order derivative of the function $v(\Delta^{(1)})$ is a linear combination of the expression

$$B_\rho = D^\rho_{\Delta(1)}\left(E + \frac{\partial a(\Delta(\Delta^{(1)}))}{\partial \Delta}\right)^{-1} D^r_{\Delta(1)} \frac{\partial a(\Delta(\Delta^{(1)}))}{\partial \Delta_\alpha}, \tag{18.26}$$

where $0 \leqslant |\rho| \leqslant s_1$, $|r| = s_1 - |\rho|$.

The expression (18.26) contains derivatives of order not exceeding s_1 of the function $v(\Delta^{(1)})$, so that in evaluating this it is possible, by manipulating the inequality (18.24), to show that

$$|B_\rho| \leqslant \left| D^\rho_{\Delta(1)}\left(E + \frac{\partial a}{\partial \Delta}\right)^{-1}\right| \left| D^r_{\Delta(1)} \frac{\partial a}{\partial \Delta_\alpha}\right| \leqslant$$

$$\leqslant c_1 \max(1, M^{|\rho|+1}\delta_0)\, M^{|r|+1}\delta_0 =$$

$$= c_1 \max(M^{|r|+1}\delta_0,\; M^{s_1+2}\delta_0^2) < c_1 M^{s_1+1}\delta_0. \tag{18.27}$$

The inequalities (18.27) imply (18.25) and, consequently, also the estimate (18.6) of the function $v(\Delta^{(1)})$.

Now, consider the function $u(\theta, \Delta^{(1)})$ defined by (18.18). What has been proved above implies the inequality

$$|\Delta(\Delta^{(1)})|_{s_1} = |\Delta^{(1)} + v(\Delta^{(1)})|_{s_1} < c_2 \quad (0 \leqslant s_1 \leqslant s_0 + 1).$$

So $u(\theta, \Delta^{(1)})$ satisfies the estimate

$$| u(\theta, \Delta^{(1)}) |_{s_1} = | \tilde{T}f(\theta, \Delta(\Delta^{(1)})) |_{s_1} \leqslant c_2 | \tilde{T}f(\theta, \Delta) |_{s_1}.$$

Using the properties of the operator «~» and T, we can write

$$| \tilde{T}f(\theta, \Delta) |_{s_1} \leqslant c_2 \max_{0 \leqslant |r| + |\rho| \leqslant s_1} | D_\theta^{2m+|r|+1} D_\Delta^\rho Tf | \leqslant$$

$$\leqslant c_2 \max_{0 \leqslant |r| + |\rho| \leqslant s_1} N^{2m+|r|+1} M^{|\rho|} |f|_0 < c_2 M^{s_1} N^{2m+1} \delta_0. \quad (18.28)$$

Taking note of the relation between the parameters M, N and δ_0, we see that (18.28) yields the estimate (18.6) of $u(\theta, \Delta^{(1)})$,

$$| u(\theta, \Delta^{(1)}) |_{s_0+1} < N^{-1}.$$

Finally, consider the function $f^{(1)}(\theta, \Delta^{(1)})$. Since the inequality (18.6) has already been established, this can be invoked to yield the estimates

$$\left| \left(E + \frac{\partial \tilde{T}f}{\partial \theta} \right)^{-1} \right| < c_3, \quad \left| \left(E + \frac{\partial \tilde{T}f}{\partial \theta} \right)^{-1} - E \right| = \left| \frac{\partial \tilde{T}f}{\partial \theta} \left(E + \frac{\partial \tilde{T}f}{\partial \theta} \right)^{-1} \right| < c_3,$$

$$\left| \frac{\partial \tilde{T}f}{\partial \theta} \right| < \bar{c}_3 N^{2(m+1)} \delta_0,$$

implying

$$| f^{(1)}(\theta, \Delta^{(1)}) | < \bar{c}_3 N^{2(m+1)} \delta_0 M^{-1} + c_3 [| f(\theta + \tilde{T}f, \Delta) - Tf(\theta + \tilde{T}f, \Delta) |_0 +$$

$$+ | Tf(\theta + \tilde{T}f, \Delta) - Tf(\theta, \Delta) |_0]. \quad (18.29)$$

Furthermore, from the properties of smoothing operators by inequalities (14.3) and (14.7), we get

$$| f(\varphi, \Delta) - Tf(\varphi, \Delta) |_0 < C \sup_{|\rho| + |r| = l} N^{-|\rho|} M^{-|r|} | D_\varphi^\rho D_\Delta^r f(\varphi, \Delta) |_0 \leqslant$$

$$\leqslant C N^{-l} N_0^l = C \left(\frac{N_0}{N} \right)^l = C N_0^{(1-\alpha) l}, \quad (18.30)$$

$$| Tf(\theta + \tilde{T}f, \Delta) - Tf(\theta, \Delta) |_0 < m \left| \frac{\partial Tf(\varphi, \Delta)}{\partial \varphi} \right|_0 | \tilde{T}f |_0 \leqslant$$

$$\leqslant c_4 N \delta_0 | \tilde{T}f |_0 \leqslant \bar{c}_4 N^{2(m+1)} \delta_0^2. \quad (18.31)$$

The inequalities (18.29) through (18.31) imply the estimate (18.7) for $|f^{(1)}(\theta, \Delta^{(1)})|$:

$$|f^{(1)}(\theta, \Delta^{(1)})| < c_5 (N^{2(m+1)} M^{-1}\delta_0 + N^{(1-\alpha)\,l} + N^{2\,(m+1)}\delta_0^2) \leqslant$$

$$\leqslant c_5 \left(N^{2(m+1)} M_0^{\left(1-\frac{1}{\alpha}\right)\beta-\alpha} + N_0^{(1-\alpha)\,l+\alpha\beta\nu} + \right.$$

$$\left. + N^{2(m+1)} M_0^{-\beta\,(2-\alpha)} \right) \delta_0^{\alpha} < \delta^0.$$

The procedure for estimating the l-th order derivative of the function $f^{(1)}(\theta, \Delta^{(1)})$ is as follows: Replace the variables θ, $\Delta^{(1)}$ by θ_1, Δ_1, setting $\theta_1 = N\theta$, $\Delta_1 = M\Delta^{(1)}$ and $(\bar{\Delta}) = M\Delta$ in the expression

$$f^{(1)}(\theta, \Delta^{(1)}) = \left(E + \frac{\partial u(\theta, \Delta^{(1)})}{\partial\theta}\right)^{-1} \left[-\frac{\partial u(\theta, \Delta^{(1)})}{\partial\theta}\Delta^{(1)} + \right.$$

$$+ f(\theta + u(\theta, \Delta^{(1)}), \Delta^{(1)} + v(\Delta^{(1)}))$$

$$\left. - Tf(\theta, \Delta^{(1)} + v(\Delta^{(1)}))\right]. \tag{18.32}$$

Then (18.32) assumes the form

$$F^{(1)}(\theta_1, \Delta_1) = \left(E + \frac{\partial U}{\partial\theta_1}\right)^{-1} \left[-\frac{\partial u}{\partial\theta_1}\Delta_1 \frac{N}{M} + F(\theta_1 + U, \Delta_1 + V)\right.$$

$$\left. - Tf(\theta_1, \Delta_1 + V)\right] \qquad (|\Delta_1| < 1); \tag{18.33}$$

the notations introduced here are

$$V(\Delta_1) = Mv\left(\frac{\Delta_1}{M}\right), \qquad U(\theta_1, \Delta_1) = Nu\left(\frac{\theta_1}{N}, \frac{\Delta_1}{M}\right),$$

$$F(\theta_1, \bar{\Delta}) = Nf\left(\frac{\theta_1}{N}, \frac{\bar{\Delta}}{M}\right), \quad F^{(1)}(\theta_1, \Delta_1) = Nf^{(1)}\left(\frac{\theta_1}{N}, \frac{\Delta_1}{M}\right). \tag{18.34}$$

The identity

$$D_\theta^\rho D_{\Delta^{(1)}}^r f^{(1)}(\theta, \Delta^1) = N^{-1} N^{|\rho|} M^{|r|}$$

is obvious and implies the inequality (18.7) for the l-th order derivative of the function $f^{(1)}(\theta, \Delta^{(1)})$, provided that

$$\left| D_{\theta_1}^\rho D_{\Delta_1}^r F^{(1)}(\theta_1, \Delta^{(1)}) \right| < c_6 \quad \text{for } |\rho| + |r| = l, \tag{18.35}$$

where c_6 is a constant not depending on N and M.

Thus, to complete the proof of the inductive theorem, it remains to verify the inequality (18.35). It is seen from (18.33) that the derivatives of order upto and including l of the function $F^{(1)}$ are definable by means of the derivatives of order not exceeding l of the functions $V, U, \partial U/\partial\theta$ and F. Consequently, to establish (18.35) it is sufficient to show that the norms $V|_l$, $|U|_{l+1}$, $|F|_l$ are bounded constants not depending on N and M. Taking account of (18.19) and (18.22), we obtain the inequality

$$\left| D^{\rho}_{\Delta^{(1)}} v(\Delta^{(1)}) \right| < c_7 M^{|\rho|} \delta_0 \qquad (0 \leqslant |\rho| \leqslant l). \tag{18.36}$$

Similarly, for the derivatives of the function $u(\theta, \Delta^{(1)})$, we have

$$\left| D^r_\theta D^{\rho}_{\Delta^{(1)}} u(\theta, \Delta^{(1)}) \right| < c_7 N^{|r|+2m+1} M^{|\rho|} \delta_0 \qquad (0 \leqslant |\rho| + |r| \leqslant l+1). \tag{18.37}$$

For $|V|_l$ and $|U|_{l+1}$ the inequalities (18.36) and (18.37) lead to the estimates

$$|V|_l = \max_{0\leqslant|\rho|\leqslant l} \left| D^{\rho}_{\Delta_1} V(\Delta_1) \right| = \max_{0\leqslant|\rho|\leqslant l} M^{-|\rho|+1} \left| D^{\rho}_{\Delta^{(1)}} v(\Delta^{(1)}) \right| \leqslant c_7 M \delta_0 \leqslant 1, \tag{18.38}$$

$$\begin{aligned} |U|_{l+1} &= \max_{0\leqslant|r|+|\rho|\leqslant l+1} \left| D^r_{\theta_1} D^{\rho}_{\Delta_1} U(\theta_1, \Delta_1) \right| = \\ &= \max_{0\leqslant|r|+|\rho|\leqslant l+1} N^{-|r|+1} M^{-|\rho|} \left| D^r_\theta D^{\rho}_{\Delta^{(1)}} u(\theta, \Delta^{(1)}) \right| \leqslant \\ &\leqslant c_7 N^{2m+2} \delta_0 < 1. \end{aligned}$$

Consider the norm $|F(\theta, \bar{\Delta})|_l$. Make use of the inequalities (18.3) to obtain, for $|r| + |\rho| = l$,

$$\begin{aligned} |D^r_{\theta_1} D^{\rho}_{\bar\Delta} F(\theta_1, \bar\Delta)| &\leqslant N^{1-|r|} M^{-|\rho|} \left| D^r_\theta D^{\rho}_\Delta f(\theta, \Delta) \right| \leqslant \\ &\leqslant \max_{|r|+|\rho|\leqslant l} N \left(\frac{N_0}{N} \right)^{-|r|} \left(\frac{M_0}{M} \right)^{-|\rho|} \leqslant \\ &\leqslant N N_0^{(1-\alpha)\, l} \leqslant 1. \end{aligned} \tag{18.39}$$

Since the function $F(\theta_1, \bar\Delta)$ satisfies, in addition to (18.39), the inequality

$$|F(\theta_1 \bar\Delta)| \leqslant N |f(\theta, \Delta)| \leqslant N\delta_0 < 1, \tag{18.40}$$

t is sufficient for evaluating the norm $|F|_l$ to manipulate one of the relations connecting $|F|_0$, $|F|_l$ and $|F|_{l'}$ for $0 \leqslant l' \leqslant l$. Making use of the properties of smoothing operators for this purpose, we get

$$|F(\theta_1, \bar{\Delta})|_{l'} = \max_{0 \leqslant |\rho|+|r| \leqslant l'} \left| D^r_{\theta_1} D^\rho_{\bar{\Delta}} F(\theta_1, \bar{\Delta}) \right| \leqslant$$

$$\leqslant \max_{0 \leqslant |\rho|+|r| \leqslant l'} \left| D^r_{\theta_1} D^\rho_{\bar{\Delta}} F(\theta_1, \bar{\Delta}) - T_{11} D^r_{\theta_1} D^\rho_{\bar{\Delta}} F(\theta_1, \bar{\Delta}) \right| +$$

$$+ \left| T_{11} D^r_{\theta_1} D^\rho_{\bar{\Delta}} F(\theta_1, \bar{\Delta}) \right| \leqslant$$

$$\leqslant 2c \quad \text{for } |\bar{\Delta}| \leqslant \frac{M}{M_0} - 1,$$

which completes the proof.

§ 19. Proof of the Theorem on the Reducibility of Equations on a Torus

It is not difficult to see that the hypotheses of Theorem 12 for parameters N_0, M_0, δ_0 verify the conditions (18.3), if δ_0 is so small that $\delta_0 < \delta^0 = \min(\varepsilon^\beta, c_0^{-\beta\nu})$. Hence, applying the inductive theorem to the equation

$$\frac{d\varphi}{dt} = \mu + \Delta + f(\varphi), \tag{19.1}$$

we find that the substitutions

$$\begin{aligned} \varphi &= \varphi^{(1)} + u^{(1)}(\varphi^{(1)}, \Delta^{(1)}) = \varphi^{(1)} + \overline{T_{N_1 M_1} f(\varphi^{(1)}, \Delta(\Delta^{(1)}))} \\ \Delta &= \Delta(\Delta^{(1)}) = \Delta^{(1)} + v^{(1)}(\Delta^{(1)}) \end{aligned} \tag{19.2}$$

reduce eqn. (19.1) to the form

$$\frac{d\varphi^{(1)}}{dt} = \mu + \Delta^{(1)} + f^{(1)}(\varphi^{(1)}, \Delta^{(1)}), \tag{19.3}$$

where $u^{(1)}(\varphi^{(1)}, \Delta^{(1)})$, $v^{(1)}(\Delta^{(1)})$, $f^{(1)}(\varphi^{(1)}, \Delta^{(1)})$ are 2π-periodic functions in $\varphi^{(1)}$ and satisfy the inequalities

$$\begin{aligned} &|v^{(1)}(\Delta^{(1)})| \leqslant M_1^{-1} = M_0^{-\alpha}, \quad |v^{(1)}(\Delta^{(1)})|_{s_0+1} \leqslant N_1^{-1} = N_0^{-\alpha}, \\ &|u^{(1)}(\varphi^{(1)}, \Delta^{(1)})|_{s_0+1} \leqslant N_1^{-1}, \quad |f^{(1)}(\varphi^{(1)}, \Delta^{(1)})| \leqslant \delta_1 = \delta_0^\alpha, \\ &\left| D^r_{\varphi^{(1)}} D^\rho_{\Delta^{(1)}} f^{(1)}(\varphi^{(1)}, \Delta^{(1)}) \right| \leqslant N_1^{|r|} M_1^{|\rho|} \end{aligned} \tag{19.4}$$

with $|r| + |\rho| = l$, for $|\Delta^{(1)}| \leqslant M_1^{-1}$.

The inequalities (19.4) assure the applicability of the inductive theorem to eqn. (19.3). Hence, it is possible to find the substitutions

$$\varphi^{(1)} = \varphi^{(2)} + u^2(\varphi^{(2)}, \Delta^{(2)}) = \varphi^{(2)} + \overline{T_{N_2M_2} f^{(1)}}(\varphi^{(2)}, \Delta^{(1)}(\Delta^{(2)})),$$
$$\Delta^{(1)} = \Delta^{(1)}(\Delta^{(2)}) = \Delta^{(2)} + v^{(2)}(\Delta^{(2)}), \tag{19.5}$$

which reduce eqn. (19.3) to the form

$$\frac{d\varphi^{(2)}}{dt} = \mu + \Delta^{(2)} + f^{(2)}(\varphi^{(2)}, \Delta^{(2)}), \tag{19.6}$$

where $u^{(2)}(\varphi^{(2)}, \Delta^{(2)})$, $v^{(2)}(\Delta^{(2)})$, $f^{(2)}(\varphi^{(2)}, \Delta^{(2)})$ are 2π-periodic functions in $\varphi^{(2)}$, which satisfy the inequalities

$$v^{(2)}(\Delta^{(2)}) | \leqslant M_2^{-1} = M_1^{-\alpha}, \quad | v^{(1)}(\Delta^{(2)}) |_{s_0+1} \leqslant N_2^{-1} = N_1^{-\alpha},$$
$$u^{(2)}(\varphi^{(2)}, \Delta^{(2)}) |_{s_0+1} \leqslant N_2^{-1}, \qquad | f^{(2)}(\varphi^{(2)}, \Delta^{(2)}) | \leqslant \delta_2 = \delta_1^{\alpha}, \tag{19.7}$$
$$| D^r_{\varphi^{(2)}} D^{\rho}_{\Delta^{(2)}} f^{(2)}(\varphi^{(2)}, \Delta^{(2)}) | \leqslant N_2^{|r|} M_2^{|\rho|} \qquad \text{for } |r| + |\rho| = l$$

with $| \Delta^{(2)} | \leqslant M_2^{-1}$.

The inequalities (19.7) ensure the applicability of the inductive theorem to eqn. (19.6), etc.

Thus, for an arbitrary s ($s = 1, 2, \ldots$), we can find the substitutions

$$\varphi^{(s-1)} = \varphi^{(s)} + u^{(s)}(\varphi^{(s)}, \Delta^{(s)}) = \varphi^{(s)} + \overline{T_{N_sM_s} f^{(s-1)}}(\varphi^{(s)}, \Delta^{(s-1)}(\Delta^{(s)})),$$
$$\Delta^{(s-1)} = \Delta^{(s-1)}(\Delta^{(s)}) = \Delta^{(s)} + v^{(s)}(\Delta^{(s)}), \tag{19.8}$$

which reduce the equation for $\varphi^{(s-1)}$ to the form

$$\frac{d\varphi^{(s)}}{dt} = \mu + \Delta^{(s)} + f^{(s)}(\varphi^{(s)}, \Delta^{(s)}), \tag{19.9}$$

where $u^{(s)}(\varphi^{(s)}, \Delta^{(s)})$, $v^{(s)}(\Delta^{(s)})$, $f^{(s)}(\varphi^{(s)}, \Delta^{(s)})$ are 2π-periodic functions in $\varphi^{(s-1)}$ satisfying the inequalities

$$| v^{(s)}(\Delta^{(s)}) | \leqslant M_s^{-1} = M_{s-1}^{-\alpha}, \quad | v^{(s)}(\Delta^{(s)}) |_{s_0+1} \leqslant N_s^{-1} = N_{s-1}^{-\alpha},$$
$$| u^{(s)}(\varphi^{(s)}, \Delta^{(s)}) |_{s_0+1} \leqslant N_s^{-1}, \quad | f^{(s)}(\varphi^{(s)}, \Delta^{(s)}) | \leqslant \delta_s = \delta_{s-1}^{\alpha}, \tag{19.10}$$
$$\left| D^r_{\varphi^{(s)}} D^{\rho}_{\Delta^{(s)}} f^{(s)}(\varphi^{(s)}, \Delta^{(s)}) \right| \leqslant N_s^{|r|} M_s^{|\rho|}$$

with $|r| + |\rho| = l$ for $| \Delta^{(s)} | \leqslant M_s^{-1}$.

It is evident that, when combined, the substitutions (19.2), (19.5) and (19.8) reduce to the form

$$\varphi = \varphi^{(s)} + \Phi^{(s)}(\varphi^{(s)}, \Delta^{(s)}) = \theta + \Phi^{(s)}(\theta, \Delta^{(s)}),$$
$$\Delta = A^{(s)}(\Delta^{(s)}) = \Delta^{(s)} + \Psi^{(s)}(\Delta^{(s)}), \tag{19.11}$$

which relate φ, Δ to $\varphi^{(s)} = \theta$, $\Delta^{(s)}$ and reduce the original equation (19.1) to the form (19.9). Now, if we assume that $|\Delta^{(s)}| \leqslant M_s^{-1} \to 0$ and that $|f^{(s)}(\theta, \Delta^{(s)})| \leqslant \delta_s \to 0$ as $s \to \infty$, then what remains in proving the reducibility theorem is to show the convergence of the numerical sequence $A^{(s)}(0)$ and the uniform convergence of the sequence of functions $\Phi^{(s)}(\theta, 0)$ and their derivatives up to order s_0.

Before proceeding to prove the convergence of the above sequences we note that, by the change-of-variable formulas (19.8) and (19.11), the relations

$$A^{(s+1)}(\Delta^{(s+1)}) = A^{(s)}(\Delta^{(s)}(\Delta^{(s+1)})),$$
$$\Phi^{(s+1)}(\theta, \Delta^{(s+1)}) = u^{(s+1)}(\theta, \Delta^{(s+1)}) + \Phi^{(s)}(\theta + u^{(s+1)}(\theta, \Delta^{(s+)}), \Delta^{(s)}(\Delta^{(s+1)})) \tag{19.12}$$

hold for the functions $A^{(s+1)}(\Delta^{(s+1)})$ and $\Phi^{(s+1)}(\theta, \Delta^{(s+1)})$.

Let $|\Delta^{(s+1)}| \leqslant M_{s+1}^{-1}$, then (19.8) and (19.10) imply that the values $\Delta^{(j)}(\Delta^{(j+1)}(\ldots\Delta^{(s+1)}))$ do not leave the domain $|\Delta^{(j)}| \leqslant M_j^{-1}$ for every $j = 1, 2, \ldots, s$. Hence differentiating the expression $\Delta^{(j)}(\Delta^{(j+1)}(\ldots\Delta^{(s+1)}))$, it is possible to exploit the estimate (19.10) for the function $v^{(j)}(\Delta^{(j)})$ and obtain the system of inequalities

$$\sum_{|\rho|=1} |D^{\rho}_{\Delta^{(1)}}\Delta| \leqslant 1 + mN_1^{-1}, \ldots, \sum_{|\rho|=1} |D^{\rho}_{\Delta^{(s)}}\Delta^{(s-1)}| \leqslant 1 + mN_s^{-1},$$

as a result of which, the inequalities

$$\sum_{|\rho|=1} |D^{\rho}_{\Delta^{(s)}}A^{(s)}(\Delta^{(s)})| \leqslant (1 + mN_1^{-1}) \ldots (1 + mN_s^{-1}) \leqslant \leqslant \prod_{1\leqslant p<\infty}\left(1 + mN_0^{-\alpha^p}\right) = c(N_0) \leqslant 1 \tag{19.13}$$

hold for the first derivatives of the function $A^{(s)}(\Delta^{(s)})$ with $|\Delta^{(s)}| < M_s^{-1}$. Here $c(N_0) \to 0$ as $N_0 \to \infty$.

We use (19.13) to obtain

$$|A^{(s+1)}(0) - A^{(s)}(0)| = |A^{(s+1)}(\Delta^{(s)}(0)) - A^{(s)}(0)| \leqslant |\Delta^{(s)}(0)| < M_s^{-1},$$

which implies the well-known test for convergence to be satisfied by the sequence $\{A^{(s)}(0)\}$:

$$|A^{(s+k_0)}(0) - A^{(s)}(0)| \leqslant \sum_{p=0}^{\infty} M_{s+p}^{-1} \quad (k_0 \geqslant 1). \tag{19.14}$$

Thus, the limit

$$\lim_{s\to\infty} A^{(s)}(0) = A^{(\infty)}(0) \tag{19.15}$$

exists and satisfies the inequality

$$|A^{(\infty)}(0)| \leqslant \lim_{k_0\to\infty} |A^{(1+k_0)}(0) - A^{(1)}(0)| + |A^{(1)}(0)| \leqslant$$

$$\leqslant \sum_{p=0}^{\infty} M_{1+p}^{-1} + \overline{|T_{N_1M_1}f(\varphi)|} \leqslant \sum_{p=0}^{\infty} M_p^{-1}. \tag{19.16}$$

The uniform convergence of the sequence of functions $\Phi^s(\theta, 0)$ has still to be shown. For this, we differentiate the second identity in (19.12) and evaluate the first derivatives of the function $\Phi^{(s)}(\theta, \Delta^{(s)})$.

We introduce the notation

$$x = (x_1, x_2) = (\theta, \Delta),$$

$$w^{(s)} = (w_1^{(s)}, w_2^{(s)}) = (u^{(s)}, v^{(s)}), \tag{19.17}$$

$$y = (y_1, y_2), \quad y = x + w^{(s+1)}(x),$$

which allows us to reduce the second identity in (19.12) to the simplified form

$$\Phi^{(s+1)}(x) = u^{(s+1)}(x) + \Phi^{(s)}(x + w^{(s+1)}(x)). \tag{19.18}$$

We differentiate (19.18) to obtain

$$\frac{\partial \Phi^{(s+1)}(x)}{\partial x_\alpha} = \frac{\partial u^{(s+1)}(x)}{\partial x_\alpha} + \sum_{\beta=1}^{2m} \frac{\partial \Phi^{(s)}(y)}{\partial y_\beta}\left(\delta_{\alpha\beta} + \frac{\partial w_\beta^{(s+1)}(x)}{\partial x_\alpha}\right), \tag{19.19}$$

where

$$\delta_{\alpha\beta} = \begin{cases} 0 \text{ for } \alpha \neq \beta, \\ 1 \text{ for } \alpha = \beta. \end{cases}$$

Evaluating the derivatives of the function $\Phi^{(s+1)}(x)$ with regard to (19.10) and (19.19), we obtain the inequality

$$\sum_{\alpha=1}^{2m} \left|\frac{\partial \Phi^{(s+1)}(x)}{\partial x_\alpha}\right|_0 \leqslant 2mN_{s+1}^{-1} + \sum_{\beta=1}^{2m} \left|\frac{\partial \Phi^{(s)}(y)}{\partial y_\beta}\right|_0 (1 + 2mN_{s+1}^{-1}). \tag{19.20}$$

We introduce the notation

$$z_s = \sum_{\alpha=1}^{2m} \left|\frac{\partial \Phi^{(s+1)}(x)}{\partial x_\alpha}\right|_0,$$

and rewrite (19.20) in the form

$$z_{s+1} \leqslant 2mN_{s+1}^{-1} + z_s(1 + 2mN_{s+1}^{-1}). \tag{19.21}$$

Since $z_0 = 0$, it is easy, starting from (19.21), to obtain the inequality

$$z_s \leqslant \prod_{1 \leqslant p \leqslant s} (1 + 2mN_p^{-1}) - 1 \leqslant e^{\sum_{p=0}^{\infty} 2mN_p^{-1}} - 1 \leqslant 1,$$

whence

$$\left| \sum_{\alpha=1}^{2m} \frac{\partial \Phi^{(s)}(x)}{\partial x_\alpha} \right|_0 = \sum_{|\rho|+|r|=1} | D_\theta^\rho D_{\Delta^{(s)}}^r \Phi^{(s)}(\varphi, \Delta^{(s)}) |_0 \leqslant 1. \tag{19.22}$$

We manipulate (19.22) to obtain

$$| \Phi^{(s+1)}(\theta, 0) - \Phi^{(s)}(\theta, 0) | \leqslant | u^{(s+1)}(\theta, 0) | + | \Phi^{(s)}(\theta + u^{(s+1)}(\theta, 0), \Delta^{(s)}(0))$$

$$- \Phi^{(s)}(\theta, 0) | \leqslant N_{s+1}^{-1} = 2N_{s+1}^{-1},$$

which assures the uniform convergence of the sequence of functions $\Phi^{(s)}(\theta, 0)$:

$$| \Phi^{(s+k_0)}(\theta, 0) - \Phi^{(s)}(\theta, 0) | \leqslant 2 \sum_{p=1}^{\infty} N_{s+p}^{-1} \quad (k_0 \geqslant 1). \tag{19.23}$$

Thus, the limit

$$\lim_{s \to \infty} \Phi^{(s)}(\theta, 0) = \Phi^{(\infty)}(\theta, 0) \tag{19.24}$$

exists uniformly with respect to θ, and satisfies the inequality

$$| \Phi^{(\infty)}(\theta, 0) | \leqslant \lim_{k_0 \to \infty} | \Phi^{(1+k_0)}(\theta, 0) - \Phi^{(1)}(\theta, 0) | + | \Phi^{(1)}(\theta, 0) | \leqslant$$

$$\leqslant 2 \sum_{p=1}^{\infty} N_{1+p}^{-1} + | u^{(1)}(\theta, 0) | \leqslant 2 \sum_{p=1}^{\infty} N_p^{-1}. \tag{19.25}$$

Since the function $\Phi^{(s)}(\theta, 0)$ is continuous, the convergence of (19.24) implies the continuity of the limit function $\Phi^{(\infty)}(\theta, 0)$. To show that it is s_0-times differentiable, we differentiate (19.19) to obtain

$$\frac{\partial^2\Phi^{(s+1)}(x)}{\partial x_{\alpha_2}\,\partial x_\alpha} = \frac{\partial^2 u^{(r+1)}(x)}{\partial x_{\alpha_2}\,\partial x_\alpha} + \sum_{\beta=1}^{2m}\sum_{\beta_2=1}^{2m} \frac{\partial^2\Phi^{(s)}(y)}{\partial y_{\beta_2}\,\partial y_\beta}\left(\delta_{\alpha_2\beta_2} + \frac{\partial w_{\beta_2}^{(s-1)}(x)}{\partial x_{\alpha_2}}\right)\times$$

$$\times\left(\delta_{\alpha\beta} + \frac{\partial w_\beta^{(s+1)}(x)}{\partial x_\alpha}\right) + \sum_{\beta=1}^{2m}\frac{\partial\Phi^{(s)}(y)}{\partial y_\beta}\cdot\frac{\partial^2 w_\beta(x)}{\partial x_{\alpha_2}\,\partial x_\alpha} \tag{19.26}$$

for every $\alpha_2 = 1, 2, \ldots, 2m$.

Taking account of the estimates obtained for the functions $u^{(s+1)}(x)$, $w^{(s+1)}(x)$ and their first and second derivatives, and also of the estimate of the first derivatives of the functions $\Phi^{(s)}(x)$, we infer from the inequality (19.22) that

$$\sum_{\alpha_2,\alpha=1}^{2m}\left|\frac{\partial^2\Phi^{(s+1)}(x)}{\partial x_{\alpha_2}\,\partial x_\alpha}\right|_0 \leqslant 2\sum_{\alpha_2,\alpha=1}^{2m} N_{s+1}^{-1} + \sum_{\beta_2,\beta=1}^{2m}\left|\frac{\partial^2\Phi^{(s)}(y)}{\partial y_{\beta_2}\,\partial y_\beta}\right|_0 (1 + 2mN_{s+1}^{-1})^2 =$$

$$= 2(2m)^2 N_{s+1}^{-1} + \sum_{\beta_2,\beta=1}^{2m}\left|\frac{\partial^2\Phi^{(s)}(y)}{\partial y_{\beta_2}\,\partial y_\beta}\right|_0 (1 + 2mN_{s+1}^{-1})^2.$$

If we write

$$z_s = \sum_{\alpha_2,\alpha=1}^{2m}\left|\frac{\partial^2\Phi^{(s)}(x)}{\partial x_{\alpha_2}\,\partial x_\alpha}\right|_0,$$

then we can express the last inequality in the form

$$z_{s+1} \leqslant 2(2m)^2 N_{s+1}^{-1} + z_s(1 + 2m\,N_{s+1}^{-1})^2. \tag{19.27}$$

The desired solution of the inequality (19.27) in the form of a product is

$$z_s = y_s \prod_{0\leqslant p\leqslant s}(1 + 2m\,N_p^{-1})^2, \tag{19.28}$$

where y_s is given by

$$y_{s+1} \leqslant \frac{2(2m)^2 N_{s+1}^{-1}}{\prod_{0\leqslant p\leqslant s}(1 + 2mN_p^{-1})^2} + y_s \leqslant 2(2m)^2 N_{s+1}^{-1} + y_s.$$

For $y_0 = 0$ this implies the inequality

$$y_s \leqslant 2\,(2m)^2 \sum_{p=0}^{\infty} N_{p+1}^{-1},$$

which leads to

$$\sum_{|\rho|+|r|=2} |D_\theta^\rho D_{\Delta^{(s)}}^r \Phi^{(s)}(\theta, \Delta^{(s)})| = \sum_{\alpha_2,\,\alpha=1}^{2m} \left|\frac{\partial^2 \Phi^{(s)}(x)}{\partial x_{\alpha_2}\,\partial x_\alpha}\right|_0 \leqslant$$

$$\leqslant 2\,(2m)^2 \prod_{0\leqslant p<\infty} (1 + 2mN_p^{-1})^2 \sum_{p=0}^{\infty} N_{p+1}^{-1} \leqslant$$

$$\leqslant 2\,(2m)^2\, e^{2\sum_{p=0}^{\infty} 2mN_p^{-1}} \sum_{p=0}^{\infty} N_{p+1}^{-1} \leqslant 1. \tag{19.29}$$

Making use of the relations (19.19), (19.22) and the estimate of the difference

$\left|\dfrac{\partial \Phi^{(s+1)}(\theta, 0)}{\partial \theta_\alpha} - \dfrac{\partial \Phi^{(s)}(\theta, 0)}{\partial \theta_\alpha}\right|$, we find

$$\sum_{\alpha=1}^{m} \left|\frac{\partial \Phi^{(s+1)}(\theta, 0)}{\partial \theta_\alpha} - \frac{\partial \Phi^{(s)}(\theta, 0)}{\partial \theta_\alpha}\right|_0 \leqslant$$

$$\leqslant \sum_{\alpha=1}^{m} \left|\frac{\partial u^{(s+1)}(\theta, 0)}{\partial \theta_\alpha}\right|_0 +$$

$$+ \sum_{\alpha=1}^{m} \left|\frac{\partial \Phi^{(s)}(\theta + u^{(s+1)}(\theta, 0), \Delta^{(s)}(0))}{\partial \theta_\alpha} - \frac{\partial \Phi^{(s)}(\theta, 0)}{\partial \theta_\alpha}\right|_0 +$$

$$+ \sum_{\beta=1}^{m} \left|\frac{\partial \Phi^{(s)}(\varphi, \Delta^{(s)}(0))}{\partial \varphi_\beta}\right|_0 \sum_{\alpha=1}^{m} \left|\frac{\partial u_\beta^{(s+1)}(\theta, 0)}{\partial \theta_\alpha}\right|_0 \leqslant$$

$$\leqslant 2 \cdot 2mN_{s+1}^{-1} + \sum_{\alpha_2,\,\alpha=1}^{m} \left|\frac{\partial^2 \Phi^{(s)}(x)}{\partial x_{\alpha_2},\,\partial x_\alpha}\right|_0 |w^{(s+1)}|_0 \leqslant$$

$$\leqslant 4mN_{s+1}^{-1} + N_{s+1}^{-1} = (1 + 4m)\,N_{s+1}^{-1}, \tag{19.30}$$

from which we deduce immediately the uniform convergence of the sequence

$$\sum_{\alpha=1}^{m}\left|\frac{\partial\Phi^{(s+k_0)}(\theta, 0)}{\partial\theta_\alpha}-\frac{\partial\Phi^{(s)}(\theta, 0)}{\partial\theta_\alpha}\right|_0 \leqslant (1+4m)\sum_{p=1}^{\infty} N_{s+p}^{-1} \quad (k_0 \geqslant 1). \tag{19.31}$$

Thus, the function $\Phi^{(\infty)}(\theta, 0)$ has the first-order derivatives

$$\frac{\partial\Phi^{(\infty)}(\theta, 0)}{\partial\theta_\alpha}=\lim_{s\to\infty}\frac{\partial\Phi^{(s)}(\theta, 0)}{\partial\theta_\alpha}, \tag{19.32}$$

which satisfy

$$\left|\frac{\partial\Phi^{(\infty)}(\theta, 0)}{\partial\theta_\alpha}\right| \leqslant \lim_{k\to\infty}\left|\frac{\partial\Phi^{(1+k)}(\theta, 0)}{\partial\theta_\alpha}-\frac{\partial\Phi^{(1)}(\theta, 0)}{\partial\theta_\alpha}\right|+\left|\frac{\partial\Phi^{(1)}(\theta, 0)}{\partial\theta_\alpha}\right| \leqslant$$

$$\leqslant (1+4m)\sum_{p=1}^{\infty} N_{1+p}^{-1}+N_1^{-1} \leqslant (1+4m)\sum_{p=1}^{\infty} N_p^{-1}. \tag{19.33}$$

Differentiating (19.26) and taking into account the estimates of the functions $u^{(s+1)}(x)$, $w^{(s+1)}(x)$ and their derivatives upto and including third order, as well as of the estimates for first and second derivatives of the functions $\Phi^{(s)}(x)$ by (19.22) and (19.29), we verify that the criterion for uniform convergence is satisfied by the sequence of second derivatives of the functions $\Phi^{(s)}(\theta, 0)$:

$$\sum_{\alpha_2,\alpha=1}^{m}\left|\frac{\partial^2\Phi^{(s+k_0)}(\theta, 0)}{\partial\theta_{\alpha_2}\,\partial\theta_\alpha}-\frac{\partial^2\Phi^{(s)}(\theta, 0)}{\partial\theta_{\alpha_2}\,\partial\theta_\alpha}\right|_0 \leqslant c_2\sum_{p=1}^{\infty} N_{s+p}^{-1} \quad (k_0 \geqslant 1).$$

This proves the existence of the second order continuous derivatives of the functions $\Phi^{(\infty)}(\theta, 0)$ and gives for them the estimate

$$\left|\frac{\partial^2\Phi^{(\infty)}(\theta, 0)}{\partial\theta_{\alpha_2}\,\partial\theta_\alpha}\right| \leqslant c_2\sum_{p=1}^{\infty} N_p^{-1}.$$

Repeating the above arguments, we can prove the s_0-times differentiability of the functions $\Phi^{(\infty)}(\theta, 0)$ and also obtain the estimate

$$|\Phi^{(\infty)}(\theta, 0)|_{s_0} \leqslant c_{s_0}\sum_{p=1}^{\infty} N_p^{-1} \tag{19.34}$$

for the norm of s_0-th derivative.

The inequality (19.34) concludes the proof of Theorem 12, because by the change-of-variable formula

$$\varphi = \theta + \Phi^{(\infty)}(\theta, 0) \tag{19.35}$$

the original eqn. (19.1) reduces, for $\Delta = A^{(\infty)}(0)$, to the form

$$\frac{d\theta}{dt} = \mu,$$

and this substitution satisfies all the hypotheses of Theorem 12†, as does the constant $A^{(\infty)}(0)$.

Example. Let $f(\varphi)$ be an analytic function of the parameter γ in some domain ϑ and satisfy everywhere in this domain, the hypotheses of the reducibility Theorem 12 with fixed constants δ_0, c_0.

Then $A^{(s)}(0)$ and $\Phi^{(s)}(\theta, 0)$ are analytic functions of γ on ϑ and, since the convergence of the sequences of $A^{(s)}(0)$ and $\Phi^{(s)}(\theta, 0)$ is defined by the quantities δ_0, c_0 not depending on γ, the limit functions $A^{(\infty)}(0)$ and $\Phi^{(\infty)}(\theta, 0)$ are also analytic in γ on this domain.

†See Appendix 6.

Chapter 5

LINEAR SYSTEMS WITH QUASI-PERIODIC COEFFICIENTS

§ 20. Reducibility Theorem

Consider a system of differential equations

$$\frac{dx}{dt} = Ax + P(\varphi)x,$$
$$\frac{d\varphi}{dt} = \omega, \tag{20.1}$$

where A is a constant, $P(\varphi)$ is an $n \times n$ matrix, 2π-periodic in $\varphi = (\varphi_1, \ldots, \varphi_m)$ and real for real φ ; $\omega = (\omega_1, \ldots, \omega_m)$ are frequencies of the matrix $P(\omega t)$; $x = (x_1, \ldots, x_n)$ is an n-dimensional vector and t is time.

For the system (20.1), we seek the change of variable

$$x = \Phi(\varphi)y \tag{20.2}$$

with $\Phi(\varphi)$ a non-singular marix, periodic and real for real φ, reducing the system (20.1) to the following system with constant coefficients :

$$\frac{dy}{dt} = A_0 y.$$
$$\frac{d\varphi}{dt} = \omega. \tag{20.3}$$

This problem has been considered by several authors. For the linear systems with periodic coefficients, i.e. for the system (20.1) with $m = 1$, the results of Floquet and Lyapunov, establishing the existence of the substitution (20.2) are well known.

The problems of the construction of solutions and of the reducibility of equations with quasi-periodic coefficients of form (20.1) and even more general form have been investigated by Artem'ev [5], Erugin [18, 19], Shtokalo [67], Kolmogorov [27-29], Belaga [6], Gel'man [22], Andrianova [1], Mitropoliskii [41], Mitropoliskii and Samoilenko [48-50], Samoilenko [66], Blinov [7], Harashal [24] and others. However, the problem of the reducibility of a system with quasi-periodic coefficients is yet to be solved completely.

The present chapter is devoted to the problem of finding a solution of the system (20.1), when the vector function $P(\varphi)$ is small, by constructing the transformation matrix $\Phi(\varphi)$ by the methods set forth earlier.

It may be remarked that for the system (20.1) with small $P(\varphi)$, Shtokalo [67] constructed asymptotic solutions invoking asymptotic methods of nonlinear mechanics and examined their stability. However, the expansions obtained by him were asymptotic expansions and because of the presence of small divisors were, in general, divergent. The reducibility of system (20.1) to a system with constant coefficients of the form (20.3) could thus be considered only in an asymptotic sense.

By means of the method outlined above, it will emerge from the subsequent discussion that the reducibility of system (20.1) to the system (20.3) with constant coefficients is effected by the transformation matrix $\Phi(\varphi)$, expressed as a rapidly convergent series.

System (20.1), the main object of investigation in this chapter is, generally speaking, a special case of the system of equations (7.4) with $F(h, \varphi) = P(\varphi)\, h$. In spite of the fact that in constructing the solutions it is the method and results of Chap. 2 that shall be used, it has been considered expedient to examine the system (20.1) in a separate chapter. This is fully warranted by the fact that systems of linear equations with quasi-periodic coefficients occupy a central place in the theory of differential equations. In addition, for the system of eqns. (20.1), apart from constructing the transformation matrix $\Phi(\varphi)$, it is intended to study transformation properties connected with the measure of the system.

After these preliminary remarks, we now proceed to formulate and prove the following theorem on the reducibility of a system of the form (20.1).

Theorem 14. *Suppose that the right-hand sides of the system of equations*

$$\frac{dx}{dt} = Ax + P(\varphi)x, \qquad \frac{d\varphi}{dt} = \omega, \tag{20.4}$$

satisfy the following conditions:

(i) *The matrix $P(\varphi)$ is periodic in $\varphi = (\varphi_1, \dots, \varphi_m)$ and analytic on the domain*

$$|\operatorname{Im} \varphi| = \sup_{\alpha} |\operatorname{Im} \varphi_\alpha| \leqslant \rho_0 \quad (\rho_0 > 0), \tag{20.5}$$

and real for real φ.

(ii) *For some positive ε and d, the inequality*

$$|(k, \omega)| \geqslant \varepsilon |k|^{-d} \quad (|k| \neq 0) \tag{20.6}$$

holds for every integral-valued vector $k = (k_1, \dots, k_m)$.

(iii) *The eigenvalues $\lambda = (\lambda_1, \dots, \lambda_m)$ of the matrix A have distinct real parts.*

Then a sufficiently small positive constant M_0 can be found, such that for

$$|P(\varphi)| = \sum_{i,j=1}^{n} |P_{ij}(\varphi)| \leqslant M_0, \tag{20.7}$$

the system (20.4) *reduces to the form*

$$\frac{dy}{dt} = A_0 y, \qquad \frac{d\varphi}{dt} = \omega \tag{20.8}$$

with A_0 a constant matrix, by means of the non-singular change of variable

$$x = \Phi(\varphi)\, y \tag{20.9}$$

with $\Phi(\varphi)$ a marix, 2π-periodic in φ, analytic and analytically invertible in the domain

$$|\operatorname{Im} \varphi| \leqslant \frac{\rho_0}{2}, \tag{20.10}$$

and real for $\operatorname{Im} \varphi = 0$.

Proof. By this theorem, the fundamental matrix of the solutions of system (20.1) has the form

$$X = \Phi(\omega t + \varphi_0)\, e^{A_0 t}, \quad \varphi = \omega t + \varphi_0, \tag{20.11}$$

where $\Phi(\omega t + \varphi_0)$ is a non-singular matrix, with the frequency basis $\omega = (\omega_1, \ldots, \omega_m)$, is quasi-periodic in t and real for real φ_0.

Making use of the well-known Lagrange formula expressing the analytic function $f(A)$ of a matrix as a polynomial in A and of the facts that the matrix $P(\varphi)$ is small and the eigenvalues $\lambda_1^0, \ldots, \lambda_n^0$ of the matrix A_0 are close to the eigenvalues $\lambda_1, \ldots, \lambda_n$ of the matrix A, i.e. also have distinct real parts, we see that the fundamental matrix (20.11) can be expressed in the form

$$X = \Phi(\omega t + \varphi_0) \sum_{k_0=1}^{n} \frac{(A_0 - \lambda_1^0) \ldots (A_0 - \lambda_{k_0-1}^0)(A_0 - \lambda_{k_0+1}^0) \ldots (A_0 - \lambda_n^0)}{(\lambda_{k_0}^0 - \lambda_1^0) \ldots (\lambda_{k_0}^0 - \lambda_{k_0-1}^0)(\lambda_{k_0}^0 - \lambda_{k_0+1}^0) \ldots (\lambda_{k_0}^0 - \lambda_n^0)}\, e^{\lambda_{k_0}^0 t}. \tag{20.12}$$

§ 21. Solution of the Auxiliary Equation

Theorem 14 will be proved by actually constructing the transformation matrix $\Phi(\varphi)$. In this an important role is played by the periodic solution of the equation

$$\left(\frac{\partial U}{\partial \varphi}, \omega\right) + UA = AU + P(\varphi), \tag{21.1}$$

where A, U and $P(\varphi)$ are matrices, and

$$\left(\frac{\partial U}{\partial \varphi}, \omega\right) = \sum_{\alpha=1}^{m} \frac{\partial U}{\partial \varphi_\alpha} \omega_\alpha$$

is the scalar product of a matrix by a vector.

With the object of demonstrating the solubility of equation (21.1) with the matrix U and establishing some properties of its periodic solutions, we express the matrix $P(\varphi)$ in the form of the Fourier series

$$P(\varphi) = \sum_{(k)} P_k e^{i(k,\varphi)}, \tag{21.2}$$

and seek the solution $U(\varphi)$ also in the form of the series

$$U(\varphi) = \sum_{(k)} Y_k e^{i(k,\varphi)}. \tag{21.3}$$

To determine the coefficients Y_k, we substitute (21.2) and (21.3) in (21.1), to obtain the matrix equation for the coefficients:

$$Y_k (A + i(k, \omega)) - AY_k = P_k. \tag{21.4}$$

Eqn. (21.4) is representable by a matrix equation of the type

$$XA - BX = C, \tag{21.5}$$

(where A, B and C are the given matrices), which has been studied extensively.

It is known [21] that, if the matrices A and B do not have common eigenvalues, then equation (21.5) has a solution for an arbitrary matrix C and this solution is unique. In the present case $\lambda_1, \ldots, \lambda_n$ and $\lambda_1 + i(k, \omega), \ldots, \lambda_n + i(k, \omega)$ are respectively the eigenvalues of the matrices A and B. Thus, for those $k = (k_1, \ldots, k_n)$ for which

$$|\lambda_\alpha - \lambda_\beta + i(k, \omega)| \neq 0 \quad (\alpha, \beta = 1, 2, \ldots, n), \tag{21.6}$$

equation (21.4) has a unique solution. To determine this solution, we reduce matrix A to the normal Jordan form:

$$A = CJC^{-1}, \tag{21.7}$$

where

$$J = \{J_{\rho_1}(\lambda_1), \ldots, J_{\rho_l}(\lambda_l)\},$$
$$J_{\rho\alpha}(\lambda_\alpha) = \lambda_\alpha E_{\rho\alpha} + Z_{\rho\alpha}, \tag{21.8}$$

with E_{ρ_α} a $\rho_\alpha \times \rho_\alpha$ identity matrix and Z_{ρ_α} a $\rho_\alpha \times \rho_\alpha$ matrix whose first diagonal element is unity and the rest are zero. Substituting the expression for A in (21.4) and carrying out simple transformations, we obtain the equation

$$X_k (J + i(k, \omega)) - JX_k = Q_k, \tag{21.9}$$

where

$$X_k = C^{-1}Y_kC, \qquad Q_k = C^{-1}P_kC. \tag{21.10}$$

To solve the equation (21.9), we partition the matrices X_k and Q_k into blocks: $X_k = \{X_{jr}^{(k)}\}$, $Q_k = \{Q_{jr}^{(k)}\}$ consisting of ρ_j rows and ρ_r columns, where $j, r = 1, 2, \ldots, l$. Equation (21.9) then decomposes into the system of independent equations

$$X_{jr}^{(k)}(J_{\rho_r}(\lambda_r) + i(k, \omega)) - J_{\rho_j}(\lambda_j) X_{jr}^{(k)} = Q_{jr}^{(k)} \quad (j, r = 1, \ldots, l). \tag{21.11}$$

We denote by $x_\alpha^{(jr)}$ and $q_\alpha^{(jr)}$ the vectors forming the α-th rows of the matrices $X_{jr}^{(k)}$ and $Q_{jr}^{(k)}$. Then, taking account of the form of the Jordan cell $J_{\rho_j}(\lambda_j)$, the system of equations (21.11) can be expressed as

$$x_\alpha^{(jr)} J_{\rho_r}(\lambda_r - \lambda_j + i(k, \omega)) = x_{\alpha-1}^{(jr)} + q_\alpha^{(jr)}, \tag{21.12}$$

where

$$x_0^{(jr)} = 0, \qquad \alpha = 1, \ldots, \rho_j, \qquad j, r = 1, 2, \ldots, l.$$

Equations (21.12) uniquely define the vector

$$x_\alpha^{(jr)} = (x_{\alpha-1}^{(jr)} + q_\alpha^{(jr)}) J^{(jr)}(\lambda_r - \lambda_j + i(k, \omega)), \tag{21.13}$$

where $\bar{J}^{(jr)}(\lambda)$ denotes the $\rho_r \times \rho_r$ matrix, given by

$$\bar{J}^{(jr)}(\lambda) = \frac{E_{\rho_r}}{\lambda} - \frac{Z_{\rho_r}}{\lambda^2} + \ldots + (-1)^{\rho_r - 1} \frac{Z_{\rho_r}^{\rho_r - 1}}{\lambda^{\rho_r}}. \tag{21.14}$$

Thus, in the given case, the solution of equation (21.4) is defined by the formulas (21.10), (21.13) and (21.14). For those k, for which inequality (21.6) is violated even for a single α, β, equation (21.4) is either inconsistent or has infinitely many solutions. For the latter case to hold, we need to show, as is plain from equation (21.12), that the components of matrix P_k satisfy

$$Q_{\alpha\beta}^{(k)} = 0 \tag{21.15}$$

for every k, α, and β, for which

$$\lambda_\alpha - \lambda_\beta + i(k, \omega) = 0. \tag{21.16}$$

Subject to (21.15) being fulfilled, the solution of equation (21.4) is uniquely defined by the formulas (21.10), (21.13) and (21.14), for those blocks whose indices satisfy inequality (21.6), and by the equation

$$X_{\alpha\beta}^{(k)} = 0 \tag{21.17}$$

for the remaining blocks. Having determined Y_k from system (21.4), by the same token we obtain a normal periodic solution of equation (21.1). This solution, indeed, satisfies equation (21.1), if the series (21.3) converges and has first derivatives in φ. To make the conditions for the convergence of series (21.3) explicit, we must estimate the norm of the matrix Y_k. For this, we estimate $|x_\alpha^{(jr)}|$ by manipulating (21.13) and (21.14), getting

$$|x_\alpha^{(jr)}| \leqslant c_1^\alpha (|q_1^{(jr)}| + \ldots + |q_\alpha^{(jr)}|)\,|\lambda_r - \lambda_j + i(k, \omega)|^{-\rho_r^0 \alpha},$$

$$(\alpha = 1, 2, \ldots, \rho_j). \tag{21.18}$$

For the matrix Y_k we now obtain

$$|Y_k| \leqslant |C|\,|C^{-1}|\,|X_k|\,\bar{c}_2\,|Q_k| \max_{r,j} |\lambda_r - \lambda_j + i(k, \omega)|^{-d_1} \leqslant$$

$$\leqslant |c_2 \max_{r,j} |\lambda_r - \lambda_j + i(k, \omega)|^{-d_0}\,|P_k|, \tag{21.19}$$

where c_2 is a constant not depending on $|k|$,

$$d_0 = \begin{cases} 1 & \text{for } |\lambda_r - \lambda_j + i(k, \omega)| \geqslant 1, \\ \max\limits_j \rho_j^2 & \text{for } |\lambda_r - \lambda_j + i(k, \omega)| < 1. \end{cases} \tag{21.20}$$

The inequality (21.19) motivates the following proposition.

Lemma 7′. *Suppose that the matrix $P(\varphi)$ satisfies the hypotheses of Theorem 14. In addition, suppose that the eigenvalues of the matrix A satisfy the inequality*

$$|\lambda_\alpha - \lambda_\beta + i(k, \omega)| \geqslant \varepsilon |k|^{-d} \quad (|k| \neq 0) \tag{21.21}$$

for every $\alpha, \beta = 1, \ldots, n$, any integral-valued vector $k = (k_1, \ldots, k_m)$ and some $\varepsilon > 0$, $d > 0$.

Then, if

$$|P(\varphi)| \leqslant M, \tag{21.22}$$

and if

$$P_0 = \bar{P}(\varphi) = \frac{1}{(2\pi)^m} \int_0^{2\pi} \cdots \int_0^{2\pi} P(\varphi)\, d\varphi_1 \ldots d\varphi_m$$

satisfies the condition (21.15), *then the series* (21.3), *defining the periodic solution of equation* (21.1), *converges uniformly in the domain*

$$|\operatorname{Im} \varphi| \leqslant \rho_0 - 2\delta, \quad 0 < 2\delta < \rho_0, \tag{21.23}$$

and represents a matrix analytic in this domain and real for real φ, satisfying

$$|U(\varphi) - \bar{U}| \leqslant c_3 \frac{M}{\delta^{d_1+m}},$$

$$\left|\frac{\partial U}{\partial \varphi}\right| = \sum_{\alpha=1}^{m} \left|\frac{\partial U}{\partial \varphi_\alpha}\right| \leqslant c_3 \frac{M}{\delta^{d_1+m+1}}, \tag{21.24}$$

with c_3 a constant depending only on ε, d and m, $d_1 = dd_0$, d_0 a constant defined by (21.20), *and $\bar{U}$ the mean of the matrix $U(\varphi)$.*

Proof. Since the matrix $P(\varphi)$ is analytic in the strip $|\operatorname{Im} \varphi| \leqslant \rho_0$, its Fourier coefficients P_k satisfy the inequalities

$$|P_k| < |P|_0 e^{-\rho_0 |k|} \leqslant M e^{-\rho_0 |k|}, \tag{21.25}$$

which together with (21.21) permit the extension of the bound (21.19) for the coefficients Y_k to the form

$$|Y_k| \leqslant c_2 \varepsilon^{-d_0} |k|^{d_0 d} |P_k| \leqslant c_2 \varepsilon^{-d_0} |k|^{d_1} e^{-\rho_0 |k|} M \quad (|k| \neq 0).$$

Hence, it is possible to set

$$|U(\varphi) - \bar{U}| \leqslant \sum_{|k| \neq 0} |Y_k| e^{|\operatorname{Im} \varphi|\, |k|} \leqslant c_2 \varepsilon^{-d_0} \Big(\sum_{|k| \neq 0} |k|^{d_1} e^{(-\rho_0 + |\operatorname{Im} \varphi|)\, |k|} \Big) M \leqslant$$

$$\leqslant c_2 \varepsilon^{-d_0} \sum_{|k| \neq 0} |k|^{d_1} e^{-2\delta |k|} M \tag{21.26}$$

for $| \operatorname{Im} \varphi | \leqslant \rho_0 - 2\delta$. Taking account of the estimate (2.13) for the sum $\sum_{|k| \neq 0} | k |^{d_1} e^{-2\delta|k|}$ obtained in § 2, we deduce that

$$| U(\varphi) - \bar{U} | \leqslant c_2 \varepsilon^{-d_0} \left(\frac{d_1}{e} \right)^{d_1} (1 + e)^m \delta^{-(d_1+m)} M \leqslant$$

$$\leqslant c_3(\varepsilon) \frac{M}{\delta^{-(d_1+m)}} \tag{21.27}$$

for $| \operatorname{Im} \varphi | \leqslant \rho_0 - 2\delta$.

If we differentiate the series (21.3) term by term and estimate its coefficients, we obtain for the derivatives of function $U(\varphi)$, the bound

$$\left| \frac{\partial U}{\partial \varphi} \right| \leqslant \sum_{\alpha=1}^{m} \left| \frac{\partial U}{\partial \varphi_\alpha} \right| \leqslant \sum_{\alpha=1}^{m} \sum_{|k| \neq 0} | k_\alpha | \, | Y_k | e^{|\operatorname{Im} \varphi| \, |k|} \leqslant$$

$$\leqslant m \sum_{|k| \neq 0} | k | \, | Y_k | e^{|\operatorname{Im} \varphi| \, |k|} \leqslant$$

$$\leqslant c_2 m \varepsilon^{-d_0} \sum_{|k| \neq 0} |k|^{d_1+1} e^{-2\delta|k|} M \leqslant$$

$$\leqslant c_3(\varepsilon) \frac{M}{\delta^{-(d_1+m+1)}} \tag{21.28}$$

for $| \operatorname{Im} \varphi | \leqslant \rho_0 - 2\delta$.

It follows from (21.27) and (21.28) that the function $U(\varphi)$ is defined and analytic on the domain $| \operatorname{Im} \varphi | \leqslant \rho_0 - 2\delta$ and satisfies (21.24). To prove that the matrix $U(\varphi)$ is real for real φ, it is enough to remark that its coefficients Y_k and Y_{-k} as solutions of equation (21.4) are complex conjugates. This is a consequence of the fact that the Fourier coefficients P_k and P_{-k} of the matrix $P(\varphi)$, which is real for $\operatorname{Im} \varphi = 0$, are conjugates.

§ 22. Proof of Reducibility Theorem

Suppose that the hypotheses of Theorem 14 of § 20 are satisfied for the system of equations

$$\frac{dx}{dt} = Ax + P(\varphi) x,$$

$$\frac{\varphi d}{dt} = \omega. \tag{22.1}$$

Then the matrix A, as indicated above, can be expressed in its Jordan form

$J = J(\lambda_1, \ldots, \lambda_n)$ by

$$A = CJC^{-1}, \tag{22.2}$$

C a non-singular real matrix. The change of variable $x = Cy$ reduces (22.1) to the system

$$\frac{dy}{dt} = Jy + C^{-1} P(\varphi) Cy,$$
$$\frac{d\varphi}{dt} = \omega. \tag{22.3}$$

Let the matrix $C^{-1}P(\varphi)C$ be expressed as a sum of two matrices: $C^{-1}P(\varphi)C = D + Q$, one of which, $D = \{d_1, \ldots, d_n\}$, is a diagonal matrix whose entries are the diagonal elements of the matrix $C^{-1}P(\varphi)C$. Now, effect a change of variables in (22.3), by putting

$$y = (E + U^{(1)}(\varphi))\, y^{(1)}, \tag{22.4}$$

where $U^{(1)}(\varphi)$ is a periodic solution of the equation

$$\left(\frac{\partial U^{(1)}}{\partial \varphi}, \omega\right) + U^{(1)}y = JU^{(1)} + C^{-1}P(\varphi)\, C - \bar{D},$$
$$\bar{D} = \frac{1}{(2\pi)^m} \int_0^{2\pi} \cdots \int_0^{2\pi} D(\varphi)\, d\varphi_1 \ldots d\varphi_m. \tag{22.5}$$

Since the matrix J is diagonal and, by the hypotheses of the theorem, its diagonal elements $\lambda_1, \ldots, \lambda_n$ are real and distinct, it follows from the lemma proved above that equation (22.5) is always solvable; moreover, its solution is an analytic function of φ for every φ belonging to the domain

$$|\operatorname{Im} \varphi| < \rho_0 - 2\delta_0 = \rho_1, \tag{22.6}$$

it is real for real φ and satisfies the inequality

$$|U^{(1)}(\varphi)| \leqslant |Y_0| + \sum_{|k| \neq 0} |Y_k e^{i(k,\varphi)}| \leqslant |Y_0| + c_3(\varepsilon) \frac{M_0}{\delta_0^{d_1+m}} \leqslant$$
$$\leqslant \left[c_3\left(\min_{\alpha \neq \beta} |\lambda_\alpha - \lambda_\beta|\right) + c_3(\varepsilon)\, \delta_0^{-(d_1+m)}\right] M_0 \leqslant$$
$$\leqslant c_4(r_0, \varepsilon) \frac{M_0}{\delta_0^{d_1+m}}, \tag{22.7}$$

where

$$r_0 = \min_{\alpha \neq \beta} | \lambda_\alpha - \lambda_\beta |, \quad c_4(r_0, \varepsilon) = c_3(r_0) + c_3(\varepsilon).$$

Since the function $U^{(1)}(\varphi)$ satisfies equation (22.5), the substitution (22.4) reduces system (22.1) to the form

$$\begin{aligned} \frac{dy^{(1)}}{dt} &= (J + \bar{D})\, y^{(1)} + P^{(1)}(\varphi)\, y^{(1)}, \\ \frac{d\varphi}{dt} &= \omega, \end{aligned} \tag{22.8}$$

where we are using the notation

$$P^{(1)}(\varphi) = (E + U^{(1)}(\varphi))^{-1} (C^{-1} P(\varphi)\, CU^{(1)}(\varphi) - U^{(1)}(\varphi)\, \bar{D}). \tag{22.9}$$

We choose $M_0 = M_0(r_0, \delta_0)$ so small that the inequality

$$4nc_4(r_0, \varepsilon) \frac{M_0^{2-\varkappa}}{\delta_0^{d_1+m}} \leqslant 1, \tag{22.10}$$

with $\varkappa$ fixed in the interval $1 < \varkappa < 2$, is satisfied. Subject to such a choice of the constant M_0, inequality (22.7) assumes the form

$$| U^{(1)}(\varphi) | \leqslant \frac{M_0^{\varkappa-1}}{4}.$$

The bound of the function $P^{(1)}(\varphi)$ for φ belonging to domain (22.6) is given, in the notation (22.9), by

$$\begin{aligned} P^{(1)}(\varphi) | &\leqslant | (E + U^{(1)}(\varphi))^{-1} | (| C^{-1} | | P(\varphi) | | C | | U^{(1)}(\varphi) | + \\ &+ | U^{(1)}(\varphi) | | \bar{D} |) \leqslant \\ &\leqslant \left(| E | + \sum_{\alpha=1}^{\infty} | U^{(1)}(\varphi) |^\alpha \right) \left(| C^{-1} | | M_0 | C | \frac{M_0^{\varkappa-1}}{4n | C | | C^{-1} |} \right. \\ &\left. + \frac{M_0^{\varkappa-1}}{4n | C | | C^{-1} |} | C | | C^{-1} | M_0 \right) \leqslant \\ &\leqslant \left(n + \frac{M_0^{\varkappa-1}}{4n - M_0^{\varkappa-1}} \right) \frac{M_0^{\varkappa}}{2n} \leqslant M_0^{\varkappa} = M_1. \end{aligned}$$

We denote by J_1 the matrix $J_1 = J + \overline{D}$. Then, (22.8) assumes the form

$$\frac{dy^{(1)}}{dt} = J_1 y^{(1)} + P^{(1)}(\varphi)\, y^{(1)},$$
$$\frac{d\varphi}{dt} = \omega. \tag{22.11}$$

The matrix $J_1 = \{\lambda_1^{(1)}, \ldots, \lambda_n^{(1)}\}$ is diagonal, its entries being real and satisfying the inequality

$$\min_{j \neq r} |\lambda_j^{(1)} - \lambda_r^{(1)}| \geqslant \min_{j \neq r} |\lambda_j - \lambda_r| - \max_{j \neq r} |d_j - d_r| \geqslant r_0 - |C||C^{-1}| M_0 = r_1.$$

For small M_2, in particular for $|C||C^{-1}| M_0 < r_0/2$, the constant r_1 is positive. Consequently, the constants $\lambda_1^{(1)}, \ldots, \lambda_n^{(1)}$ are not only real but distinct.

The transformed system (22.11) has the same properties as the original system (22.3). Hence, it admits the same transformations as the system (22.3).

Suppose that the indicated substitution process is carried out s-times. The possibility of making the $(s+1)$-th substitution is assured by the following inductive theorem.

Theorem 15. *Suppose that the right-hand sides of the system of equations*

$$\frac{dx}{dt} = Jx + P(\varphi)\, x,$$
$$\frac{d\varphi}{dt} = \omega \tag{22.12}$$

satisfy the following conditions:

(i) *The matrix $P(\varphi)$, 2π-periodic in φ, is analytic in φ on the domain*

$$|\operatorname{Im} \varphi| \leqslant \rho_s$$

is real for real φ, and is bounded by a constant M_s:

$$|P(\varphi)| \leqslant M_s.$$

(ii) *The matrix $J = (\lambda_1, \ldots, \lambda_n)$ is diagonal and real, with distinct entries which satisfy the inequalities*

$$\min_{j \neq r} |\lambda_j - \lambda_r| \geqslant r_s. \tag{22.13}$$

Then a positive constanl M^0 not depending on s can be found, such that for every $M_0 < M^0$ and for all integers $s \geqslant 0$ there exists a change of variable

$$x = (E + U(\varphi))\, y$$

reducing the system of equations (22.12) *to the form*

$$\frac{dy}{dt} = J_1 y + P_1(\varphi)\, y,$$

$$\frac{d\varphi}{dt} = \omega,$$

such that

(*a*) *The matrices* $U(\varphi)$ *and* $P_1(\varphi)$ *are* 2π*-periodic in* φ, *analytic on* φ *in the domain*

$$|\operatorname{Im}\varphi| \leqslant \rho_{s+1}, \tag{22.14}$$

real for real φ, *and are bounded respectively by the constants* $M_s^{-\varkappa}/4n$ *and* M_{s+1}, *so that*

$$|P_1(\varphi)| \leqslant M_{s+1}; \quad |U(\varphi)| < \frac{M_s^{\varkappa-1}}{4n}; \tag{22.15}$$

(*b*) *The matrix* $J_1 = (\lambda_1^1, \ldots, \lambda_n^1)$ *is real and diagonal, with distinct entries which satisfy the inequalities*

$$\min_{j\neq r} |\lambda_j^1 - \lambda_r^1| \leqslant r_{s+1}. \tag{22.16}$$

Moreover, the positive constants M_{s+1}, ρ_{s+1}, r_{s+1} *are related to the constants* M_s, ρ_s *and* r_s *by*

$$M_{s+1} = M_s^{\varkappa}, \quad \rho_{s+1} = \rho_s = 2\delta_0^s, \quad r_{s+1} = r_s - M_s^{\varkappa-1} > 0 \quad (s \geqslant 0), \tag{22.17}$$

where

$$1 < \varkappa < 2, \quad \delta_0 \leqslant \frac{\rho_0}{4+\rho_0}.$$

Proof. We make a change of variable in system (22.12), by setting

$$x = (E + U(\varphi))\, y, \tag{22.18}$$

where $U(\varphi)$ is a periodic solution of the equation

$$\left(\frac{\partial U}{\partial \varphi}, \omega\right) + UJ = JU + P(\varphi) - \bar{D}, \tag{22.19}$$

and $\bar{D}$ is an integral mean of the diagonal matrix $D=\{p_{11}, \ldots, p_{nn}\}$ having as its entries the diagonal elements of the matrix $P(\varphi)$. By the assertions of Lemma 7 of § 21 it follows that the periodic solution $U(\varphi)$ of equation (22.19) has the form

$$U(\varphi) = \sum Y_k e^{i(k,\varphi)},$$

is analytic on the domain (22.14), is real for $\operatorname{Im}\varphi = 0$ and satisfies the inequality

$$| U(\varphi) | \leqslant c_4(r_s, \varepsilon) \frac{M_s}{\delta_0^{(d_1+m)s}}.$$

Choose M_0 so small that, in addition to condition (22.10), the inequalities

$$4nc_4\left(r_0 - \sum_{k_0=0}^{s-1} M_0^{(\varkappa-1)\varkappa^{k_0}}, \varepsilon\right)\frac{M_0^{(2-\varkappa)\varkappa^s}}{\delta_0^{(d_1+m)s}} \leqslant 1,$$

$$r_0 - \sum_{k_0=0}^{\infty} M_0^{(\varkappa-1)\varkappa^{k_0}} > 0 \tag{22.20}$$

are satisfied for all integers $s \geqslant 0$. Given these inequalities, the matrix $U(\varphi)$ evidently has the bound (22.15).

Since $U(\varphi)$ is a solution of equation (22.19), the substitution (22.18) reduces (22.12) to the form

$$\frac{dy}{dt} = (J + \bar{D})\, y + (E + U(\varphi))^{-1}(P(\varphi)\, U(\varphi) - U(\varphi)\, \bar{D})\, y. \tag{22.21}$$

Setting

$$J_1 = J + \bar{D}, \quad P_1(\varphi) = (E + U(\varphi))^{-1}(P(\varphi)\, U(\varphi) - U(\varphi)\, \bar{D}),$$

we see the system (22.21) assumes the form required for the proof of Theorem 15; it remains to establish the estimate (22.15) for the norm of $P_1(\varphi)$ and the estimate (22.16) for the diagonal elements of the matrix J_1.

We have

$$| P_1(\varphi) | \leqslant | (E + U(\varphi))^{-1} | \, (| P(\varphi) | \, | U(\varphi) | + | U(\varphi) | \, | \bar{D} |) \leqslant$$

$$\leqslant \left(n + \frac{M_{s-1}^{\varkappa-1}}{4n - M_{s-1}^{\varkappa-1}}\right)\left(M_{s-1}\frac{M_{s-1}^{\varkappa-1}}{4n} - \frac{M_{s-1}^{\varkappa-1}}{4n} M_{s-1}\right) \leqslant$$

$$\leqslant M_{s-1}^{\varkappa} = M_s.$$

Moreover, since

$$J_1 = (\lambda_1' = \lambda_1 + \bar{p}_{11}, \ldots, \lambda_n' = \lambda_1 + \bar{p}_{nn}), \tag{22.22}$$

we have

$$\min_{j\neq r} |\lambda'_j - \lambda'_r| = \min_{j\neq r} |\lambda_j + \bar{p}_{jj} - \lambda_r - p_{rr}| >$$

$$> \min_{j\neq r} |\lambda_j - \lambda_r| - \max_{j\neq r} |\bar{p}_{jj} - \bar{p}_{rr}| =$$

$$= r_s - \max_{j\neq r} |\bar{p}_{jj} - \bar{p}_{rr}|_0 > r_0 - 2M_s > r_s - M_s^{\varkappa - 1} =$$

$$= r_{s+1} > r_0 - \sum_{k_0=0}^{\infty} M_0^{(\varkappa-1)\varkappa^{k_0}} > 0.$$

Thus, taking those values of M_0 which satisfy (22.10) and (22.20), the inductive theorem is proved for every positive $M_0 \leqslant M^0$.

The transformation of system (22.3) can be continued by means of the substitutions which began with (22.4). This leads to the fact that the repetition of s substitutions of this form, i.e. the change of variable

$$y = (E + U^{(1)}(\varphi))(E + U^{(r)}(\varphi)) \dots (E + U^{(s)}(\varphi))\, y^{(s)}, \qquad (22.23)$$

permits to transform the original system of equations into the form

$$\frac{dy^{(s)}}{dt} = J_s y^{(s)} + P^{(s)}(\varphi)\, y^{(s)}. \qquad (22.24)$$

Here, the matrices $U^{(\alpha)}(\varphi)$ and $P^{(s)}(\varphi)$ are 2π-periodic in φ, analytic in φ on the domain $|\operatorname{Im}\varphi| < \rho_s$, real for real φ and bounded respectively by the constants $M_\alpha^{\varkappa-1}/4n$ and M_s, with

$$|U^{(\alpha)}(\varphi)| \leqslant \frac{M_\alpha^{\varkappa-1}}{4n}, \quad |P^{(s)}(\varphi)| \leqslant M_s. \qquad (22.25)$$

Set

$$\Phi^{(s)}(\varphi) = (E + U^{(1)}(\varphi)) \dots (E + U^{(s)}(\varphi)) = \prod_{\alpha=1}^{s} (E + U^{(\alpha)}(\varphi)), \quad (22.26)$$

to show that the sequence (22.26) converges uniformly to a matrix, analytic on the domain

$$|\operatorname{Im}\varphi| < \frac{\rho_0}{2}. \qquad (22.27)$$

By the definition of the constants ρ_s it is plain that $\rho_s > \rho_{s+1} > \dots > \rho_0/2$.

Consequently, every matrix $\Phi(\varphi)$ is defined and analytic on the domain (22.27). What is more, because of inequalities (22.25), we have the estimate

$$|\Phi^{(s+1)}(\varphi) - \Phi^{(s)}(\varphi)| = |\Phi^{(s)}(\varphi)|\,|U^{(s+1)}(\varphi)| \leqslant$$

$$\leqslant \left|\prod_{\alpha=1}^{s}\left(E + \frac{M_{\alpha-1}^{\varkappa-1}}{4n}\right)\right| \frac{M_s^{\varkappa-1}}{4n} \leqslant$$

$$\leqslant \frac{n}{4}\prod_{\alpha=1}^{s}\left(1 + \frac{M_{\alpha-1}^{\varkappa-1}}{4}\right) M_s^{\varkappa-1} \leqslant$$

$$\leqslant \frac{n}{4}\prod_{\alpha=1}^{\infty}\left(1 + \frac{M_{\alpha-1}^{\varkappa-1}}{4}\right) M_s^{\varkappa-1} = c_5 M_s^{\varkappa-1},$$

whence the criterion for uniform convergence of the sequence (22.26) is satisfied:

$$|\Phi^{(s+k_0)}(\varphi) - \Phi^{(s)}(\varphi)| < c_5 \sum_{\alpha=s}^{s+k_0-1} M_\alpha^{\varkappa-1} < c_5 \sum_{\alpha=s}^{\infty} M_\alpha^{\varkappa-1}. \tag{22.28}$$

We denote by $\Phi(\varphi)$ the limit matrix of sequence (22.26). The matrix $\Phi(\varphi)$ is 2π-periodic in φ, real for $\operatorname{Im}\varphi = 0$ and analytic on the domain (22.27), for it is the limit of a uniformly convergent sequence of matrices analytic for $|\operatorname{Im}\varphi| < \rho_0/2$ and real for $\operatorname{Im}\varphi = 0$. Taking the limit in (22.23) as $s \to \infty$, we verify that the change of variable

$$y = \Phi(\varphi)\, z$$

reduces the system (22.3) to the form

$$\frac{dz}{dt} = J^0 z,$$

$$\frac{d\varphi}{dt} = \omega,$$

using the notation

$$J^0 = \lim_{s\to\infty} J_s = \lim_{s\to\infty} (J + \bar{D}_1 + \ldots + \bar{D}_s).$$

Finally, to show that the matrix $\Phi(\varphi)$ is non-singular, we estimate the difference $|\Phi^{(s)}(\varphi) - E|$, getting

$$|\Phi^{(s)}(\varphi) - E| = \left| \prod_{\alpha=1}^{s} (E + U^{(\alpha)}(\varphi)) - E \right| \leqslant$$

$$\leqslant \left| \left(\frac{M_0^{\varkappa-1}}{4n} + \ldots + \frac{M_{s-1}^{\varkappa-1}}{4n} \right) I + \right.$$

$$+ \left(\frac{M_0^{\varkappa-1}}{4n} \cdot \frac{M_1^{\varkappa-1}}{4n} n + \ldots + \frac{M_{s-r}^{\varkappa-1}}{4n} \cdot \frac{M_{s-1}^{\varkappa-1}}{4n} n \right) I +$$

$$\left. + \ldots + \frac{M_0^{\varkappa-1}}{4n} \cdot \frac{M_1^{\varkappa-1}}{4n} \cdots \frac{M_{s-1}^{\varkappa-1}}{4n} n^{s-1} I \right| \leqslant$$

$$\leqslant \left(\prod_{\alpha=1}^{s} \left(1 + \frac{M_{\alpha-1}}{4} \right) - 1 \right) |I|,$$

where

$$I = \begin{pmatrix} 1 & 1 & \ldots & 1 \\ \vdots & \vdots & & \vdots \\ 1 & 1 & \ldots & 1 \end{pmatrix}.$$

For small M_0, the last inequality implies the estimate $\Phi^{(s)}(\varphi) - E| \leqslant c_6 < 1$, which, in the limit, yields $|\Phi(\varphi) - E| \leqslant c_6 < 1$, implying that the series $\sum_{\alpha=0}^{\infty} (E - \Phi(\varphi))^{\alpha}$ converges and defines the matrix $\Phi^{-1}(\varphi)$, the inverse of the matrix $\Phi(\varphi)$.

§ 23. Construction of a Fundamental Matrix of Solutions

The process of constructing the transformation matrix employed in § 22 for proving the reducibility theorem can serve as a device for the construction of a fundamental matrix of solutions of system (20.1). The main practical difficulty encountered in this connection is related to the representation of matrix A in terms of its Jordan form J, by

$$A = CJC^{-1}. \tag{23.1}$$

If, however, the matrices C and J are found, then the determination of the solutions of the equations considered reduces to simple algebraic operations. Thus, in the first approximation, it is possible to set

$$x = C(E + U^{(1)}(\varphi))\, y^{(1)},$$
$$\varphi = \omega t + \varphi_0, \tag{23.2}$$

where $U^{(1)}(\varphi)$ is a periodic solution of the equations

$$\left(\frac{\partial U^{(1)}}{\partial \varphi}, \omega\right) + U^{(1)}J = JU^{(1)} + P^0(\varphi) - \bar{D}_0, \tag{23.3}$$

$$P^0(\varphi) = C^{-1}P(\varphi)\,C = \{p^0_{\alpha\beta}(\varphi)\}, \quad \bar{D}_0 = \{\overline{p^0_{11}(\varphi)}, \overline{p^0_{22}(\varphi)}, \ldots, \overline{p^0_{nn}(\varphi)}\}, \tag{23.4}$$

$$y^{(1)} = e^{(J+\bar{D}_0)t}\, y_0 = e^{J_1 t}\, y_0 = \left\{e^{(\lambda_1+\bar{p}^0_{11})\,t}\, y_1^0, \ldots, e^{(\lambda_n+\bar{p}^0_{nn})t}\, y_n^0\right\},$$

φ_0, y_0, arbitrary constants.

For the second approximation, we can take the expression

$$x = C(E + U^{(1)}(\varphi))(E + U^{(2)}(\varphi))\, y^{(2)},$$
$$\varphi = \omega t + \varphi_0, \tag{23.5}$$

where $U^{(2)}(\varphi)$ is a periodic solution of the equations

$$\left(\frac{\partial U^{(2)}}{\partial \varphi}, \omega\right) + U^{(2)}J_1 = J_1 U^{(2)} + P^{(1)}(\varphi) - \bar{D}_1, \tag{23.6}$$

$$P^{(1)}(\varphi) = (E + U^{(1)}(\varphi))^{-1}(P^{(0)}(\varphi)\,U^{(1)}(\varphi) - U^{(1)}(\varphi)\,\bar{D}_0) = \{p^{(1)}_{\alpha\beta}(\varphi)\},$$

$$\bar{D}_1 = \{\overline{p^{(1)}_{11}(\varphi)}, \ldots, \overline{p^{(1)}_{nn}(\varphi)}\}, \tag{23.7}$$

$$y^2 = e^{(J+\bar{D}_0+\bar{D}_1)t} y_0 = e^{J_2 t} y_0 = \left\{e^{(\lambda_1+\bar{p}^{(0)}_{11}+\bar{p}^{(1)}_{11})t}\, y_1^0, \ldots, e^{(\lambda_n+\bar{p}^0_{nn}+\bar{p}^{(1)}_{nn})t}\, y_n^0\right\}.$$

Each of the succeeding approximations, say the s-th approximation, admits

$$x = C(E + U^{(1)}(\varphi)) \ldots (E + U^{(s)}(\varphi))\, y^{(s)} = C\Phi^{(s)}(\varphi),$$
$$\varphi = \omega t + \varphi_0, \tag{23.8}$$

where $U^{(\alpha)}(\varphi)$ $(\alpha = 1, 2, \ldots, s)$ is a periodic solution of the equations

$$\left(\frac{\partial U^{(\alpha)}}{\partial \varphi}, \omega\right) + U^{(\alpha)} J_{\alpha=1} = J_{\alpha-1} U^{(\alpha)} + P^{(\alpha-1)}(\varphi) - \bar{D}_{\alpha-1}, \tag{23.9}$$

$$P^{(\alpha-1)}(\varphi) = (E + U^{(\alpha-1)}(\varphi))^{-1} (P^{(\alpha-2)}(\varphi) U^{(\alpha-1)}(\varphi) - U^{(\alpha-1)}(\varphi) \bar{D}_{\alpha-2}) =$$

$$= \{p_{\alpha\beta}^{(\alpha-1)}(\varphi)\},$$

$$\bar{D}_{\alpha-1} = \left\{ \overline{p_{11}^{(\alpha-1)}(\varphi)}, \ldots, \overline{p_{nn}^{(\alpha-1)}(\varphi)} \right\}, \tag{23.10}$$

$$y^{(s)} = e^{(J+\bar{D}_0+\bar{D}_1+\ldots+\bar{D}_{s-1})t} y_0 = e^{J_s t} y_0 =$$

$$= \left\{ e^{(\lambda_1+\overline{p_{11}^0}+\ldots+\overline{p_{11}^{(s-1)}})t} y_1^0, \ldots, e^{(\lambda_n+\overline{p_{nn}^0}+\ldots+\overline{p_{nn}^{s-1}})t} y_n^0 \right\}.$$

The periodic solutions of equations (23.9) are sought by the method described in § 21 and $U^{(\alpha)}(\varphi)$ is obtained in the form of the series expansion

$$U^{(\alpha)}(\varphi) = \sum Y_k^{(\alpha)} e^{i(k,\varphi)}, \tag{23.11}$$

where the $Y_k^{(\alpha)}$ are solutions of the equations

$$Y_k^{(\alpha)} [J_{\alpha-1} + i(k, \omega)] - J_{\alpha-1} Y_k^{(\alpha)} = P_k^{(\alpha-1)}, \tag{23.12}$$

in which $P_k^{(\alpha-1)}$ are the Fourier coefficients of the matrix $P^{(\alpha-1)}(\varphi) - D_{\alpha-1}$.

As shown in the preceding section, the sequences of matrices $\Phi^{(s)}(\varphi)$ and J_s converge rapidly, so that even the use of a few initial approximations assures a good degree of accuracy in determining the general solution of the system of equations (20.1). The degree of accuracy is also easy to estimate. Indeed, the expression (22.28) yields

$$|\Phi^{(\infty)}(\varphi) - \Phi^{(s)}(\varphi)| \leqslant C \sum_{\alpha=s}^{\infty} M_\alpha^{\varkappa-1} \leqslant 2CM_s^{\varkappa-1} =$$

$$= \frac{n}{2} \prod_{\alpha=1}^{\infty} \left(1 + \frac{M_{\alpha-1}^{\varkappa-1}}{4}\right) M_s^{\varkappa-1} \leqslant \tag{23.13}$$

$$\leqslant nM_s^{\varkappa-1} = nM_0^{(\varkappa-1)\varkappa^s},$$

whenever M_0 is so small that

$$\prod_{\alpha=1}^{\infty} \left(1 + \frac{M_{\alpha-1}^{\varkappa-1}}{4}\right) \leqslant 2.$$

A similar inequality is obtained for the difference $| J_\infty - J_s |$ by

$$| J_\infty - J_s | \leqslant \sum_{\alpha=s}^{\infty} | D_\alpha | \leqslant \sum_{\alpha=s}^{\infty} | P^\alpha (\varphi) | \leqslant \sum_{\alpha=s}^{\infty} M_s \leqslant 2M_s = 2M_0^{\varkappa^s}, \qquad (23.14)$$

whenever $\sum_{\alpha=s}^{\infty} M_\alpha M_s^{-1} \leqslant 2$. The estimates derived characterise the difference

$$| X(t) - \Phi^{(s)} (\omega t + \varphi_0) e^{J_s t} | = | \Phi^{(\infty)} (\omega t + \varphi_0) e^{J_\infty t} - \Phi^{(s)} (\omega t + \varphi_0) e^{J_s t} |,$$

where $X(t)$ is the fundamental matrix of solutions of the linear system (20.1), and $X_s(t) = \Phi^{(s)} (\omega t + \varphi_0) e^{J_s t}$ is its s-th approximation.

The scheme conceived above for determining the general solution of system (20.1) is singularly convenient for a system of two first order equations. In this case the matrices C and J are easy to determine, permitting $\Phi^{(s)}\ \varphi$ and $J^{(s)}$ to be expressed explicitly. Aiming at a deeper investigation, we consider the system of equations

$$\frac{dx}{dt} = Ax + P(\varphi)\, x, \qquad \frac{d\varphi}{dt} = \omega, \qquad (23.15)$$

assuming $x = (x_1, x_2)$ to be a two-dimensional vector. For this system, we write out explicit expressions for the matrices $\Phi^{(1)}(\varphi)$ and $J^{(1)}$ and construct first approximations. Let

$$A = \begin{pmatrix} a & b \\ c & d \end{pmatrix}, \quad P(\varphi) = \begin{pmatrix} p_1 & p_2 \\ p_3 & p_4 \end{pmatrix}. \qquad (23.16)$$

The eigenvalues λ_1, λ_2 of the matrix A are defined by the expressions

$$\lambda_1 = \frac{\sigma(A) + r_0}{2}, \quad \lambda_2 = \frac{\sigma(A) - r_0}{2}, \qquad (23.17)$$

where $\sigma(A) = a + d$ is the trace of the matrix A, $r_0^2 = \sigma^2(A) - 4\Delta(A)$, and $\Delta(A) = ad - bc$ is the determinant of the matrix A; λ_1 and λ_2 are distinct provided that

$$r_0^2 > 0. \qquad (23.18)$$

Assuming that the condition (23.18) is satisfied we seek the matrix C. For this, we solve the equation

$$AC = CJ, \qquad (23.19)$$

where $J = (\lambda_1, \lambda_2)$, relative to C. As a result,

$$C = \begin{pmatrix} \dfrac{b}{\lambda_1 - a} & \dfrac{b}{\lambda_2 - a} \\ 1 & 1 \end{pmatrix}. \tag{23.20}$$

(It is natural to choose one of the possible solutions of equation (23.19) and also assume that $b \neq 0$ which is always plausible whenever A is not a diagonal matrix.) An inverse of the matrix C has the form

$$C^{-1} = \frac{1}{\lambda_2 - \lambda_1} \begin{pmatrix} \dfrac{(\lambda_1 - a)(\lambda_2 - a)}{b} & -(\lambda_1 - a) \\ -\dfrac{(\lambda_1 - a)(\lambda_2 - a)}{b} & \lambda_2 - a \end{pmatrix}. \tag{23.21}$$

The formulas (23.20) and (23.21) allow us to express the product $C^{-1} P(\varphi) C$ in the form

$$C^{-1}P(\varphi)\, C =$$

$$= \begin{pmatrix} q_1 & q_2 \\ q_3 & q_4 \end{pmatrix} = \frac{1}{\lambda_2 - \lambda_1} \times$$

$$\times \left[\begin{matrix} p_1(\lambda_2 - a) + \dfrac{p_2}{b}(\lambda_1 - a)(\lambda_2 - a) - (\lambda_1 - a)\left(\dfrac{p_3 b}{\lambda_1 - a} + p_4\right), \\ -p_1(\lambda_2 - a) - \dfrac{p_2}{b}(\lambda_1 - a)(\lambda_2 - a) + (\lambda_2 - a)\left(\dfrac{p_3 b}{\lambda_1 - a} + p_4\right), \end{matrix} \right.$$

$$\left. \begin{matrix} p_1(\lambda_1 - a) + \dfrac{p_2}{b}(\lambda_1 - a)(\lambda_2 - a) - (\lambda_1 - a)\left(\dfrac{p_3 b}{\lambda_2 - a} + p_4\right) \\ -p_1(\lambda_1 - a) - \dfrac{p_2}{b}(\lambda_1 - a)(\lambda_2 - a) + (\lambda_2 - a)\left(\dfrac{p_3 b}{\lambda_2 - a} + p_4\right) \end{matrix} \right]. \tag{23.22}$$

Suppose that

$$Q_k = \begin{pmatrix} q_1^{(k)}, & q_2^{(k)} \\ q_3^{(k)}, & q_4^{(k)} \end{pmatrix} = \frac{1}{(2\pi)^m} \int_0^{2\pi} \dots \int_0^{2\pi} C^{-1}P(\varphi)\, Ce^{-i(k,\varphi)}\, d\varphi_1 \dots d\varphi_m$$

are the Fourier coefficients of the matrix $C^{-1} P(\varphi) C$ such that

$$C^{-1}P(\varphi)\, C = \overline{Q} + \sum_{|k| \neq 0} Q_k e^{i(k,\varphi)}, \quad \overline{Q} = Q_0 = \begin{pmatrix} \overline{q_1}, & \overline{q_2} \\ \overline{q_3}, & \overline{q_4} \end{pmatrix}.$$

By (23.2), the solution of system (23.15) in first approximation is expressible in the form

$$x = C\,(E + U^{(1)}(\varphi))\,\{e^{(\lambda_1+\bar{q}_1)\,t}c_1,\ e^{(\lambda_2+\bar{q}_1)\,t}c_2\},$$
$$\varphi = \omega t + \varphi_0, \tag{23.23}$$

where

$$U^{(1)}(\varphi) = Y_0 + \sum_{|k|\neq 0} Y_k e^{i(k,\varphi)}$$

is a periodic matrix, whose coefficients Y_k are defined by (23.12), and c_1, c_2 are arbitrary constants. Solving (23.12), we find that the Y_k are given by

$$Y_0 = \begin{pmatrix} 0 & \dfrac{\bar{q}_2}{\lambda_2 - \lambda_1} \\ \dfrac{\bar{q}_3}{\lambda_1 - \lambda_3} & 0 \end{pmatrix},$$

$$Y_k = \begin{pmatrix} \dfrac{q_1^{(k)}}{i\,(k,\omega)} & \dfrac{q_2^{(k)}}{\lambda_2 - \lambda_1 + i(k,\omega)} \\ \dfrac{q_3^{(k)}}{\lambda_1 - \lambda_2 + i\,(k,\omega)} & \dfrac{q_4^{(k)}}{i\,(k,\omega)} \end{pmatrix}. \tag{23.24}$$

The relations (23.24) permit us to express the solution of the system of equations (23.15) in the first approximation, as

$$x_1 = \left\{\frac{1}{\lambda_1 - a} - \frac{\bar{q}_3}{(\lambda_2 - a)(\lambda_1 - \lambda_2)} + \sum_{|k|\neq 0} \times \right.$$
$$\left.\times \left[\frac{q_1^{(k)}}{i(\lambda_1 - a)(k,\omega)} + \frac{q_3^{(k)}}{(\lambda_2 - a)(\lambda_1 - \lambda_2 + i\,(k,\omega))}\right] e^{i\,(k,\varphi)}\right\} \times$$
$$\times\, e^{(\lambda_1+\bar{q}_1)t}c_1 + \left\{\frac{1}{\lambda_2 - a} + \frac{\bar{q}_2}{\lambda_2 - \lambda_1} + \sum_{|k|\neq 0} \times \right.$$
$$\left.\times \left[\frac{q_4^{(k)}}{i\,(\lambda_2 - a)(k,\omega)} + \frac{q_2^{(k)}}{(\lambda_1 - a)(\lambda_2 - \lambda_1 + i\,(k,\omega))}\right] e^{i\,(k,\varphi)}\right\} e^{(\lambda_2+\bar{q}_4)t}c_2,$$
$$\tag{23.25}$$
$$x_2 = \left\{1 + \frac{\bar{q}_3}{\lambda_1 - \lambda_2} + \sum_{|k|\neq 0}\left[\frac{q_1^{(k)}}{i(k,\omega)} + \frac{q_3^{(k)}}{\lambda_1 - \lambda_2 + i(k,\omega)}\right] e^{i(k,\varphi)}\right\} e^{(\lambda_1+\bar{q}_1)t}c_1 +$$
$$+ \left\{1 + \frac{\bar{q}}{\lambda_2 - \lambda_1} + \sum_{|k|\neq 0}\left[\frac{q_4^{(k)}}{i(k,\omega)} + \frac{q_2^{(k)}}{\lambda_2 - \lambda_1 + i(k,\omega)}\right] e^{i(k,\varphi)}\right\} e^{(\lambda_2+\bar{q}_4)t}c_2.$$

Thus, for example, for the equation

$$\frac{d^2x}{dt^2} + \mu \frac{dx}{dt} + \omega_0^2 x = \varepsilon \left(\sum_{|k| \leqslant N} \alpha_k e^{i(k,\omega)t} \frac{dx}{dt} + \sum_{|k| \leqslant N} \beta_k e^{i(k,\omega)t} x \right), \tag{23.26}$$

with μ and ω_0 satisfying the inequality

$$\mu^2 > 4\omega_0^2, \tag{23.27}$$

manipulation of (23.25) yields in the first approximation:

$$x = \left\{ \frac{1}{\lambda_1} - \varepsilon \frac{\lambda_1 \alpha_0 + \beta_0}{\mu^2 - 4\omega_0^2} + \varepsilon \sum_{0 < k \leqslant N} \left[\frac{\lambda_1 \alpha_k + \beta_k}{\lambda_1 (\lambda_2 - \lambda_1)(\lambda_1 - \lambda_2 - i(k,\omega))} \right. \right.$$

$$\left. \left. - i \frac{\lambda_1 \alpha_k + \beta_k}{\lambda_1 (\lambda_2 - \lambda_1)(k,\omega)} \right] e^{i(k,\varphi)} \right\} e^{\left(\lambda_1 + \varepsilon \frac{\lambda_1 \alpha_0 + \beta_0}{\lambda_2 - \lambda_1} \right) t} c_1 +$$

$$+ \left\{ \frac{1}{\lambda_2} - \varepsilon \frac{\lambda_1 (\lambda_2 \alpha_0 + \beta_0)}{\lambda_2 (\mu^2 - 4\omega_0^2)} - \varepsilon \sum_{0 < |k| \leqslant N} \left[\frac{\lambda_2 \alpha_k + \beta_k}{\lambda_2 (\lambda_2 - \lambda_1)(\lambda_2 - \lambda_1 + i(k,\omega))} \right. \right. +$$

$$\left. \left. + i \frac{\lambda_2 \alpha_k + \beta_k}{\lambda_2 (\lambda_2 - \lambda_1)(k,\omega)} \right] e^{i(k,\varphi)} \right\} e^{\left(\lambda_2 + \varepsilon \frac{\lambda_2 \alpha_0 - \beta_0}{\lambda_2 - \lambda_1} \right) t} c_2, \tag{23.28}$$

where

$$\lambda_1 = \frac{-\mu + \sqrt{\mu^2 - 4\omega_0^2}}{2}, \quad \lambda_2 = \frac{-\mu - \sqrt{\mu^2 - 4\omega_0^2}}{2}.$$

In case an explicit representation of the matrix A in the form (23.1) is inconvenient, the s-th approximation of solutions of the system of equations

$$\frac{dx}{dt} = Ax + P(\varphi)x, \qquad \frac{d\varphi}{dt} = \omega \tag{23.29}$$

can be found in the form

$$x = (E + U^{(1)}(\varphi)) \ldots (E + U^{(s)}(\varphi)) y^{(s)} = \Phi^{(s)}(\varphi) y^{(s)}, \qquad \varphi = \omega t + \varphi_0, \tag{23.30}$$

where $U^{(\alpha)}(\varphi)$ $(\alpha = 1, 2, \ldots, s)$ is a periodic solution of the equations

$$\left(\frac{\partial U^{(\alpha)}}{\partial \varphi}, \omega \right) + U^{(\alpha)} A_{\alpha-1} = A_{\alpha-1} U^{(\alpha)} + P^{(\alpha-1)}(\varphi) - \overline{P^{(\alpha-1)}(\varphi)}, \tag{23.31}$$

$$P^{(\alpha-1)}(\varphi) = (E + U^{(\alpha)}(\varphi))^{-1}(P^{(\alpha-2)}(\varphi)\,U^{(\alpha-1)}(\varphi) - U^{(\alpha-1)}(\varphi)\,\overline{P^{(\alpha-2)}(\varphi)}),$$

$$P^{(0)}(\varphi) = P(\varphi),$$

$$A_{\alpha-1} = \sum_{\beta=0}^{\alpha-2} \overline{P^{(\beta)}(\varphi)} + A, \quad A_0 = A, \tag{23.32}$$

$$y^{(s)} = e^{A_{s-1}t} y_0.$$

It is easy to see that the approximations (23.30) tend to exact solutions of the system of equations (23.29):

$$|\,\Phi^{(\infty)}(\varphi) - \Phi^{(s)}(\varphi)\,| \leqslant nM_0^{(\varkappa-1)\varkappa^s},$$
$$(23.33)$$
$$|\,A_\infty - A_s\,| \leqslant 2\,M_0^{\varkappa^s} \qquad (s = 1, 2, \ldots).$$

The fundamental matrix of solutions of system (23.29) can now be expressed in the form

$$X = \Phi^{(\infty)}(\varphi_t)\,e^{A_\infty t}, \tag{23.34}$$

where $\varphi_t = \omega t + \varphi_0$ and its s-th approximation X_s is given by

$$X_s = \Phi^{(s)}(\varphi_t)\,e^{A_s t}. \tag{23.35}$$

The representation of the fundamental matrix of solutions of system (23.29) with the aid of formula (23.34) is especially convenient in solving the stability problem of the trivial solution $x = 0$, when one of the eigenvalues of the matrix A is zero and all the rest are negative.

We can now determine the stability conditions in this critical case. For simplicity we assume that the matrix $P(\varphi)$ is of the form $\varepsilon P(\varphi)$, where ε is a small positive parameter. From the formula (23.34) defining the fundamental matrix of solutions of the system (23.29) it is plain that the question of stability for the trivial solution of the system of equations (23.29) is solved by knowing the eigenvalues $\lambda_0(\varepsilon)$ of the matrix $A_\infty(\varepsilon)$ which vanish when $\varepsilon = 0$: $\lambda(0) = 0$. The inequality (23.33) implies that

$$|\,A_\infty(\varepsilon) - A_1(\varepsilon)\,| \leqslant 2\varepsilon^{\varkappa}\,|\,P\,|_0 = c_1\varepsilon^{\varkappa} \qquad (1 < \varkappa < 2). \tag{23.36}$$

Thus, the sign of $\lambda(\varepsilon)$ is defined by the sign of the eigenvalue $\lambda_1(\varepsilon)$ of the matrix $A_1(\varepsilon)$, for which $\lambda_1(0) = 0$. Hence, for the purpose of determining $\lambda_1(\varepsilon)$ we express the characteristic equation of the matrix $A_1(\varepsilon) = A + \varepsilon\overline{P}$ by

$$\lambda^n - \sigma(A + \varepsilon\overline{P})\,\lambda^{n-1} + \ldots + \rho(A + \varepsilon\overline{P})\,\lambda + (-1)^n\Delta(A + \varepsilon\overline{P}) = 0,$$
$$(23.37)$$

where $\sigma(A_1)$ is the trace of the matrix A_1 and $\Delta(A_1)$ is its determinant.

The eigenvalue $\lambda_1(\varepsilon)$ is an analytic function of the parameter ε when it is small and, consequently,

$$\lambda_1(\varepsilon) = \varepsilon\lambda_1^0 + \varepsilon^2 \ldots . \tag{23.38}$$

Substituting the series (23. 38) into equation (23.37), we get λ_1^0 as a solution of the equation

$$\rho(A)\lambda_1^0 + (-1)^n \lim_{\varepsilon\to 0} \frac{\Delta(A+\varepsilon\bar{P})}{\varepsilon} = 0. \tag{29.39}$$

However,

$$\lim_{\varepsilon\to 0} \frac{\Delta(A+\varepsilon\bar{P})}{\varepsilon} = \sum_{i=1}^{n} \Delta(\ldots, p_i, \ldots), \tag{23.40}$$

where $\Delta(\ldots, \bar{p}, \ldots)$ denotes the determinant of the matrix obtained from A on replacing the i-th row of A by the i-th column of the matrix $\bar{P}$. Solving equation (23.39) by using (23.40) and the fact that $\rho(A) \neq 0$, we obtain

$$\lambda_1^0 = \frac{(-1)^{n-1} \sum_{i=1}^{n} \Delta(\ldots, \bar{p}_i, \ldots)}{\rho(A)}. \tag{23.41}$$

Since, by (23.36), the characteristic equation of the matrix $A_\infty(\varepsilon)$ coincides, to within a magnitude of order $\varepsilon^\varkappa$, with the characteristic equation of the matrix $A_1(\varepsilon)$, $\lambda_0(\varepsilon)$ has the expansion

$$\lambda_0(\varepsilon) = \varepsilon\lambda_1^0 + \varepsilon^2 \ldots, \tag{23.42}$$

with λ_1^0, defined by (23.41). Thus, if $\lambda_1^0 \neq 0$, then there is an $\varepsilon_0 > 0$ such that for all positive $\varepsilon < \varepsilon_0$ the sign of the root $\lambda_0(\varepsilon)$ coincides with that of the constant λ_1^0. However, for the matrix having one zero root, $(-1)^{n-1}\rho(A)$ is the product of all remaining non-zero roots, which being negative leads to

$$\operatorname{sign} \rho(A) = \operatorname{sign} (-1)^{(n-1)^2}, \tag{23.43}$$

and, consequently,

$$\operatorname{sign} \lambda_1^0 = \operatorname{sign} \sum_{i=1}^{n} \Delta(\ldots, \bar{p}_i, \ldots)(-1)^{(n-1)}. \tag{23.44}$$

Summing up the above assertions we are led to the next theorem.

Theorem 16. *Suppose that the right-hand sides of the system of equations*

$$\frac{dx}{dt} = Ax + \varepsilon P(\varphi)\, x, \qquad \frac{d\varphi}{dt} = \omega \tag{23.45}$$

satisfy the hypotheses of Theorem 14.

Then, if one of the eigenvalues of the matrix A is zero and all the rest are negative, the trivial solution of the system (23.45) *is asymptotically stable for sufficiently small* ε, *if*

$$(-1)^{n-1} \sum_{i=1}^{n} \Delta(\ldots, \overline{p}_i, \ldots) < 0, \tag{23.46}$$

and unstable, if

$$(-1)^{n-1} \sum_{i=1}^{n} \Delta(\ldots, \overline{p}_i, \ldots) > 0. \tag{23.47}$$

In particular, for the system of two equations (23.15), satisfying

$$\Delta(A) = 0, \quad \sigma(A) < 0,$$

the trivial solution is asymptotically stable, if

$$\overline{p}_1 d - \overline{p}_2 c + a\overline{p}_4 - b\overline{p}_3 > 0, \tag{23.48}$$

and unstable if the inequality is reversed.

Remark. The stability conditions of the trivial solution of system (23.45), similar to (23.46), can also be found easily when the eigenvalues of the matrix A have multiple zeros.

§ 24. The Measure of Reducible Systems. Statement of the Problem

As above, we consider the system

$$\frac{dx}{dt} = Ax + P(\varphi)\, x, \qquad \frac{d\varphi}{dt} = \omega \tag{24.1}$$

assuming that the matrix $P(\varphi)$ is 2π-periodic in $\varphi = (\varphi_1, \ldots, \varphi_m)$, analytic for

$$|\operatorname{Im} \varphi| \leqslant \rho_0 \tag{24.2}$$

and real for real φ, and the frequencies $\omega = (\omega_1, \ldots, \omega_m)$ satisfy the inequality

$$|(k, \omega)| \geqslant \varepsilon |k|^{-d} \qquad (|k| \neq 0) \tag{24.3}$$

for every integer $k = (k_1, \ldots, k_m)$ and some $\varepsilon > 0, d > 0$.

Denote by $\mathfrak{B}$ the space of all real $n \times n$ matrices $A = \{a_{\alpha\beta}\}$ and by S_R the set of matrices A in the space $\mathfrak{B}$, satisfying

$$|A| = \max_{\alpha, \beta} |a_{\alpha\beta}| \leqslant R. \tag{24.4}$$

Let $\mathfrak{M}_{r_0}^R$ be a set of matrices A contained in S_R, having real eigenvalues $\lambda = (\lambda_1, \ldots, \lambda_m)$, and also such that

$$\min_{\alpha \neq \beta} |\lambda_\alpha - \lambda_\beta| > r_0 \qquad (r_0 > 0). \tag{24.5}$$

Then, the reducibility Theorem 16 implies the proposition: If R, r_0, ε are given, it is always possible to find $M_0 = M_0(R, r_0, \varepsilon) > 0$, such that for all matrices $P(\varphi)$ whose norms are bounded by a constant M_0 and all the matrices A of the set $\mathfrak{M}_{r_0}^R$ the system of equations (24.1) reduces (in the sense of Lyapunov) to the system with constant coefficients

$$\frac{dy}{dt} = A_0 y, \qquad \frac{d\varphi}{dt} = \omega. \tag{24.6}$$

The set $\mathfrak{M}_{r_0}^R$ obviously does not exhaust all the matrices A of the set S_R, for which the reducibility assertion holds, posing the problem: how *many* of the matrices A of the set S_R ensure the reducibility of system (24.1) to the system (24.6) for every fixed matrix $P(\varphi)$?

Before taking up this problem, we must give the expression 'how many' a more precise sense. For this, we introduce into the space $\mathfrak{B}$ the usual notions of measurability and the measure of a set. We set up a homeomorphism Γ of the space $\mathfrak{B}$ onto an n^2-dimensional Euclidean space E_{n^2}, by assigning to a matrix $A = \{a_{ij}\}$ the point

$$a = \{a_{11}, \ldots, a_{1n}, a_{21}, \ldots, a_{2n}, \ldots, a_{nn}\}.$$

It is said that the set $\mathfrak{M}$ of matrices is measurable in $\mathfrak{B}$ and that its Lebesgue measure is μ : mes $\mathfrak{M} = \mu$, if $\Gamma\mathfrak{M}$ is measurable in the space E_{n^2} and the Lebesgue measure of $\Gamma\,\mathfrak{M}$ is μ.

We denote by $\mathfrak{M}(P)$ the set of all those matrices A of the set S_R, for which the system (24.1), with the matrix $P(\varphi)$ given, can be reduced to the system with constant coefficients (24.6) by means of the non-singular transformation

$$x = \Phi(\varphi)\, y, \tag{24.7}$$

2π-periodic in φ and analytic in some strip $|\operatorname{Im}\varphi| \leqslant \rho_1$. The problem set forth above can now be formulated in the form: What is the measure of set $\mathfrak{M}(P)$?

In the sections below it will be shown that

$$\lim_{|P|\to 0} = \frac{\underline{\text{mes}}\, \mathfrak{M}(P)}{\text{mes}\, S_R} = 1, \tag{24.8}$$

where $\underline{\text{mes}}$ denotes the interior measure of $\mathfrak{M}(P)$. This, in turn, will show that for every sufficiently small matrix $P(\varphi)$ the reducibility of system (24.1) holds for the set of matrices A of almost total Lebesgue measure.

§ 25. A Generalized Reducibility Theorem

Suppose that the required matrix A_0 is real, constant and that its eigenvalues $\lambda = (\lambda_1^0, \ldots \lambda_n^0)$ satisfy the inequality

$$|\lambda_\alpha^0 - \lambda_\beta^0 + i(k, \omega)| \geqslant \varepsilon |k|^{-d} \quad (|k| \neq 0, \quad \alpha, \beta = 1, 2, \ldots, n) \tag{25.1}$$

for every integer $k = (k_1, \ldots, k_m)$ and some $\varepsilon > 0$, $d > 0$.

As in Chap. 2, in the investigation of the system (24.1) we introduce small corrections $\xi = (\xi_{\alpha\beta})$ representing the difference

$$\xi = A - A_0 \tag{25.2}$$

between the matrix A of the system of equations (24.1) and some matrix A_0 with the property (25.1). This facilitates the passage from system (24.1) to another system equivalent to it, given by

$$\begin{aligned} \frac{dx}{dt} &= A_0 x + [P(\varphi, \xi) + \xi]\, x, \\ \frac{d\varphi}{dt} &= \omega. \end{aligned} \tag{25.3}$$

It is now required to determine those values of ξ for which the system (25.3) is reducible to (24.6) by means of the substitution (24.7). The existence of such ξ is established in the following generalized reducibility theorem.

Theorem 17. *For the system of equations* (25.3), *suppose that the matrix* $P(\varphi) = P(\varphi, \xi)$ *is* 2π*-periodic in* φ, *analytic in* φ *and* ξ *on the domain*

$$|\operatorname{Im} \varphi| \leqslant \rho_0, \qquad |\xi| = \sum_{\alpha,\beta} |\xi_{\alpha\beta}| < \sigma \tag{25.4}$$

and real for real φ *and* ξ; *suppose also that the eigenvalues* $\lambda^0 = (\lambda_1^0, \ldots, \lambda_n^0)$ *of the matrix* A_0 *satisfy the inequality* (25.1).

Then, it is possible to find a small $M_0 > 0$ *and a real constant matrix* ξ^0, $|\xi^0| \leqslant 2M_0$ *such that, subject to the condition*

$$|P(\varphi, \xi)| \leqslant M_0, \tag{25.5}$$

there exists a change of variable

$$x = \Phi(\varphi)\, y \tag{25.6}$$

with $\Phi(\varphi)$ *a non-singular matrix,* 2π*-periodic in* φ, *analytic on the domain*

$$|\operatorname{Im} \varphi| \leqslant \frac{\rho_0}{2} \tag{25.7}$$

and real for real φ, *which reduces the system of equations* (25.3) *for* $\xi = \xi^0$ *to the system*

$$\frac{dy}{dt} = A_0 y, \qquad \frac{d\varphi}{dt} = \omega. \tag{25.8}$$

Proof. Let

$$U^{(1)}(\varphi) = \sum_{|k| \neq 0} Y_k e^{i(k,\varphi)} \tag{25.9}$$

be a periodic solution of the equation

$$\left(\frac{\partial U^{(1)}}{\partial \varphi}, \omega\right) + U^{(1)} A_0 = A_0 U^{(1)} + P(\varphi, \xi) - \overline{P}(\xi), \tag{25.10}$$

where

$$\overline{P}(\xi) = \frac{1}{(2\pi)^m} \int_0^{2\pi} \cdots \int_0^{2\pi} P(\varphi, \xi)\, d\varphi_1 \ldots d\varphi_m.$$

By Lemma 7 of § 21 such a solution exists, is an analytic function of φ on the domain

$$| \operatorname{Im} \varphi | \leqslant \rho_1 = \rho_0 - 2\delta_0 \qquad (0 < 2\delta_0 < \rho_0 \leqslant 1), \tag{25.11}$$

is real for real φ, and satisfies the inequality

$$| U^{(1)} (\varphi) | \leqslant c_3 (\varepsilon) \frac{M_0}{\delta_0^{d_1+m}}. \tag{25.12}$$

The change of variable

$$x = (E + U^{(1)} (\varphi)) x^{(1)} \tag{25.13}$$

reduces the system (25.3) to the form

$$\begin{aligned} \frac{dx^{(1)}}{dt} &= A_0 x^{(1)} + [P_1 (\varphi, \xi) + \xi^{(1)}] x^{(1)}, \\ \frac{d\varphi}{dt} &= \omega, \end{aligned} \tag{25.14}$$

with the notation

$$\begin{aligned} P (\varphi, \xi) = (E + U^{(1)} \varphi))^{-1} [P (\varphi) U^{(1)} (\varphi) - U^{(1)} (\varphi) \bar{P} (\xi) + \\ + \xi U^{(1)} (\varphi) - U^{(1)} (\varphi) \xi], \end{aligned} \tag{25.15}$$

$$\xi^{(1)} = \xi + \bar{P} (\xi). \tag{25.16}$$

Set $\delta_0 = \rho_0/(\rho_0 + 4)$ and choose M_0 so small that for some $1 < \varkappa < 2$ the inequality

$$\frac{c_3 (\varepsilon) M_0}{\delta_0^{d_1+m}} < \frac{M_0^{\varkappa-1}}{8n} < \frac{1}{2} \tag{25.17}$$

holds. Then the inequality (25.12) leads to the estimate

$$| U^{(1)} (\varphi) | \leqslant \frac{M_0^{\varkappa-1}}{8n}. \tag{25.18}$$

It is further required that M_0 should ensure also the solvability of equation (25.16) relative to ξ on the domain

$$| \xi^{(1)} | < M_0, \qquad | \xi | \leqslant \frac{\sigma}{2}. \tag{25.19}$$

Since the Cauchy estimate holds for the derivatives of the matrix $\bar{P} (\xi)$ whenever

$|\xi| \leqslant \sigma/2$, i.e.

$$\sum_{\alpha,\beta} \left| \frac{\partial \overline{P}(\xi)}{\partial \xi_{\alpha\beta}} \right| \leqslant \frac{M_0}{\sigma - \frac{\sigma}{2}} n^2 = \frac{2M_0 n^2}{\sigma}, \tag{25.20}$$

it is sufficient for the solvability of (25.16) that M_0 satisfies the condition

$$\frac{2M_0 n^2}{\sigma} \leqslant \frac{1}{2}. \tag{25.21}$$

Subject to such a choice of M_0, the solution of equation (25.16)

$$\xi = \xi(\xi^{(1)}) \tag{25.22}$$

is an analytic function of $\xi^{(1)} = \{\xi^{(1)}_{\alpha\beta}\}$ for $|\xi^{(1)}| < M_0$ and satisfies the inequality

$$|\xi(\xi^{(1)})| \leqslant M_0 + |\overline{P(\xi)}|_0 \leqslant 2M_0. \tag{25.23}$$

Further, differentiating the identity

$$\xi^{(1)} = \xi(\xi^{(1)}) + \overline{P(\xi(\xi^{(1)}))},$$

we see that

$$\left| \frac{\partial \xi_{k_0 q}}{\partial \xi_{\alpha\beta}} \right| \leqslant \left| \frac{\partial \xi^{(1)}_{k_0 q}}{\partial \xi^{(1)}_{\alpha\beta}} \right| + \sum_{i,j} \left| \frac{\partial \overline{p}_{k_0 q}}{\partial \xi_{ij}} \right| \left| \frac{\partial \xi_{ij}}{\partial \xi^{(1)}_{\alpha\beta}} \right| \leqslant 1 + 2n^2 \frac{M_0}{\sigma} \max_{k_0, q} \left| \frac{\partial \xi_{k_0 q}}{\partial \xi^{(1)}_{\alpha\beta}} \right|,$$

which implies the estimate

$$\max_{k_0, q} \sum_{\alpha,\beta} \left| \frac{\partial \xi_{k_0 q}(\xi^{(1)})}{\partial \xi^{(1)}_{\alpha\beta}} \right| \leqslant \frac{1}{1 - 2n^2 \frac{M_0}{\sigma}} \leqslant 1 + 4n^2 \frac{M_0}{\sigma}. \tag{25.24}$$

Taking account of (25.18) and (25.24), we see that the estimate for the norm of the matrix $P_1(\varphi, \xi(\xi^{(1)}))$ is given by

$$|P_1(\varphi, \xi(\xi^{(1)}))| \leqslant 2n\,[\,|P(\varphi, \xi)| + |\xi(\xi^{(1)})| + |\xi^{(1)}|\,]\,|U^{(1)}(\varphi)| \leqslant$$

$$\leqslant 8nM_0 \frac{M_0^{\varkappa-1}}{8n} = M_0^{\varkappa} = M_1. \tag{25.25}$$

Thus, it has been proved that the change of variable (25.13) reduces the system of equations (25.3) for $\xi = \xi(\xi^{(1)})$ to the system

$$\frac{dx^{(1)}}{dt} = A_0 x^{(1)} + [P^{(1)}(\varphi, \xi^{(1)}) + \xi^{(1)}]\, x^{(1)},$$

$$\frac{d\varphi}{dt} = \omega, \tag{25.26}$$

where

$$P^{(1)}(\varphi, \xi^{(1)}) = P_1(\varphi, \xi(\xi^{(1)})).$$

In addition, the matrices $\xi(\xi^{(1)})$, $U^{(1)}(\varphi) = U^{(1)}(\varphi, \xi(\xi^{(1)}))$ and $P^{(1)}(\varphi, \xi^{(1)})$ are analytic on the domain

$$|\operatorname{Im} \varphi| \leqslant \rho_1 = \rho_0 - 2\delta_0, \quad |\xi^{(1)}| \leqslant M_0, \tag{25.27}$$

2π-periodic in φ, real for real φ and satisfy the inequalities

$$|\xi(\xi^{(1)}) - \xi^{(1)}| \leqslant M_0, \quad |U^{(1)}(\varphi, \xi(\xi^{(1)}))| \leqslant \frac{M_0^{\varkappa-1}}{8},$$

$$|P^{(1)}(\varphi, \xi^{(1)})| \leqslant M_0^{\varkappa} = M_1, \tag{25.28}$$

$$\max_{k_0, q} \sum_{\alpha,\beta} \left| \frac{\partial \xi_{k_0 q}(\xi^{(1)})}{\partial \xi^{(1)}_{\alpha\beta}} \right| \leqslant 1 + 4n^2 \frac{M_0}{\sigma},$$

whenever M_0 is so small that the conditions (25.17) and (25.21) are fulfilled.

Since (25.26) satisfies all the hypotheses of Theorem 17 with the constants ρ_1, M_0, M_1 in place of ρ_0, σ, M_0, it follows that for this system the change of variables

$$\begin{aligned} x^{(1)} &= (E + U^{(2)}(\varphi, \xi^{(1)}))\, x^{(2)}, \\ \xi^{(1)} &= \xi^{(1)}(\xi^{(2)}) \end{aligned} \tag{25.29}$$

exists and reduces (25.26) to the system

$$\begin{aligned} \frac{dx^{(2)}}{dt} &= A_0 x^{(2)} + [P^{(2)}(\varphi, \xi^{(2)}) + \xi^{(2)}]\, x^{(2)}, \\ \frac{d\varphi}{dt} &= \omega. \end{aligned} \tag{25.30}$$

For M_1, so small that the inequalities

$$\frac{c_3(\varepsilon) M_1}{\delta_1^{d_1+m}} < \frac{M_1^{\varkappa-1}}{8n} \leqslant \frac{1}{2}, \quad \frac{2M_1 n^2}{M_0} \leqslant \frac{1}{2} \tag{25.31}$$

are satisfied, the matrices $\xi^{(1)}(\xi^{(2)})$, $U^{(2)}(\varphi, \xi^{(1)}(\xi^{(2)}))$ and $P^{(2)}(\varphi, \xi^{(2)})$ are analytic on the domain

$$|\operatorname{Im} \varphi| \leqslant \rho_1 - 2\delta_1 = \rho_2, \quad |\xi^{(2)}| \leqslant M_1, \tag{25.32}$$

2π periodic in φ, real for real φ and satisfy the inequalities

$$| \xi^{(1)} (\xi^{(2)}) - \xi^{(2)} | \leqslant M_1, \qquad | U^{(2)} (\varphi, \xi^{(1)}, (\xi^{(2)})) | \leqslant \frac{M_1^{\varkappa-1}}{8},$$

$$| P^{(2)} (\varphi, \xi^{(2)}) | \leqslant M_1^{\varkappa} = M_2, \tag{25.33}$$

$$\max_{k_0,q} \sum_{\alpha,\beta} \left| \frac{\partial \xi^{(1)}_{k_0 q} (\xi^{(2)})}{\partial \xi^{(2)}_{\alpha\beta}} \right| \leqslant 1 + 4n^2 M_0^{\varkappa-1}.$$

Setting $\delta_1 = \delta_0^2$, we verify that inequalities (25.17) and (25.21) are satisfied, provided that

$$\frac{2M_0 n^2}{\sigma} \leqslant 2M_0^{\varkappa-1} n^2 < 1, \qquad \frac{c_3(\varepsilon) M_0^{(2-\varkappa)\varkappa^{\nu}-1}}{\delta_0^{\nu(d_1+m)}} < \frac{1}{8n} \qquad (\nu = 1, 2). \tag{25.34}$$

Continuing the argument, we finally conclude that for each $s \geqslant 1$ there exists the change of variables

$$x^{(s-1)} = (E + U^{(s)} (\varphi, \xi^{(s-1)})) x^{(s)},$$
$$\xi^{(s-1)} = \xi^{(s-1)} (\xi^{(s)}), \tag{25.35}$$

which reduces the system of equations for $x^{(s-1)}$ to the system

$$\frac{dx^{(s)}}{dt} = A_0 x^{(s)} [P^{(s)} (\varphi, \xi^{(s)}) + \xi^{(s)}] x^{(s)},$$
$$\frac{d\varphi}{dt} = \omega. \tag{25.36}$$

In addition, the matrices $\xi^{(s-1)} (\xi^{(s)})$, $U^{(s)} (\varphi, \xi^{(s-1)} (\xi^{(s)}))$ and $P^{(s)} (\varphi, \xi^{(s)})$ are analytic on the domain

$$| \operatorname{Im} \varphi | \leqslant \rho_{s-1} - 2\delta_{s-1} = \rho_{s-1} - 2\delta_0^s = \rho_s, \quad | \xi^{(s)} | \leqslant M_{s-1}, \tag{25.37}$$

2π-periodic in φ, real for real φ, and satisfy the inequalities

$$| \xi^{(s-1)} (\xi^{(s)}) - \xi^{(s)} | \leqslant M_{s-1}, \qquad | U^{(s)} (\varphi, \xi^{(s-1)} (\xi^{(s)})) | \leqslant \frac{M_{s-1}^{\varkappa-1}}{8n},$$

$$| P^{(s)} (\varphi, \xi^{(s)}) | \leqslant M_{s-1}^{\varkappa} = M_s, \tag{25.38}$$

$$\max_{k_0, q} \sum_{\alpha, \beta} \left| \frac{\partial \xi^{(s-1)}_{k_0 q} (\xi^{(s)})}{\partial \xi^{(s)}} \right| \leqslant 1 + 4n^2 M_{s-2}^{\varkappa-1},$$

whenever M_0 satisfies (25.34) for $\nu = 1, 2, \ldots, s$.

Since the numerator of the second inequality in (25.34) decreases faster than the denominator as $\nu \to \infty$, the inequalities (25.34) are consistent; hence $M^0 > 0$ can be found, such that for every $M_0 < M^0$ and any ν these inequalities are satisfied. From what has been stated it follows that for $M_0 < M^0$ the system of equations (25.4) reduces to a system of the form (25.36) by the iteration of substitutions of the form (25.13), . . . , (25.35), i.e. by the substitution

$$x = [E + U^{(1)}(\varphi, \xi)]\,[E + U^{(2)}(\varphi, \xi^{(1)})] \ldots [E + U^{(s)}(\varphi, \xi^{(s-1)})]\, x^{(s)} =$$
$$= \Phi^{(s)}(\varphi, \xi^{(s)})\, x^{(s)}, \qquad (25.39)$$
$$\xi = \xi(\xi^{(1)}), \quad \xi^{(1)} = \xi^{(1)}(\xi^{(2)}), \ldots, \xi^{(s-1)} = \xi^{(s-1)}(\xi^{(s)}).$$

As demonstrated in § 22, the estimates (25.38) ensure the uniform convergence of the sequence of functions $\Phi^{(s)} = \Phi^{(s)}(\varphi, 0)$ in the domain

$$|\operatorname{Im} \varphi| \leqslant \rho_0 - 2 \sum_{s=1}^{\infty} \delta_0^s = \frac{\rho_0}{2},$$

the analyticity of limit function

$$\Phi(\varphi) = \Phi^{(\infty)}(\varphi, 0) = \lim_{s \to \infty} \Phi^{(s)}(\varphi, 0) \qquad (25.40)$$

in the same domain, the non-degeneracy of matrix $\Phi(\varphi)$ and its property of being real for real φ.

Differentiating $\xi = \xi(\xi^{(1)} \ldots \xi^{(s)}))$ for $|\xi^{(s)}| \leqslant M_{s-1}$ and taking account of the inequalities (25.28), (25.33), . . . , (25.38) for the function $\xi^{(\alpha)}(\xi^{(\alpha+1)})$ and its derivatives, we get the estimate

$$\max_{k_0,\, q} \sum_{\alpha, \beta} \left| \frac{\partial \xi_{k_{0}q}(\xi^{(s)})}{\partial \xi^{(s)}_{\alpha\beta}} \right| \leqslant \left(1 + 4n^2 \frac{M_0}{\sigma}\right)(1 + 4n^2 M_0^{\varkappa-1}) \ldots (1 + 4n^2 M_{s-2}^{\varkappa-1}) \leqslant$$
$$\leqslant 2 \prod_{0 \leqslant p < \infty} (1 + 4n^2 M_p^{\varkappa-1}) \leqslant 2e^{\sum_{p=0}^{\infty} 4n^2 M_p^{\varkappa-1}} = c, \qquad (25.41)$$

from which we deduce that

$$|\xi(\xi^{(1)} \ldots (\xi^{(s-2)}(0)) - \xi(\xi^{(1)} \ldots \xi^{(s-2)}(0)))| \leqslant$$
$$\leqslant c\, |\xi^{(s-1)}(0)| \leqslant c M_{s-1}. \qquad (25.42)$$

The inequality (25.42) ensures the convergence of the sequence of matrices

$$\xi(0),\ \xi(\xi^{(1)}(0)), \ldots, \xi(\xi^{(1)}(\ldots \xi^{(s-1)}(0))) \ldots \qquad (25.43)$$

as $s \to \infty$.

Thus, if in the system of equations (25.3) we set

$$\xi = \xi^0 = \lim_{s\to\infty} \xi\,(\xi^{(1)} \ldots \xi^{(s-1)}\,(0)), \tag{25.44}$$

we can reduce it to the system (25.8) by the change of variables $x = \Phi(\varphi)y$ with the matrix $\Phi\,(\varphi) = \Phi^{(\infty)}(\varphi, 0)$ defined by (25.40). Furthermore, from the inequality (25.23) we deduce that

$$|\,\xi\,(\xi^{(1)}\,(\ldots\,\xi^{(s-1)}\,(0)))\,| \leqslant 2M_0 \tag{25.45}$$

for every $s \geqslant 1$. This inequality establishes the estimate $|\,\xi^0\,| \leqslant 2M_0$ and, consequently, completes the proof of the generalized reducibility theorem.

The existence of the matrix $\xi = \xi^0$, ensuring the reducibility of the system (25.3), is thus proved and we now look for an equation satisfied by this matrix. In the first place, we observe that the inequalities (25.38) ensure the existence of the limits

$$\xi_0^{(\alpha)} = \lim_{s\to\infty} \xi^{(\alpha)}\,(\xi^{(\alpha+1)}\,(\ldots\,\xi^{(s-1)}\,(0))) \tag{25.46}$$

for $\alpha = 0, 1, 2, \ldots$, where ξ^0 and $\xi^0\,(\xi^{(1)}\,(\ldots))$ denote the matrices ξ^0 and $\xi\,(\xi^{(1)}\,(\ldots))$, respectively.

The relations (25.46) with regard to the continuity of functions $\xi^{(\alpha)}(\xi^{(\alpha+1)})$ imply

$$\xi_0^{(\alpha)} = \xi_0^{(\alpha)}(\xi_0^{(\alpha+1)}) \qquad (\alpha = 0, 1, 2, \ldots). \tag{25.47}$$

The expression (25.47) yields the representation

$$\Phi^{(s)}\,(\varphi, \xi_0^{(s)}) = (E + U^{(1)}\,(\varphi, \xi_0^{(0)}))\,(E + U^{(2)}\,(\varphi, \xi_0^{(1)}))\ldots(E + U^{(s)}\,(\varphi, \xi_0^{(s-1)})).$$

As $s \to \infty$, $\Phi^{(s)}\,(\varphi, \xi_0^{(s)})$ behaves in the same way as the matrix $\Phi^{(s)}\,(\varphi, 0)$ and, consequently,

$$\lim_{s\to\infty} \Phi^{(s)}\,(\varphi, \xi_0^{(s)}) = \Phi^{(s)}\,(\varphi, 0) = \prod_{\alpha=1}^{\infty} (E + U^{(\alpha)}\,(\varphi, \xi_0^{(\alpha-1)})). \tag{25.48}$$

Introducing the notation

$$P\,(\varphi, \xi^{(0)}) + \xi^{(0)} = P_0^{(0)}\,(\varphi, \xi^{(0)}), \tag{25.49}$$

we can write the system of equations (25.3) for $\xi = \xi^0$ as

$$\begin{aligned} \frac{dx}{dt} &= A_0x + P_0^{(0)}\,(\varphi, \xi_0^{(0)})\,x, \\ \frac{d\varphi}{dt} &= \omega. \end{aligned} \tag{25.50}$$

It is trivial to verify that the recurrence relations defining the matrices $U^{(\alpha)}(\varphi, \xi_0^{(\alpha-1)})$ are expressed by the equations

$$\left(\frac{\partial U^{(\alpha)}}{\partial \varphi}, \omega\right) + U^{(\alpha)} A_0 = A_0 U^{(\alpha)} + P_0^{(\alpha-1)}(\varphi) - \overline{P_0^{(\alpha-1)}(\varphi)},$$

$$P_0^{(\alpha)}(\varphi, \xi) = (E + U^{(\alpha)}(\varphi, \xi))^{-1} (\overline{P_0^{(\alpha-1)}(\xi)}\, U^{(\alpha)}(\varphi, \xi) - U^{(\alpha)}(\varphi, \xi)\, \overline{P_0^{(\alpha-1)}(\xi)}), \tag{25.51}$$

and that the substitution

$$x = \prod_{\alpha=1}^{\infty} (E + U^{(\alpha)}(\varphi, \xi_0))\, y \tag{25.52}$$

reduces the system of equations (25.50) to the system

$$\frac{dy}{dt} = \left(A_0 + \sum_{\alpha=0}^{\infty} \overline{P_0^{(\alpha)}(\xi)}\right) y, \qquad \frac{d\varphi}{d} = \omega. \tag{25.53}$$

Starting from the expression (25.48), we conclude that

$$\sum_{\alpha=0}^{\infty} \overline{P_0^{(\alpha)}(\xi)} = 0,$$

i.e. that ξ^0 is a solution of the equation

$$\xi^0 + \overline{P(\varphi, \xi^0)} + \sum_{\alpha=1}^{\infty} \overline{P_0^{(\alpha)}(\varphi, \xi^0)} = 0. \tag{25.54}$$

Thus, if the system of equations (24.1) reduces to the system (24.6), where the eigenvalues of the matrix A_0 satisfy the inequality (25.1), then for small $P(\varphi)$ the reduction can be realized by means of the substitution (25.52), and the correction $\xi = \xi^0$ can be determined from equation (25.54). In addition, it is obvious that $A_0 = A - \xi^0$.

§ 26. Metric Propositions

It has been shown in the preceding section that a real matrix $A = A_0 + \xi^0$ can be associated with every real matrix A_0, whose eigenvalues satisfy the inequality

$$| \lambda_\alpha - \lambda_\beta + i(k, \omega) | \geqslant \varepsilon | k |^{-d} \quad (| k | \neq 0, \quad \alpha, \beta = 1, \ldots, n) \tag{26.1}$$

for every integral vector $k = (k_1, \ldots, k_m)$, so that the system of equations (24.1) is reducible to a system with constant coefficients; in addition, the transformation matrix $\Phi(\varphi)$ turns out to be such that $A \in \mathfrak{M}(P)$.

Hence a mapping $F : \mathfrak{M}_\varepsilon \to \mathfrak{M}(P)$ can be regarded as defined on a subset $\mathfrak{M}_\varepsilon$ of S_R consisting of matrices A_0, whose eigenvalues satisfy the inequality (26.1). If this mapping is measure-preserving and the measure of $\mathfrak{M}_\varepsilon$ turns out to be large, then by measuring $\mathfrak{M}_\varepsilon$ the measure of $\mathfrak{M}(P)$ can be ascertained. Before we take up this project, we shall outline some results from the theory of Lebesgue measure and prove some propositions concerning measure and sets of the sorts being considered.

We suppose that M is a set belonging to a Euclidean space E_n and contained in a bounded open set Ω_0.

The interior measure of a set M is the least upper bound of the measures of closed sets F contained in M, and the exterior measure of M is the greatest lower bound of the measures of the open sets Ω containing M.

A set M is called measurable if its interior ($\underline{\text{mes}}$) and exterior ($\overline{\text{mes}}$) measures are equal, and the measure of M is defined by

$$\text{mes}\, M = \sup_{F \subset M} \text{mes}\, F = \inf_{\Omega \supset M} \text{mes}\, \Omega. \tag{26.2}$$

The measure of an open set Ω is defined by

$$\text{mes}\, \Omega = \int_\Omega dx = \lim_{k_0 \to \infty} \int_{\Phi_{h_{k_0}}} dx; \tag{26.3}$$

and that of a closed set F by

$$\text{mes}\, F = \int_F dx = \text{mes}_{\Omega \supset F} \Omega - \text{mes}\, (\Omega \setminus F), \tag{26.4}$$

where $\Phi_{h_{k_0}}$ is a system of interior nets which completely cover the set Ω (cubical h_{k_0}-net for the set Ω, see [71]).

The measure of sets in the space E_n introduced by (26.2) — (26.4) is called the Lebesgue measure.

We shall draw upon the following well-known properties of the Lebesgue measure of sets:

(i) Let $\Omega_1 \supseteq \Omega_2 \ldots \supseteq \Omega_{k_0} \supseteq \ldots$ be a nested sequence of bounded open (closed) sets and let their intersection $\Omega_0 = \bigcap_{\alpha=1}^{\infty} \Omega_\alpha$ be either open or closed. Then

$$\operatorname{mes} \Omega_0 = \lim_{k_0 \to \infty} \operatorname{mes} \Omega_{k_0}. \tag{26.5}$$

(ii) Let F be a closed bounded set and let $f_{k_0}(x)$ ($k_0 = 1, 2, \ldots$) be a sequence of continuous mappings of F into E_n, uniformly convergent to $f_0(x)$. Then

$$\operatorname{mes} F_0 = \operatorname{mes} f_0\{F\} \geqslant \overline{\lim} \operatorname{mes} f_{k_0}\{F\} = \overline{\lim} \operatorname{mes} F_{k_0}, \tag{26.6}$$

where $f\{F\}$ denotes an image of the set F under the mapping f.

(iii) Let $F_1 \supseteq F_2 \supseteq \ldots \supseteq F_{k_0} \supseteq \ldots$ be a nested sequence of closed subsets of E_n, $[F_{k_0}]$ the boundary of F_{k_0} and let $F_0 = \bigcap_{\alpha=1}^{\infty} F_\alpha$ be an intersection of F_{k_0}. If f is a continuous and one-to-one mapping of F_1 into E_n, then

$$\operatorname{mes} f\{F_0\} = \lim_{k_0 \to \infty} \operatorname{mes} f\{F_{k_0}\} \geqslant \lim_{k_0 \to \infty} \operatorname{mes} f\{F_{k_0} \setminus [F_{k_0}]\}. \tag{26.7}$$

(iv) Let M be a measurable set of the cube $\Omega\,\{0 < x_i < 1,\ 0 < y_j < 1,\ i = 1, \ldots, n,\ j = 1, \ldots, m\}$; let M_y be the projection of M on the manifold $y_1, \ldots, y_m$ and $M(y)$ a cut of this set by the manifold $y_1 = \text{const}$, $y_2 = \text{const}, \ldots, y_m = \text{const}$. Then, the function

$$\psi_M(y) = \begin{cases} 0 & \text{for } y \,\bar{\in}\, M_y, \\ \displaystyle\int_{M(y)} dx & \text{for } y \in M_y \end{cases}$$

is measurable, and

$$\operatorname{mes} M = \int_{0<y_i<1} \psi_M(y)\, dy. \tag{26.8}$$

A function $f(x) = \{f_1(x), \ldots, f_m(x)\}$ defined on an open set Ω is said to be uniformly differentiable on this set, if $f(x)$ has continuous derivatives $(\partial f/\partial x) = \{\partial f_\alpha/\partial x_\beta\}$ at each point of Ω.

An important characteristic of mappings which are uniformly differentiable on the set Ω is stated in the following lemma.

Lemma 8. *Suppose that Ω is an open subset of E_n and that f is a continuous one-to-one mapping of Ω into E_n. If $f(x)$ is uniformly differentiable on the set Ω and if*

$$0 < m \leqslant \left| \det \left\{ \frac{\partial f}{\partial x} \right\} \right| \leqslant M \qquad (x \in \Omega), \tag{26.9}$$

then

$$\operatorname{mes} f\{\Omega\} \geqslant \frac{\operatorname{mes} \Omega}{M}. \tag{26.10}$$

Proof. It is required to establish the inequality (26.10). Since Ω is an open set, it has a system of interior nets $\Phi_{h_{k_0}}$ that cover it. For every fixed k_0, $\Phi_{h_{k_0}}$ consists of a finite number of cubic nets and since it is closed, it decomposes into a finite number (l_{k_0}) of closed disjoint domains: $\Phi_{h_{k_0}} = \bigcup_{\alpha=1}^{l_{k_0}} \Phi_{h_{k_0}}^{(\alpha)}$.

Obviously, the boundary of $\Phi_{h_{k_0}}^{(\alpha)}$ is piece-wise smooth. The uniform differentiability of the mapping f on the set Ω implies the continuity of $f(x)$ on this set and the continuous differentiability of $f(x)$ on each of the domains $\Phi_{h_{k_0}}^{(\alpha)}$ ($\alpha = 1, \ldots, l_{k_0}$; $k_0 = 1, 2, \ldots$). Since the mapping f is one-to-one on the set Ω, the disjointness of the sets $\Phi_{h_{k_0}}^{(\alpha)}$ ($\alpha = 1, \ldots, l_{k_0}$) implies the disjointness of their images:

$$f\left\{ \Phi_{h_{k_0}}^{(\alpha_1)} \right\} \cap f\left\{ \Phi_{h_{k_0}}^{(\alpha_2)} \right\} = \Phi \qquad \text{for } \alpha_1 \neq \alpha_2,$$

where Φ is an empty set. Hence, for every k_0, we can set

$$\operatorname{mes} f\{\Omega\} \geqslant \operatorname{mes} \{\Phi_{h_{k_0}}\} = \sum_{\alpha=1}^{l_{k_0}} \operatorname{mes} f\left\{\Phi_{h_{k_0}}^{(\alpha)}\right\} = \sum_{\alpha=1}^{l_{k_0}} \int_{f\left\{ \Phi_{h_{k_0}}^{(\alpha)} \right\}} dx.$$

Consider a system of functions $x = f(y)$ defined in the closed domain $\Phi_{h_{k_0}}^{(\alpha)}$ of the space E_n with coordinates $y_1, \ldots, y_n$. Since $f(y)$ is a continuously differentiable function of y in the domain $\Phi_{h_{k_0}}^{(\alpha)}$ and, by hypothesis, its Jacobian $J(y)$ satisfies the inequality

$$0 < m \leqslant \left| \det\left\{ \frac{\partial f}{\partial y} \right\} \right| = |\, \mathrm{J}(y) \,| \leqslant M \qquad \text{for } y \in \Phi_{h_{k_0}}^{(\alpha)},$$

the system considered is uniquely solvable:

$$y = f^{-1}(x) \quad \text{for } x \in f\left\{\Phi_{h_{k_0}}^{(\alpha)}\right\},$$

$f^{-1}(x)$ being a continuously differentiable function of x in the domain $f\left\{\Phi_{h_{k_0}}^{(\alpha)}\right\}$ and its Jacobian satisfying the inequality

$$\frac{1}{M} \leqslant \frac{1}{\left| \det\left\{ \dfrac{\partial f(y)}{\partial y} \right\} \right|} = \left| \det\left\{ \frac{\partial f^{-1}(x)}{\partial x} \right\} \right| = |\, J(x) \,| \leqslant \frac{1}{m}.$$

The integral $\int\limits_{f\left\{\Phi^{(\alpha)}_{h_{k_0}}\right\}} dx$ is the usual multiple Riemann integral and since the boundary of the set $f\left\{\Phi^{(\alpha)}_{h_{k_0}}\right\}$ is piece-wise smooth, we may make the change of variable $x = f^{-1}(y)$ in it, to obtain

$$\int\limits_{f\left\{\Phi^{(\alpha)}_{h_{k_0}}\right\}} dx = \int\limits_{\Phi^{(\alpha)}_{h_{k_0}}} \left| \det \left\{ \frac{\partial f^{-1}(y)}{\partial y} \right\} \right| dy \geqslant \frac{1}{M} \operatorname{mes} \Phi^{(\alpha)}_{h_{k_0}}.$$

Thus

$$\operatorname{mes} f\{\Omega\} \geqslant \sum_{\alpha=1}^{l_{k_0}} \frac{\operatorname{mes} \Phi^{(\alpha)}_{h_{k_0}}}{M} = \frac{1}{M} \operatorname{mes} \left\{ \bigcup_{\alpha=1}^{l_{k_0}} \Phi^{(\alpha)}_{h_{k_0}} \right\} = \frac{\operatorname{mes} \Phi_{h_{k_0}}}{M}.$$

Taking the limit in the last inequality as $k_0 \to \infty$, we get the required estimate

$$\operatorname{mes} f\{\Omega\} \geqslant \frac{1}{M} \lim \operatorname{mes} \Phi_{h_{k_0}} = \frac{\operatorname{mes} \Omega}{M}.$$

In the following, the condition (26.1), identifying the set of matrices A belonging to $\mathfrak{M}_\varepsilon$, expressed in terms of the eigenvalues of A, will be bypassed in favour of a more convenient condition, equivalent to it, connected directly with the coefficient matrix $A = \{a_{\alpha\beta}\}$. To accomplish this, we recall the equation

$$Y_k (A + i(k, \omega)) - AY_k = P_k \qquad (|k| \neq 0). \tag{26.11}$$

Given that $A \in \mathfrak{M}_\varepsilon$, equation (26.11) can be transformed if it has a solution satisfying the inequality

$$| Y_k | \leqslant C(\varepsilon) \, | k |^{d_1} \, | P_k |. \tag{26.12}$$

Let

$$y = (y_{11}, \ldots, y_{1n}, y_{21}, \ldots, y_{nn}), \quad p = (p_{11}, \ldots, p_{1n}, p_{21}, \ldots, p_{nn})$$

be n^2-dimensional vectors. Write the matrix equation (26.11) in the form

$$[\mathfrak{A}_0 + i(k, \omega)] \, y = p, \tag{26.13}$$

where

$$\mathfrak{A}_0 = \begin{pmatrix} A' - a_{11}E, & -a_{12}E, \ldots, & -a_{1n}E \\ -a_{21}E, & A' - a_{22}E, \ldots, & -a_{2n}E \\ \vdots & \vdots & \vdots \\ -a_{n1}E, & -a_{n2}E, \ldots, & A' - a_{nn}E \end{pmatrix}, \quad A = \{a_{\alpha\beta}\}, \quad A' = \{a_{\beta\alpha}\}, \tag{26.14}$$

E an $n \times n$ identity matrix. Consequently, the matrix $\mathfrak{A}_0$ consists of $n \times n$ blocks $A' - a_{\alpha\alpha}E$ and $-a_{\alpha\beta}E$. The condition for the solvability of equation (26.13) is that the determinant of the matrix $\mathfrak{A}_0 + i(k, \omega)$ be different from zero. Let $\det[\mathfrak{A}_0 + i(k, \omega)] \neq 0$. Then, obviously, $\det[\mathfrak{A}_0 - i(k, \omega)] \neq 0$. Multiplying equation (26.13) by the matrix $\mathfrak{A}_0 - i(k, \omega)$, we get

$$[\mathfrak{A}_0^2 + (k, \omega)^2]\, y = [\mathfrak{A}_0 - i(k, \omega)]\, p, \tag{26.15}$$

which is equivalent to equation (26.11).

We denote by $F^0_{n^2}(\mathfrak{A}_0, \lambda)$ the characteristic polynomial of the matrix $\mathfrak{A}_0$. By manipulating this polynomial, we set up in the desired form the condition that the matrix A belongs to the set $\mathfrak{M}_\varepsilon$. Equation (26.15), obviously, has a solution, if

$$\det(\mathfrak{A}_0^2 + (k, \omega)^2) = F^0_{n^2}(\mathfrak{A}_0^2 - (k, \omega)^2) \neq 0. \tag{26.16}$$

Subject to inequality (26.16) being satisfied, the solution of equation (26.15) has the estimate

$$|y| \leqslant |[\mathfrak{A}_0^2 + (k, \omega)^2]^{-1}|\, |\mathfrak{A}_0 - i(k, \omega)|\, |p| \leqslant$$
$$\leqslant \frac{|D(a, (k, \omega)^2)|}{|F^0_{n^2}(\mathfrak{A}_0^2, -(k, \omega)^2)|}\, [\,|\mathfrak{A}_0| + |(k, \omega)|\, n\,]\, |p|; \tag{26.17}$$

here $D(a, (k, \omega)^2)$ is a matrix, consisting of the cofactors of the elements of the matrix $\mathfrak{A}_0^2 + (k, \omega)^2$. However,

$$|D(a, (k, \omega)^2)| \leqslant c_1 |k|^{2(n^2-1)}, \qquad |\mathfrak{A}_0| + n\,|(k, \omega)| \leqslant c_1 |k|, \tag{26.18}$$

where c_1 is a constant not depending on k. By inequality (26.18), the estimate (26.17) is representable in the form

$$y| \leqslant \frac{c_1^2 |k|^{2n^2-1}}{|F^0_{n^2}(\mathfrak{A}_0^2, -(k, \omega)^2)|},$$

whence it is easy to deduce

$$|Y_k| = \frac{c_1^2 |k|^{2n^2-1}}{|F^0_{n^2}(\mathfrak{A}_0^2, -(k, \omega)^2)|}\, |P_k| \leqslant c(\varepsilon)\, |k|^{d_1}\, |P_k|,$$

whenever

$$|F^0_{n^2}(\mathfrak{A}_0^2, -(k, \omega)^2)| \geqslant \varepsilon^{\gamma_0} |k|^{-d_0}, \tag{26.19}$$

where c_0, γ_0, d_0, are positive constants.

The set $\mathfrak{M}_\varepsilon$ of matrices A, satisfying condition (26.1) (with $d_1 \geqslant (2n^2-1)\, d_0$, $c(\varepsilon) > (c_1^2/c)\, \varepsilon^{-\gamma_0}$), is to be defined as equivalent to the set $\mathfrak{M}_\varepsilon = \mathfrak{M}_\varepsilon(\gamma_0, d_0)$ consisting of the points $a = \{a_{11}, a_{12}, \ldots, a_{1n}, \ldots, a_{nn}\}$ of the cube S_R, for which inequality (26.19) is satisfied.

For definiteness, we put $R = 1$ and prove the following proposition concerning the measure of a set $\mathfrak{M}_\varepsilon$.

Lemma 9. *If the inequality*

$$|(k, \omega)| \geqslant \varepsilon |k|^{-d} \qquad (|k| \neq 0)$$

is satisfied for every integral vector $k = (k_1, \ldots, k_m)$, then it is always possible to choose positive constants γ_0 and d_0 such that for sufficiently small ε, the inequality

$$\text{mes}\, \mathfrak{M}_\varepsilon \geqslant (1 - \varepsilon) \tag{26.20}$$

holds.

Proof. Suppose that

$$N_1 < N_2 < \ldots < N_s < \ldots \tag{26.21}$$

is an increasing sequence of integers, $N_1 > 1$. Denote by $\mathfrak{M}_\varepsilon^{(s)}$ a set consisting of the points $a \in S_1$, for which

$$|F^{(0)}(a, -(k, \omega)^2)| = |F^0_{n^2}(\mathfrak{A}_0^2, -(k, \omega)^2)| \geqslant \varepsilon^{\gamma_0} |k|^{-d_0}$$

$$\text{for} \quad 0 < |k| \leqslant N_s. \tag{26.22}$$

Since

$$\mathfrak{M}_\varepsilon^{(s)} \supset \mathfrak{M}_\varepsilon^{(s+1)}, \quad \bigcap_{s=1}^{\infty} \mathfrak{M}_\varepsilon^{(s)} = \mathfrak{M}_\varepsilon \qquad (s = 1, 2, \ldots)$$

and each of the sets $\mathfrak{M}_\varepsilon^{(s)}$ is closed, it follows that

$$\text{mes}\, \mathfrak{M}_\varepsilon = \lim_{s\to\infty} \mathfrak{M}_\varepsilon^{(s)}. \tag{26.23}$$

Let U_k be a set of the points $a \in S_1$, satisfying

$$|F^{(0)}(a, -(k, \omega^2)| \leqslant \varepsilon^{\gamma_0} |k|^{-d_0}. \tag{26.24}$$

The inclusion

$$\mathfrak{M}_\varepsilon^{(s)} \supset S_1 \setminus \bigcup_{|k|=1}^{N_s} U_k$$

holds, and so

$$\text{mes}\, \mathfrak{M}_\varepsilon^{(s)} \geqslant 1 - \text{mes} \bigcup_{|k|=1}^{N_s} U_k \geqslant 1 - \sum_{|k|=1}^{N_s} \text{mes}\, U_k. \tag{26.25}$$

The requirement now is to estimate mes U_k. Since $F^{(0)}(a, -(k, \omega)^2)$ denotes the function $F_{n^2}^{(0)}(\mathfrak{A}_0^2, -(k, \omega)^2)$, then taking account of the form of the matrix $\mathfrak{A}_0$ we see that $F^0(a, -(k, \omega)^2)$ is a polynomial of degree not exceeding $2n^2$ in each of the arguments of the function $F^{(0)}$. We now write $F^{(0)}(a, -(k, \omega)^2)$ in the form of a polynomial in a_{11}, by the equation

$$F^{(0)}(a, -(k, \omega)^2) = a_{11}^{m_1} F_0^{(1)}(a_{22}, \ldots, -(k, \omega)^2) + a_{11}^{m_1-1} F_1^{(1)} + \ldots + F_{m1}\,.$$

Let $U_k^{(1)} = U_k^{(1)}(\gamma_1, d_1, c_1)$ be a set of the points $a \in S_1$, satisfying

$$| F_0^{(1)}(a_{22}, \ldots, -(k, \omega)^2) | \leqslant c_1 \varepsilon^{\gamma_1} | k |^{-d_1}, \tag{26.26}$$

and $\overline{U}_k$ a set of the points $a \in S_1$ for which

$$| F^{(0)}(a, -(k, \omega)^2) | \leqslant \varepsilon^{\gamma_0} | k |^{-d_1},$$

$$| F_0^{(1)}(a_{22}, \ldots, -(k, \omega)^2) | \geqslant c_1 \varepsilon^{\gamma_1} | k |^{-d_1}.$$

Obviously,

$$U_k \subseteq U_k^{(1)} \cup \overline{U}_k,$$

so that

$$\text{mes}\, U_k \leqslant \text{mes}\, U_k^{(1)} + \text{mes}\, \overline{U}_k. \tag{26.27}$$

Let $M_{a'}$ be a projection of the set $\overline{U}_k$ on the manifold $a' = \{a_{22}, \ldots, a_{n-1,\, n}\}$, and $M(a')$ a cut of the set $\overline{U}_k$ by the manifold $a' =$ const. By the properties of the measure of a set deduced above,

$$\text{mes}\, \overline{U}_k = \int_{M_{a'}} \left\{ \int_{M(a')} da_{11} \right\} da',$$

and, consequently,

$$\text{mes}\, \overline{U}_k \leqslant \sup_{a' \in M_{a'}} \left| \int_{M(a')} da_{11} \right|.$$

For $a' \in M_{a'}$, $M(a')$ is a set of the points a_{11} of the segment $[-1, 1] = I$, $[(a_{11}, a') \in \overline{U}_k]$, satisfying

$$| F^{(0)}(a_{11}, a', -(k, \omega)^2) | = | a_{11}^{m_1} F_0^{(1)}(a', -(k, \omega)^2) + a_1^{m_1-1} F_1^{(1)}(a', -(k, \omega)^2) + + \ldots + F_{m_1}(a', -(k, \omega)^2) | \leqslant \leqslant \varepsilon^{\gamma_0} |k|^{-d_0}, \tag{26.28}$$

and for $a' \in M_{a'}$ the integral $\int_{M(a')} da_{11}$ is the Lebesgue measure of this set, considered as a subset $a' =$ const. of the real line. The process of estimating this measure is described in what follows.

For $a' \in M_{a'}$ fixed, we denote by $a_{11}^{(\alpha)}(a')$ ($\alpha = 1, \ldots, r$) the real zeros of functions

$$F^{(0)}(a_{11}, a', -(k, \omega)^2) \quad \text{and} \quad \frac{\partial F^{(0)}(a_{11}, a', -(k, \omega)^2)}{\partial a_{11}}.$$

Since $F_0^{(1)}(a', -(k, \omega)^2) \neq 0$ for $a' \in M_{a'}$, the number of these zeros is finite : $r \leqslant 2m_1 - 1$.

Let the point $a_{11}^{(\alpha)}(a') \in I$ be contained in the segment $I^{(\alpha)} = [a_{11}^{(\alpha)} - \varepsilon^{\overline{\gamma}} |k|^{-\overline{\delta}}, a_{11}^{(\alpha)} + \varepsilon^{\overline{\gamma}} |k|^{-\overline{\delta}}]$. We can show that for an appropriate choice of the constants $\gamma_1, \overline{\gamma}, d_1, \overline{\delta}$ and c_1, the set $M(a')$, for every $a' \in M_{a'}$ is contained in the union of segments $I^{(\alpha)}$:

$$M(a') \subseteq \bigcup_{\alpha=1}^{r} I^{(\alpha)} \qquad (a' \in M_{a'}). \tag{26.29}$$

For this purpose, we take a point $a_{11} \in I \setminus \bigcup_{\alpha=1}^{r} I^{(\alpha)}$, $(a_{11}, a') \in \overline{U}_k$, and show that $a_{11} \overline{\in} M(a')$, i.e. for this point the inequality (26.29) is not satisfied. In fact, by the choice of the segment $I^{(\alpha)}$ and the definition of the set $\overline{U}_k$ we have the inequalities

$$F^{(0)}(t, a', -(k, \omega)^2) \neq 0, \qquad \frac{dF^{(0)}(t, a', -(k, \omega)^2)}{dt} \neq 0,$$

$$\left| \frac{d^{m_1} F^{(0)}(t, a', -(k, \omega)^2)}{dt^{m_1}} \right| = m!, \qquad | F_0^{(1)}(a', -(k, \omega)^2) | \geqslant m!\, c_1 \varepsilon^{\gamma_1} |k|^{-d_1},$$

for every $t \in (a_{11} - \varepsilon^{\overline{\gamma}} |k|^{-\overline{\delta}}, a_{11} + \varepsilon^{\overline{\gamma}} |k|^{-\overline{\delta}}$. That these inequalities imply the estimate

$$| F^{(0)}(a_{11}, a', -(k, \omega)^2) | \geqslant \frac{m! c_1 \varepsilon^{\gamma_1} |k|^{-d_1}}{(2m_1)^{m_1}} \varepsilon^{m_1 \overline{\delta}} |k)^{-m_1 \overline{\delta}} \tag{26.30}$$

is a consequence of the following proposition.

Lemma 10 [38]. *Suppose that the function $f(t)$ is m-times continuously differentiable* for $a - \tau \leqslant t \leqslant a + \tau$. *Then, if*

$$f(t) \neq 0, \quad \frac{df(t)}{dt} \neq 0, \quad \left| \frac{d^m f(t)}{dt^m} \right| \geqslant \varepsilon \quad (a - \tau < t < a + \tau),$$

we have the inequality

$$| f(a) | \geqslant \frac{\varepsilon}{(2m)^m} \tau^m. \tag{26.31}$$

If γ_1, d_1, c_1 are now taken as positive constants satisfying the conditions

$$\gamma_1 + m_1 \bar{\gamma} \geqslant \gamma_0, \quad d_1 + m_1 \bar{\delta} \leqslant d_0, \quad \frac{c_1 m_1!}{(2m_1)^{m_1}} > 1, \tag{26.32}$$

then inequality (26.30) implies the inequality which is the reverse of the condition that the point a_{11} belong to the set $M(a')$, proving also the relation (26.29). This relation yields the inequality

$$\sup_{a' \in M_{a'}} \left| \int_{M(a')} da_{11} \right| \leqslant \sum_{\alpha=1}^{r} \text{mes } I^{(\alpha)} = 2\varepsilon^{\bar{\gamma}} \, | k |r^{-\bar{\delta}} \leqslant 2(2m_1 - 1) \, \varepsilon^{\bar{\gamma}} \, | k |^{-\bar{\delta}},$$

inspection of which leads to the estimate

$$\text{mes } \bar{U}_k \leqslant 2(2m_1 - 1) \, \varepsilon^{\bar{\gamma}} \, | k |^{-\bar{\delta}}. \tag{26.33}$$

Inequalities (26.27) and (26.33) imply

$$\text{mes } U_k \leqslant \text{mes } U_k^{(1)} + 2(2m_1 - 1) \, \varepsilon^{\bar{\gamma}} \, | k |^{-\bar{\delta}}. \tag{26.34}$$

We now estimate mes $U_k^{(1)}$. The function $F_0^{(1)}(a', -(k, \omega)^2)$ is a polynomial of degree not exceeding $2n^2$ in all of the arguments appearing in it. We express this function in the form

$$F_0^{(1)}(a', -(k, \omega)^2) = a_{22}^{m_2} F_0^{(2)}(a_{33}, \ldots, -(k, \omega)^2) + a_{22}^{m_2 - 1} F_1^{(2)} + \ldots + F_{m_2}^{(2)}.$$

Arguing on similar lines, we finally obtain the inequality

$$\text{mes } U_k^{(1)} \leqslant \text{mes } U_k^{(2)} + 2(2m_2 - 1) \, \varepsilon^{\bar{\gamma}} \, | k |^{-\bar{\delta}}, \tag{26.35}$$

whenever $U_k^{(2)}$ (γ_2, d_2, c_2) is a point set $a \in S_1$, satisfying

$$| F_0^{(2)} (a_{33}, \ldots, -(k, \omega)^2) | \leqslant c_2 \, \varepsilon^{\gamma_2} | k |^{-d_2}, \tag{26.36}$$

and the positive constants γ_2, d_2, c_2 obey the conditions

$$\gamma_2 + m_2\bar{\gamma} \geqslant \gamma_1, \qquad d_2 + m_2\bar{\delta} \leqslant d_1, \qquad \frac{c_2 m_2!}{(2m_2)^{m_2}} > c_1. \tag{26.37}$$

Similarly, the $(n-1)$-th step states that

$$\text{mes } U_k \leqslant \text{mes } U_k^{(n-1)} + 2\left(2 \sum_{\alpha=1}^{n-1} m_\alpha - n + 1\right) \varepsilon^{\bar{\gamma}} | k |^{-\bar{\delta}}, \tag{26.38}$$

whenever $U_k^{(n-1)}$ $(\gamma_{n-1}, d_{n-1}, c_{n-1})$ is a point set $a \in S_1$, satisfying

$$| F_0^{(n-1)} (a_{nn}, \ldots, -(k, \omega)^2) | \leqslant c_{n-1} \varepsilon^{\gamma_{n-1}} | k |^{-d_{n-1}}, \tag{26.39}$$

and the positive constants γ_α, d_α, c_α satisfy the system of inequalities

$$\gamma_\alpha + m_\alpha \bar{\gamma} \geqslant \gamma_{\alpha-1}, \quad d_\alpha + m_\alpha \bar{\delta} \leqslant d_{\alpha-1}, \quad \frac{c_\alpha m_\alpha!}{(2m_\alpha)^{m_\alpha}} > c_{\alpha-1}. \tag{26.40}$$

$$(\alpha = 1, \ldots, n-1).$$

Solving this system we verify that it is consistent, if

$$m_\alpha \neq 0, \quad \gamma_0 \geqslant \bar{\gamma} \sum_{i=1}^{n-1} m_i, \quad d_0 \geqslant \bar{\delta} \sum_{i=1}^{n-1} m_i, \tag{26.41}$$

and takes the values

$$\gamma_\alpha = \gamma_0 - \bar{\gamma} \sum_{i=1}^{\alpha} m_i, \qquad d_\alpha = d_0 - \bar{\delta} \sum_{i=1}^{\alpha} m_i,$$

$$c_\alpha = \prod_{i=1}^{\alpha} \frac{(2m_i)^{m_i}}{m_i!} c_0^\alpha \qquad \left(1 < c_0^{n-1} < \frac{3}{2}\right). \tag{26.42}$$

For a convenient choice of γ_0, d_0, the inequalities (26.41) can always be satisfied. It shall be shown below that $4m_\alpha = 4(n - \alpha)$ and that the set $U_k^{(n-1)}$ is empty for a definite choice of the constants d_0, γ_0 and c_0. For γ_{n-1}, d_{n-1} and

c_{n-1} so chosen, the measure of set U_k obviously satisfies the inequalities

$$\text{mes}\, U_k \leqslant 2(n-1)(4n-1)\, \varepsilon^{\bar{\gamma}}\, |k|^{-\bar{\delta}} \quad (|k| \neq 0),$$

so that by (26.25),

$$\text{mes}\, \mathfrak{M}_\varepsilon^{(s)} \geqslant 1 - \sum_{|k|=1}^{N_s} |k|^{-\bar{\delta}}\, 2(n-1)(4n-1) \geqslant$$
$$\geqslant 1 - 2(n-1)(4n-1)\, \varepsilon^{\bar{\gamma}} \sum_{\nu=1}^{N_s} \nu^{m-\bar{\delta}-1}. \tag{26.43}$$

We set

$$\bar{\delta} = m + 1 \tag{26.44}$$

in (26.43), to obtain for the measure of set $\mathfrak{M}_\varepsilon$ the estimate

$$\text{mes}\, \mathfrak{M}_\varepsilon = \lim_{s\to\infty} \text{mes}\, \mathfrak{M}_\varepsilon^{(s)} \geqslant 1 - \frac{2^{m+1}\pi^2}{3}(n-1)(4n-1)\,\varepsilon^{\bar{\gamma}}. \tag{26.45}$$

To determine the constants m_α, γ_{n-1}, d_{n-1} and c_{n-1}, we investigate the dependence of the function $F_0^{(\alpha)}$ on $a_{\alpha+1,\,\alpha+1}$. From its definition $F_0^{(1)}$ is the leading coefficient of degree a_{11} in the function

$$F^{(0)} = F_0^{(0)} = \det|\mathfrak{A}_0 - \lambda|\,|\det \mathfrak{A}_0 + \lambda| \qquad \text{for } \lambda = i(k, \omega).$$

Since

$$\det|\mathfrak{A}_0 - \lambda| =$$

$$= \det \begin{vmatrix} A' - (a_{11}+\lambda)E, & -a_{12}E, \dots, & -a_{1n}E \\ -a_{21}E, & A' - (a_{22}+\lambda)E, \dots, & -a_{2n}E \\ \vdots & \vdots & \vdots \\ -a_{n1}E, & -a_{n2}E, \dots, & A' - (a_{nn}+\lambda)E \end{vmatrix}, \tag{26.46}$$

a_{11} appearing in $\det|\mathfrak{A}_0 - \lambda|$ is of degree not higher than $2(n-1)$. Hence, the coefficient of a_{11} can be obtained by differentiating the determinant (26.46)

$2(n-1)$-times. It is trivial to verify that

$$\frac{1}{[2(n-1)]!}\ \frac{d^{2(n-1)}}{da_{11}^{2(n-1)}} \det|\mathfrak{A}_0-\lambda E| = \frac{2(n-1)}{[2(n-1)]!}\times$$

$$\times\left|\begin{array}{ccccccccccccc}
-\lambda & 0 & \cdots & 0 & 0 & 0 & \cdots & 0 & \cdots & 0 & 0 & \cdots & 0\\
a_{11} & 1 & \cdots & 0 & 0 & -a_{12} & \cdots & 0 & \cdots & 0 & -a_{n1} & \cdots & 0\\
a_{13} & 0 & \cdots & 0 & 0 & 0 & \cdots & 0 & \cdots & \cdot & \cdot & \cdots & \cdot\\
\vdots & \vdots & & \vdots & \vdots & \vdots & & \vdots & & \vdots & \vdots & & \vdots\\
a_{1n} & 0 & \cdots & 1 & 0 & 0 & \cdots & -a_{12} & \cdots & 0 & 0 & \cdots & -a_{n1}\\
-a_{13} & 0 & \cdots & 0 & 1 & a_{21} & \cdots & a_{n1} & \cdots & 0 & 0 & \cdots & 0\\
0 & 0 & \cdots & 0 & 0 & -\lambda & \cdots & a_{n2} & \cdots & 0 & -a_{2n} & \cdots & 0\\
\vdots & \vdots & & \vdots & \vdots & \vdots & & \vdots & & \vdots & \vdots & & \vdots\\
0 & 0 & \cdots & 0 & 0 & a_{2n} & \cdots & a_{nn}-a_{22}-\lambda & \cdots & 0 & 0 & \cdots & -a_{2n}\\
\vdots & \vdots & & \vdots & \vdots & \vdots & & \vdots & & \vdots & \vdots & & \vdots\\
-a_{n1} & 0 & \cdots & 0 & 0 & 0 & \cdots & 0 & \cdots & 1 & a_{21} & \cdots & a_{n1}\\
0 & 0 & \cdots & 0 & 0 & -a_{n2} & \cdots & 0 & \cdots & 0 & a_{22}-a_{nn}-\lambda & \cdots & a_{n2}\\
\vdots & \vdots & & \vdots & \vdots & \vdots & & \vdots & & \vdots & \vdots & & \vdots\\
0 & 0 & \cdots & 0 & 0 & 0 & \cdots & -a_{n2} & \cdots & 0 & a_{2n} & \cdots & -\lambda
\end{array}\right| =$$

$$= \frac{(-1)^n\lambda}{(2n-3)!} \det|\mathfrak{A}_0'-\lambda|, \tag{26.47}$$

where $\mathfrak{A}_0'$ is an $(n-1)^2\times(n-1)^2$ matrix, obtained from the matrix $\mathfrak{A}_0$ by deleting the first block column and the first block row, and also the first row and the first column in the remaining block matrices. Obviously, $\det|\mathfrak{A}_0'-i(k,\omega)|\not\equiv 0$, hence

$$F_0^{(1)} = -\frac{\lambda^2}{[(2n-3)!]^2}\det|\mathfrak{A}_0'-\lambda|\det|\mathfrak{A}_0'+\lambda| \tag{26.48}$$

$$\text{for}\quad \lambda = i(k,\omega),\quad m_1 = 4(n-1).$$

Since the function $F_0^{(1)}$ is similar in form to the function $F^{(0)}$, the leading coefficient $F_0^{(2)}$ of degree a_{22} in $F_0^{(1)}$ and this degree itself are defined by

$$F_0^{(2)} = \frac{\lambda^4}{[(2n-3)!]\,[(2n-5)!]}\det|\mathfrak{A}_0''-\lambda|\det|\mathfrak{A}_0''+\lambda| \tag{26.49}$$

$$\text{for}\quad \lambda = i(k,\omega),\quad m_2 = 4(n-2),$$

where $\mathfrak{A}_0''$ is an $(n-2)^2\times(n-2)^2$ matrix, obtained from $\mathfrak{A}_0'$ by deleting the first block column and first block row, and also the first row and the first column in the remaining block matrices.

Proceeding further in the same manner, it is found that

$$m_\alpha = 4(n-\alpha) \qquad \text{for } \alpha = 1, 2, \ldots, n-1, \tag{26.50}$$

$$F_0^{(n-1)} = \frac{\lambda^{2(n-1)}}{[1!\,3!\ldots(2n-3)!]^2} \det|\mathfrak{A}_0^{(n-1)} - \lambda| \det|\mathfrak{A}_0^{n-1} + \lambda|$$

$$\text{for} \quad \lambda = i(k, \omega),$$

where $\mathfrak{A}_0^{(n-1)}$ is a number given by the $2^2 \times 2^2$ matrix $\mathfrak{A}_0^{(n-2)}$ on cancellation of the first three columns and the first three rows. It is plain that $\mathfrak{A}_0^{(n-1)}$ is the last diagonal element of the matrix $\mathfrak{A}_0$, i.e. $\mathfrak{A}_0^{(n-1)} = 0$.

Consequently,

$$F_0^{(n-1)} = \frac{(k, \omega)^{2n}}{[1!\ldots(2n-3)!]^2}. \tag{26.51}$$

Thus, $U_0^{(n-1)}$ is a set of points, satisfying the inequality

$$\frac{(k, \omega)^{2n}}{[1!3!\ldots(2n-3)!]^2} < c_{n-1}\varepsilon^{\gamma_{n-1}} |k|^{-d_{n-1}} <$$

$$< \frac{3}{2} \prod_{\alpha=1}^{n-1} \frac{(8\alpha)^{4\alpha}}{(4\alpha)!} \varepsilon^{\gamma_0 - 2n(n-1)\bar{\gamma}} |k|^{-d_0 + 2n(n-1)\bar{\delta}}. \tag{26.52}$$

However, since

$$\frac{(k, \omega)^{2n}}{1!\,3!\ldots(2n-3)!} > \frac{\varepsilon^{2n}|k|^{-2nd}}{1!\,3!\ldots(2n-3)!}, \tag{26.53}$$

the inequalities (26.52) and (26.53) are inconsistent, whenever

$$\gamma_0 = 2n[(n-1)\bar{\gamma} + 1] + \bar{\delta},$$
$$d_0 = 2n[(n-1)\bar{\delta} + d] = 2n[(n-1)(m+1) + d], \tag{26.54}$$

and ε is taken so small that the inequality

$$\frac{3}{2}\varepsilon^{\bar{\gamma}} \prod_{\alpha=1}^{n-1} \frac{(8\alpha)^{4\alpha}}{(4\alpha)!} (2\alpha-1)! \leqslant 1 \tag{26.55}$$

is satisfied.

Manipulating the relations (26.54) and defining $\bar{\gamma}$ from the conditions satisfying (26.55) and the inequality

$$\frac{2^{m+1}\pi^2}{3}(n-1)(4n-1)\varepsilon^{\bar{\gamma}-1} \leqslant 1, \tag{26.56}$$

we complete the proof of Lemma 9.

§ 27. Proof of the Measure Theorem

As before, suppose that $\mathfrak{M}_\varepsilon = \mathfrak{M}_\varepsilon(\gamma_0, d_0)$ is a set of matrices $A_0 \leftrightarrow a_0$ contained in the set S_1, satisfying the condition

$$| F^{(0)}(a_0, -(k, \omega)^2) | \geqslant \varepsilon^{\gamma_0} | k |^{-d_0}, \tag{27.1}$$

and that the parameters γ_0, d_0 are fixed so that Lemma 9 holds.

The reducibility theorem proved in § 25 enables us to construct, for any matrix $A_0 \in \mathfrak{M}_\varepsilon$, a matrix ξ^0 such that for every matrix $P(\varphi) \in C^a(| \operatorname{Im} \varphi | \leqslant \rho)$, satisfying the condition $| P(\varphi) | \leqslant M_0(\varepsilon)$, the system of equations

$$\begin{aligned} \frac{dx}{dt} &= Ax + P(\varphi)\, x, \\ \frac{d\varphi}{dt} &= \omega \end{aligned} \tag{27.2}$$

with

$$A = A_0 + \xi^0 \qquad (A_0 \in \mathfrak{M}_\varepsilon), \tag{27.3}$$

reduces to the system

$$\begin{aligned} \frac{dy}{dt} &= A_0 y, \\ \frac{d\varphi}{dt} &= \omega \end{aligned} \tag{27.4}$$

by means of the non-singular matrix $\Phi(\varphi) \in C^a(| \operatorname{Im} \varphi | < \rho_0/2)$. Here $C^a(| \operatorname{Im} \varphi | \leqslant \rho)$ denotes the space of matrices analytic in the strip $| \operatorname{Im} \varphi | \leqslant \rho$, 2π-periodic in φ, and real for real φ.

Let $\mathfrak{M}(M_0) = \mathfrak{M}(M_0(\varepsilon))$ be a set of matrices A admitting the representation (27.3), which defines on $\mathfrak{M}_\varepsilon$ a mapping $F: \mathfrak{M}_\varepsilon \to \mathfrak{M}(M_0)$ by

$$a_0 \to Fa_0 = a_0 + \xi^{(0)}(a_0) \qquad (a_0 \in \mathfrak{M}_\varepsilon). \tag{27.5}$$

Investigating the properties of the mapping F, we determine the measure of set $\mathfrak{M}(M_0)$.

If this measure happens to be large, $\mathfrak{M}(M_0(\varepsilon)) \to 1$ as $\varepsilon \to 0$, i.e. if the mapping F varies the measure of set $\mathfrak{M}_\varepsilon$ by very little, the obvious inclusion

$$\mathfrak{M}(P) \supset \mathfrak{M}(M_0)$$

leads to the relation (24.8) which has to be proved. Since the set $\mathfrak{M}_\varepsilon$, on which

F is defined, is of complex structure, i.e. has the structure of the Cantor set, the process for measuring $F : \mathfrak{M}_\varepsilon$ is as follows : construct a sequence of mappings $F^{(s)}$ approximating the mapping F, each of which is defined on the set $\mathfrak{M}^{(s)}$, simpler in structure than $\mathfrak{M}_\varepsilon$; then measure $F^{(s)} : \mathfrak{M}^{(s)}$ and take the limit to obtain the desired measure of set $\mathfrak{M}(M_0)$.

To define the mappings $F^{(s)}$, we introduce the corrections ξ and express system (27.2) in the form

$$\frac{dx}{dt} = A_0 x + [P(\varphi) + \xi]\, x, \qquad \frac{d\varphi}{dt} = \omega, \tag{27.6}$$

and then make the change of variables

$$x = [E + U^{(1)}(\varphi, \xi^{(1)})]\, x^{(1)}, \quad \xi = \xi(\xi^{(1)}), \tag{27.7}$$

defining $U^{(1)}(\varphi, \xi^{(1)})$ and $\xi(\xi^{(1)})$ by the equations

$$\left(\frac{\partial U^{(1)}}{\partial \varphi}, \omega\right) + U^{(1)} A_0 = A_0 U^{(1)} + S_{N_1} P(\varphi), \quad \xi + \overline{P(\varphi)} = \xi^{(1)}, \tag{27.8}$$

where S_{N_1}, a smoothing operator considered in § 13, is defined by

$$S_{N_1} P(\varphi) = \sum_{0<|k|\leqslant N_1} P_k e^{i(k,\varphi)}, \tag{27.9}$$

with N_1 a sufficiently large integer.

The original system of equations reduces, by the substitution (27.7), to

$$\frac{dx^{(1)}}{dt} = A_0 x^{(1)} + [P^{(1)}(\varphi, \xi^{(1)}) + \xi^{(1)}]\, x^{(1)}, \qquad \frac{d\varphi}{dt} = \omega, \tag{27.10}$$

where

$$P^{(1)}(\varphi, \xi^{(1)}) = (E + U^{(1)}(\varphi, \xi^{(1)}))\, [(P(\varphi) + \xi)\, U^{(1)}(\varphi, \xi^{(1)}) - U^{(1)}(\varphi, \xi^{(1)})\, \xi^{(1)} + R_{N_1} P(\varphi)], \tag{27.11}$$

$$R_{N_1} P(\varphi) = \sum_{|k| > N_1} P_k e^{i(k,\varphi)}.$$

Since by the estimates of § 14,

$$| R_{N_1} P(\varphi) | \leqslant \left(\frac{2m}{e}\right)^m \frac{e^{-N_1\delta_0}}{\delta_0^{m+1}} | P(\varphi) |_0 \quad \text{for } | \operatorname{Im} \varphi | \leqslant \rho_0 - 2\delta_0 = \rho_1, \tag{27.12}$$

it is easy to select $M_0 = | P(\varphi) |$ and N_1 such that, for

$$| \operatorname{Im} \varphi | \leqslant \rho_1, \quad | \xi^{(1)} | \leqslant M_0, \tag{27.13}$$

the functions $U^{(1)}(\varphi)$, $P^{(1)}(\varphi, \xi^{(1)})$ and ξ are analytic in φ and $\xi^{(1)}$ and satisfy the inequalities

$$| U^{(1)}(\varphi) | \leqslant \frac{M_0^{\varkappa-1}}{8n}, \qquad | P^{(1)}(\varphi, \xi^{(1)}) | < M_1 = M_0^{\varkappa},$$

$$| \xi(\xi^{(1)}) - \xi^{(1)} | \leqslant M_0, \tag{27.14}$$

$$\sum_{q,\, l} \left| \frac{\partial \xi_{\alpha\beta}(\xi^{(1)})}{\partial \xi_{ql}^{(1)}} \right| \leqslant 2 \qquad (1 < \varkappa < 2). \tag{27.15}$$

We now consider the dependence of $U^{(1)}(\varphi)$, $P^{(1)}(\varphi, \xi^{(1)})$ and ξ on a_0. The function $\xi(\xi^{(1)})$ does not depend on a_0. The matrix

$$U^{(1)}(\varphi, \xi^{(1)}, a_0) = \sum_{0<k\leqslant N_1} Y_k(\xi(\xi^{(1)}), a^0)\, e^{i(k,\varphi)}$$

has the same properties relative to a_0 as the matrix $Y_k^{(1)}(\xi(\xi^{(1)}), a_0)$. However, by formulas (26.15) and (26.16), we see that the vector $y_k^{(1)}(a_0) = \{y_{11}^{(1)}, \ldots, y_{nn}^{(1)}\}$, made up of the elements of the matrix Y_k, is expressible in the form

$$y_k^{(1)}(a_0) = \frac{D(a_0, -(k, \omega)^2)\,[\mathfrak{A}_0 - i(k, \omega)]}{F_{n2}^{(0)}(a_0, -(k, \omega)^2)} P_k \qquad (0 < | k | \leqslant N_1) \tag{27.16}$$

and is defined on the set $\mathfrak{M}_\varepsilon^{(1)}$ of variable a_0.

Since D, $\mathfrak{A}_0$ and $F_{n2}^{(0)}$ are polynomials in $a_{\alpha\beta}^0$, the function $y_k^{(1)}$ is uniformly differentiable on the set $\Omega_\varepsilon^{(1)} = \mathfrak{M}_\varepsilon^{(1)} \setminus [\mathfrak{M}_\varepsilon^{(1)}]$, where $[\mathfrak{M}_\varepsilon^{(1)}]$ is the boundary of $\mathfrak{M}_\varepsilon^{(1)}$. Differentiating (27.16), we get

$$\left| \frac{\partial y_k^{(1)}(a_0)}{\partial a_0} \right| \leqslant \frac{1}{[F_{n2}^0]^2} \left[\frac{\partial}{\partial a_0} \{D(\mathfrak{A}_0 - i(k, \omega)) P_k\} F_{n2}^{(0)} \right.$$

$$\left. - \left(D(\mathfrak{A}_0 - i(k, \omega)) P_k, \frac{\partial F_{n2}^{(0)}}{\partial a_0} \right) \right] \leqslant$$

$$\leqslant c_1 \frac{|k|^{4n^2-1}|P_k|}{|F^{(0)}_{n^2}|} \leqslant \frac{c_1}{\varepsilon^{2\gamma_0}} |k|^{4n^2+2d_0-1}|P_k| =$$

$$= c_2(\varepsilon)|k|^{4n^2+2d_0-1}|P_k|, \quad a_0 \in \Omega^{(1)}_\varepsilon. \tag{27.17}$$

From this it is obviously easy to obtain the inequality

$$|y^{(1)}_k(a'_0) - y^{(1)}_k(a''_0)| \leqslant c_2(\varepsilon)|k|^{4n^2+2d_0-1}|P_k| \tag{27.18}$$

for $a'_0, a''_0 \in \mathfrak{M}^{(1)}_\varepsilon$. The inequalities (27.17) and (27.18) imply that the matrix

$$\Phi_1(\varphi, \xi^{(1)}, a_0) = E + U^{(1)}(\varphi, \xi(\xi^{(1)}), a_0)$$

is uniformly differentiable with respect to a_0 on the set $\Omega^{(1)}_\varepsilon$ and satisfies the inequalities

$$\sum_{k_0,q}\left|\frac{\partial\Phi_1(\varphi, \xi^{(1)}, a_0)}{\partial a^0_{k_0q}}\right| < \frac{M_0^{\varkappa-1}}{8n}, \quad a_0 \in \Omega^{(1)}_\varepsilon,$$

$$|\Phi_1(\varphi, \xi^{(1)}, a'_0) - \Phi_1(\varphi, \xi^{(1)}, a''_0)| \leqslant \frac{M_0^{\varkappa-1}}{8n}|a'_0 - a''_0|, \quad a'_0, a''_0 \in \mathfrak{M}^{(1)}_\varepsilon. \tag{27.19}$$

Consider the matrix $P^{(1)}(\varphi, \xi^{(1)}, a_0)$. Since $\Phi_1(\varphi, \xi^{(1)}, a_0)$ and its inverse matrix are uniformly differentiable on the set $\Omega^{(1)}_\varepsilon$, so also is $P^{(1)}(\varphi, \xi^{(1)}, a_0)$, being a product of matrices uniformly differentiable on $\Omega^{(1)}_\varepsilon$; in addition, the inequality

$$\left|\frac{\partial P^{(1)}(\varphi, \xi^{(1)}, a_0)}{\partial a^0_{\alpha\beta}}\right| \leqslant |\Phi_1^{-1}(\varphi, \xi^{(1)}, a_0)| \cdot \left|\frac{\partial\Phi_1(\varphi, \xi^{(1)}, a_0)}{\partial a^0_{\alpha\beta}}\right| |\Phi^{-1}(\varphi, \xi^{(1)}, a_0)| \times$$

$$\times [|P(\varphi) + \xi| \cdot |U^{(1)}(\varphi, \xi^{(1)}, a_0)| +$$

$$+ |U^{(1)}(\varphi, \xi^{(1)}, a_0)| \cdot |\xi^{(1)}| + |R_N P(\varphi)|] +$$

$$+ |\Phi_1^{-1}(\varphi, \xi^{(1)}, a_0)| \cdot \Bigg[|P(\varphi) + \xi| \times$$

$$\times \left|\frac{\partial U^{(1)}(\varphi, \xi^{(1)}, a_0)}{\partial a^0_{\alpha\beta}}\right| + \left|\frac{\partial U^{(1)}(\varphi, \xi^{(1)}, a_0)}{\partial a^0_{\alpha\beta}}\right| \cdot |\xi^{(1)}|\Bigg] \leqslant$$

$$\leqslant M_1 \qquad a \in \Omega^{(1)}_\varepsilon \tag{27.20}$$

holds, whence it is easy to obtain

$$|P^{(1)}(\varphi, \xi^{(1)}, a'_0) - P^{(1)}(\varphi, \xi^{(1)}, a''_0)| \leqslant M_1|a'_0 - a''_0|, \quad a'_0, a''_0 \in \mathfrak{M}^{(1)}_\varepsilon. \tag{27.21}$$

To define the mapping $F^{(1)} : \mathfrak{M}_\varepsilon^{(1)} \to E_{n^2}$, we denote by $\xi^{(1)}(a_0, \xi^{(2)})$, $|\xi^{(2)}| \leqslant M_1$, a solution of the equation

$$\xi^{(1)} + \overline{P^{(1)}(\varphi, a_0, \xi^{(1)})} = \xi^{(2)}, \tag{27.22}$$

and set

$$F^{(1)} a_0 = a_0 + B^{(1)}(a_0) = a_0 + \xi(\xi^{(1)}(a, 0)), \qquad a_0 \in \mathfrak{M}_\varepsilon^{(1)}. \tag{27.23}$$

Since $\overline{P^{(1)}(a_0, \xi^{(1)})}$ is a uniformly differentiable function of $a_0 \in \Omega_\varepsilon^{(1)}$, the solution $\xi^{(1)}(a_0, \xi^{(2)})$ is also uniformly differentiable on the set $\Omega_\varepsilon^{(1)}$. Hence, differentiating (27.22) we see that

$$\frac{\partial \xi^{(1)}}{\partial a_{\alpha\beta}} + \sum_{ql} \frac{\partial \overline{P^{(1)}}}{\partial \xi_{ql}^{(1)}} \cdot \frac{\partial \xi_{ql}^{(1)}}{\partial a_{\alpha\beta}} + \frac{\partial \overline{P^{(1)}}}{\partial a_{\alpha\beta}} = 0,$$

and taking account of the estimates for $\partial P^{(1)}/\partial \xi_{ql}^{(1)}$ and $\partial \overline{P^{(1)}}/\partial a_{\alpha\beta}$, that

$$\sum_{\alpha,\beta} \left| \frac{\partial \xi^{(1)}}{\partial a_{\alpha\beta}} \right| \leqslant \frac{M_1}{1 - \dfrac{2M_1}{M_0} n^2} = \frac{M_1}{1 - 2n^2 M_0^{\varkappa - 1}} \leqslant$$

$$\leqslant 4n^2 M_0^{\varkappa-1} M_1 \leqslant M_1. \tag{27.24}$$

We therefore deduce that the function $B^{(1)}(a_0)$ is uniformly differentiable on the set $\Omega_\varepsilon^{(1)}$, and satisfies the inequalities

$$\sum_{\alpha,\beta} \left| \frac{\partial B^{(1)}(a_0)}{\partial a_{\alpha\beta}} \right| \leqslant \sum_{\alpha,\beta=1}^{n} \left| \frac{\partial \xi(\xi^{(1)})}{\partial \xi_{\alpha\beta}^{(1)}} \right| \sum_{\alpha,\beta} \left| \frac{\partial \xi^{(1)}}{\partial a_{\alpha\beta}} \right| \leqslant 2n2M_1 \leqslant M_0^{\varkappa-1},$$

$$a_0 \in \Omega_\varepsilon^{(1)}, \tag{27.25}$$

$$|B^{(1)}(a_0') - B^{(1)}(a_0'')| \leqslant M_0^{\varkappa-1} |a_0' - a_0''|, \qquad a_0', a_0'' \in \mathfrak{M}_\varepsilon^{(1)}.$$

To construct the mapping $F^{(s)}$ $(s > 1)$, we make use of the following recurrence relations. Choose a sequence of numbers $N_1 > N_2 > \ldots > N_s > \ldots$ such that the inequalities

$$\left(\frac{2m}{e}\right)^m \frac{e^{-N_s \delta_0^s}}{\delta_0^{(m+s)}} M_{s-1} < \frac{M_s}{32n} \qquad (M_s = M_0^{\varkappa^s}, \quad s = 1, 2, \ldots), \tag{27.26}$$

are satisfied. Then, define the matrices $P^{(s)}(\varphi, a_0, \xi^{(s)})$ by

$$\begin{aligned} P^{(s)}(\varphi, a_0, \xi^{(s)}) = (E + U^{(s)}(\varphi, a_0, \xi^{(s)}))\,[(P^{(s-1)}(\varphi, a_0, \xi^{(s)}) + \\ + \xi^{(s-1)})\, U^{(s)}(\varphi, a_0, \xi^{(s)}) - U^{(s)}(\varphi, a_0, \xi^{(s)}) \times \\ \times (\xi^{(s-1)} + \overline{P^{(s-1)}(\varphi, a_0, \xi^{(s)})}) + R_{N_s} P^{(s-1)}(\varphi, a_0, \xi^{(s)}), \quad (27.27) \end{aligned}$$

where $U^{(s)}(\varphi, a_0, \xi^{(s)})$ are periodic solutions of the equations

$$\left(\frac{\partial U^{(s)}}{\partial \varphi}, \omega\right) + U^{(s)}A_0 = A_0 U^{(s)} + S_{N_s} P^{(s-1)}(\varphi, a_0, \xi^{(s)}), \qquad (27.28)$$

and $\xi^{(s-1)} = \xi^{(s-1)}(a_0, \xi^{(s)})$, $|\,\xi^{(s)}\,| \leqslant M_{s-1}$ are solutions of the equations

$$\xi^{(s-1)} + \overline{P^{(s-1)}(\varphi, a_0, \xi^{(s-1)})} = \xi^{(s)}. \qquad (27.29)$$

It follows from the inequalities (27.26) and (25.17) that, subject to a suitable choice of the parameters N_s and M_0, the functions $P^{(s)}(\varphi, a_0, \xi^{(s)})$ and $U^{(s)}(\varphi, a_0, \xi^{(s)})$ satisfy with respect to φ and $\xi^{(s)}$ all the hypotheses established for similar functions in § 25. To establish the properties of these functions as functions of the variable a_0, we suppose that for $s \geqslant 2$ the function $P^{(s-1)}(\varphi, a_0, \xi^{(s-1)})$ is uniformly differentiable with respect to a_0 on the set $\Omega_\varepsilon^{(s-1)} = \mathfrak{M}_\varepsilon^{(s-1)} \setminus [\mathfrak{M}_\varepsilon^{(s-1)}]$ and that it satisfies the inequalities

$$\sum_{\alpha,\beta} \left| \frac{\partial P^{(s-1)}(\varphi, a_0, \xi^{(s-1)})}{\partial a^0_{\alpha\beta}} \right| < M_{s-1}, \qquad a_0 \in \Omega_\varepsilon^{(s-1)},$$

$$(27.30)$$

$$|\, P^{(s-1)}(\varphi, a_0', \xi^{(s-1)}) - P^{(s-1)}(\varphi, a_0'', \xi^{(s-1)})\,| \leqslant M_{s-1}\,|\, a_0' - a_0''\,|,$$

$$a_0 \in \mathfrak{M}_\varepsilon^{(s-1)}.$$

We have to show that the functions $U^{(s)}(\varphi, a_0, \xi^{(s)})$, $P^{(s)}(\varphi, a_0, \xi^{(s)})$ and $\xi^{(s)}(a, \xi^{(s+1)})$ are also uniformly differentiable on the set $\Omega_\varepsilon^{(s)} = \mathfrak{M}_\varepsilon^{(s)} \setminus [\mathfrak{M}_\varepsilon^{(s)}]$. In fact, considering equation (27.22) we observe in the first place that the function $\xi^{(s-1)}(a_0, \xi^{(s)})$ is uniformly differentiable on $\Omega_\varepsilon^{(s-1)}$ and satisfies the inequalities

$$\sum_{q,l} \left| \frac{\partial \xi^{(s-1)}}{\partial a_{ql}} \right| \leqslant \frac{\left. \sum_{q,l} \left| \overline{\frac{\partial P^{(s-1)}(\varphi, a_0, \xi^{(s-1)})}{\partial a_{ql}}} \right| \right|_0}{1 - \dfrac{2M_{s-1}}{M_{s-2}}\, n^2} \leqslant 4n^2 M_{s-1}^{\varkappa-1} M_{s-1} \leqslant M_{s-1},$$

$$a_9 \in \Omega_\varepsilon^{(s-1)}, \qquad (27.31)$$

$$|\, \xi^{(s+1)}(a_0', \xi^{(s)}) - \xi^{(s+1)}(a_0'', \xi^{(s)})\,| \leqslant M_{s-1}\,|\, a_0' - a_0''\,|, \qquad a_0', a_0'' \in \mathfrak{M}_\varepsilon^{(s-1)}.$$

The matrix $U^{(s)}(\varphi, a_0, \xi^{(s)})$, a solution of equation (27.28), has the form

$$U^{(s)}(\varphi, a_0, \xi^{(s)}) = \sum_{0<|k|\leqslant N_s} Y_k^{(s)}(a_0, \xi^{(s)})\, e^{i(k,\varphi)}, \tag{27.32}$$

where $Y_k^{(s)}(a_0, \xi^{(s)})$ are vectors whose components $y_k^{(s)}$ are defined by

$$y_k^{(s)} = \frac{D(a_0 - (k, \omega)^2)\,[\mathfrak{A}_0 - i(k, \omega)]}{F_{n^2}^0(a_0, -(k, \omega)^2)} P_k^{(s-1)}(\varphi, a_0, \xi^{(s-1)} \qquad (0 < |k| \leqslant N_s).$$

The functions $y_k^{(s)}$ and, consequently, also the matrix $U^{(s)}(\varphi, a_0, \xi^{(s)})$ are defined on the set $\mathfrak{M}_\varepsilon^{(s)}$ and, obviously, are uniformly differentiable for $a_0 \in \Omega_\varepsilon^{(s)}$.

Further, from an inspection of the inequality (27.17), we deduce that

$$\left|\frac{\partial y_k^{(s)}}{\partial a_{\alpha\beta}^0}\right| \leqslant \left|\frac{\partial}{\partial a_{\alpha\beta}^0}\left\{\frac{D(\mathfrak{A}_0 - i(k, \omega))}{F_{n^2}^{(0)}}\right\} p_k^{(s-1)}\right| +$$

$$+\left|\frac{D(\mathfrak{A}_0 - i(k, \omega))}{F_{n^2}^0} \frac{\partial P_k^{(s-1)}(\varphi, a_0, \xi^{(s-1)})}{\partial a_{\alpha\beta}^0}\right| \leqslant$$

$$\leqslant c_2(\varepsilon)\,|k|^{4n^2+2d_0-1}|P_k^{(s-1)}(\varphi, a_0, \xi^{(s-1)})| + \bar{c}_2(\varepsilon)\left|\frac{\partial P_k^{(s-1)}}{\partial a_{\alpha\beta}^0}\right|.$$

However, $P_k^{(s-1)}(a_0, \xi^{(s-1)})$ and $(\partial P_k^{(s-1)}(a_0, \xi^{(s-1)}))/\partial a_{\alpha\beta}^0$, are the Fourier coefficients of the functions $P^{(s-1)}(\varphi, a_0, \xi^{(s-1)})$ and $(\partial P^{(s-1)}(\varphi, a_0, \xi^{(s-1)}))/\partial a_{\alpha\beta}^0$, which are analytic in the strip $|\operatorname{Im}\varphi| < \rho_{s-1}$; hence

$$|P_k^{(s-1)}(a_0, \xi^{(s-1)})| \leqslant M_{s-1} e^{-\rho_{s-1}|k|}, \qquad \left|\frac{\partial P_k^{(s-1)}(a_0, \xi^{(s-1)})}{\partial a_{\alpha\beta}^0}\right| \leqslant M_{s-1} e^{-\rho_{s-1}|k|}.$$

The last inequality yields the estimate

$$\left|\frac{\partial y_k^{(s)}}{\partial a_{\alpha\beta}^0}\right| \leqslant \bar{c}_3(\varepsilon)\,|k|^{4n^2+2d_0-1} M_{s-1} e^{-\rho_{s-1}|k|}, \qquad a_0 \in \Omega_\varepsilon^{(s)}, \tag{27.33}$$

implying

$$\sum_{\alpha,\beta}\left|\frac{\partial U^{(s)}(\varphi, a_0, \xi^{(s)})}{\partial a_{\alpha\beta}^0}\right| \leqslant c_3(\varepsilon)\frac{M_{s-1}}{\delta_0^{(m+4n^2+2d_0-1)s}} \leqslant \frac{M_{s-1}^{\varkappa-1}}{16n} \tag{27.34}$$

for $|\operatorname{Im}\varphi| \leqslant \rho_{s-1} - 2\overset{s}{\delta_0} = \rho_s$, $a_0 \in \Omega_\varepsilon^{(s)}$. Further, since

$$\left| \frac{\partial P^{(s-1)}(\varphi, a_0, \xi^{(s-1)})}{\partial \xi_{\alpha\beta}^{(s-1)}} \right| \leqslant \frac{M_{s-1}}{\frac{M_{s-2}}{2}} \leqslant 2M_{s-2}^{\varkappa-1} \quad \text{for } |\xi^{(s-1)}| \leqslant \frac{M_{s-2}}{2},$$

the differentiation of $y_k^{(s)}$ with respect to $\xi^{(s-1)}$ gives

$$\left| \frac{\partial y_k^{(s)}}{\partial \xi_{\alpha\beta}^{(s-1)}} \right| \leqslant \bar{c}_2(\varepsilon)\, |k|^{2n^2+d_0-1} \left| \frac{\partial P_k^{(s-1)}(\varphi, a_0, \xi^{(s-1)})}{\partial \xi_{\alpha\beta}^{(s-1)}} \right| \leqslant$$

$$\leqslant 2\bar{c}_2\, |k|^{2n^2+d_0-1} e^{-\rho_{s-1}|k|} M_{s-2}^{\varkappa-1},$$

implying that

$$\sum_{\alpha,\beta} \left| \frac{\partial U^{(s)}(\varphi, a_9, \xi^{(s-1)}(a_0, \xi^{(s)}))}{\partial \xi_{\alpha\beta}^{(s-1)}} \right| \leqslant \bar{c}_3(\varepsilon) \frac{M_{s-2}^{\varkappa-1}}{\delta_0^{(m+2n^2+d_0-1)}} \leqslant \frac{1}{16n} \tag{27.35}$$

for $|\operatorname{Im}\varphi| \leqslant \rho_s$, $|\xi^{(s)}| \leqslant \rho_s$, $a_0 \in \Omega_\varepsilon^{(s)}$.

By manipulating the relations (27.34) and (27.35), we obtain for the matrix $U^{(s)}(\varphi, a_0, \xi^{(s)})$ the inequality

$$\sum_{\alpha,\beta} \left| \frac{\partial U^{(s)}(\varphi, a_0, \xi^{(s=1)}(a_0. \xi^{(s)}))}{\partial \overset{0}{a}_{\alpha\beta}} \right| \leqslant \sum_{\alpha\beta q l} \left| \frac{\partial U^{(s)}(\varphi, a_0, \xi^{(s)})}{\partial \xi_{ql}^{(s-1)}} \right| \cdot \left| \frac{\partial \xi_{ql}^{(s-1)}}{\partial \overset{0}{a}_{\alpha\beta}} \right| +$$

$$+ \sum_{\alpha\beta} \left| \frac{\partial U^{(s)}(\varphi, a_0, \xi^{(s-1)})}{\partial \overset{0}{a}_{\alpha\beta}} \right| \leqslant$$

$$\leqslant \frac{M_{s-1}}{16n} + \frac{M_{s-1}^{\varkappa-1}}{16n} \leqslant \frac{M_{s-1}^{\varkappa-1}}{8n}, \qquad a_0 \in \Omega_\varepsilon^{(s)}. \tag{27.36}$$

The inequality (27.36) yields

$$|U^{(s)}(\varphi, a_0', \xi^{(s-1)}(a_0', \xi^{(s)})) - U^{(s)}(\varphi, a_0'', \xi^{(s-1)}(a_0'', \xi^{(s)}))| \leqslant \frac{M_{s-1}^{\varkappa-1}}{8n} |a_0' - a_0''|,$$

$$a_0', a_0'' \in \mathfrak{M}_\varepsilon^{(s)}. \tag{27.37}$$

Consider the matrix $P^{(s)}(\varphi, a_0, \xi^{(s)})$. Being the product of matrices which are uniformly differentiable on the set $\Omega_\varepsilon^{(s)}$, this matrix itself is uniformly differentiable with respect to a_0 on the set $\Omega_\varepsilon^{(s)}$. Furthermore, differentiating (27.27) with respect to a_0, we obtain an estimate for the derivatives of $P^{(s)}(\varphi, a_0, \xi^{(s)})$:

$$\left| \frac{\partial P^{(s)}(\varphi, a_0, \xi^{(s)})}{\partial a^0_{\alpha\beta}} \right| \leqslant$$

$$\leqslant |(E + U^{(s)}(\varphi, a_0, \xi^{(s)}))^{-1}| \left| \frac{\partial U^{(s)}(\varphi, a_0, \xi^{(s)})}{\partial a^0_{\alpha\beta}} \right| \times$$

$$\times |P^{(s)}(\varphi, a_0, \xi^{(s)})| + |(E + U^{(s)}(\varphi, a_0, \xi^{(s)}))^{-1}| \times$$

$$\times \left\{ 2 \left[\sum_{q,l} \left| \frac{\partial P^{(s-1)}(\varphi, a_0, \xi^{(s)})}{\partial \xi^{(s-1)}_{ql}} \right|_0 \left| \frac{\partial \xi^{(s-1)}_{ql}}{\partial a^0_{\alpha\beta}} \right|_0 + \left| \frac{\partial \xi^{(s-1)}}{\partial a^0_{\alpha\beta}} \right| \right] \times \right.$$

$$\times |U^{(s)}(\varphi, a_0, \xi^{(s)})| + 2\,[\,|P^{(s-1)}(\varphi, a_0, \xi^{(s)})|_0 + |\xi^{(s-1)}|\,] \times$$

$$\times \left| \frac{\partial U^{(s)}(\varphi, a_0, \xi^{(s)})}{\partial a^0_{\alpha\beta}} \right| + \sum_{q,l} \left| \frac{\partial R_{N_s} P^{(s-1)}(\varphi, a_0, \xi^{(s)})}{\partial \xi^{(s-1)}_{ql}} \right| \left| \frac{\partial \xi^{(s-1)}_{ql}}{\partial a^0_{\alpha\beta}} \right| +$$

$$\left. + 2 \left| \frac{\partial P^{(s-1)}(\varphi, a_0, \xi^{(s)})}{\partial a^0_{\alpha\beta}} \right|_0 + \left| \frac{\partial R_{N_s} P^{(s-1)}(\varphi, a_0, \xi^{(s)})}{\partial a^0_{\alpha\beta}} \right| \right\} \leqslant$$

$$\leqslant M^{\varkappa-1}_{s-1} M_s + (n + M^{\varkappa-1}_{s-1}) \left\{ [2\,(2n^2 M^{\varkappa-1}_{s-2} + 1) + 1] \frac{M^{\varkappa}_{s-1}}{8n} + \right.$$

$$\left. + 4 \frac{M^{\varkappa}_{s-1}}{8n} + \left(\frac{2m}{e} \right)^m \frac{e^{-N_s \delta^s_0}}{\delta_0^{s(m+1)}} [2n^2 M^{\varkappa-1}_{s-2} M_{s-1} + M_{s-1}] \right\} \leqslant$$

$$\leqslant M_s, \qquad a_0 \in \Omega^{(s)}_{\varepsilon}. \tag{27.38}$$

Hence,

$$|P^{(s)}(\varphi, a_0', \xi^{(s)}) - P^{(s)}(\varphi, a_0'', \xi^{(s)})| \leqslant M_s |a_0' - a_0''|, \quad a_0', a_0'' \in \mathfrak{M}^{(s)}_{\varepsilon}. \tag{27.39}$$

Consider the following equation in $\xi^{(s)}$:

$$\xi^{(s)} + \overline{P^{(s)}(a_0, \xi^{(s)})} = \xi^{(s+1)} \qquad (|\xi^{(s+1)}| \leqslant M_s). \tag{27.40}$$

The solution $\xi^{(s)} = \xi^{(s)}(a_0, \xi^{(s+1)})$ of this equation is obviously defined for $a_0 \in \mathfrak{M}^{(s)}_{\varepsilon}$ and is uniformly differentiable with respect to a_0 on the set $\Omega^{(s)}_{\varepsilon}$. Differentiating equation (27.40) and inspecting the estimates for the derivatives $\partial \overline{P^{(s)}(a_0, \xi^{(s)})} / \partial \xi^{(s)}_{\alpha\beta}$ and $\partial \overline{P^{(s)}(a_0, \xi^{(s)})} / \partial a^0_{\alpha\beta}$, we verify that

$$\sum_{\alpha,\beta} \left| \frac{\partial \xi^{(s)}}{\partial a^0_{\alpha\beta}} \right| \leqslant \frac{M_s}{1 - 2n^2 M^{\varkappa-2}_{s-2}} \leqslant M_s, \qquad a_0 \in \Omega^{(s)}_{\varepsilon},$$

$$|\xi^{(s)}(a_0', \xi^{(s+1)}) - \xi^{(s)}(a_0'', \xi^{(s+1)})| \leqslant M_s |a_0' - a_0''|, \qquad a_0 \in \mathfrak{M}^{(s)}_{\varepsilon}. \tag{27.41}$$

We set

$$B^{(s)}(a_0) = \xi(\xi^{(1)}(a_0, \xi^{(2)}(a_0, \ldots, \xi^{(s)}(a_0, 0)))), \qquad a_0 \in \mathfrak{M}_\varepsilon^{(s)}, \quad (27.42)$$

and define the mapping $F^{(s)} : \mathfrak{M}_\varepsilon^{(s)} \to E_{n^2}$ by the relations

$$F^{(s)} a_0 = a_0 + B^{(s)}(a_0), \qquad a_0 \in \mathfrak{M}_\varepsilon^{(s)}. \quad (27.43)$$

The structure of matrices $\xi^{(s-1)}(a_0, \xi^{(s)})$, $U^{(s)}(\varphi, a_0, \xi^{(s)})$ and $P^{(s)}(\varphi, a_0, \xi^{(s)})$ and their properties as functions of the variables φ and $\xi^{(s)}$ imply that

$$B^{(s)}(a_0) \to \xi^{(0)}(a_0), \quad \prod_{\alpha=1}^{s} (E + U^{(s)}(\varphi, a_0, 0)) \underset{s\to\infty}{\longrightarrow} \Phi(\varphi, a_0) \quad (27.44)$$

uniformly with respect to $a_0 \in \mathfrak{M}_\varepsilon$, where $\Phi(\varphi, a_0) \in C^a[|\operatorname{Im}\varphi| < (\rho_0/2)]$ is a non-singular matrix, reducing the system (27.2) to the system (27.4) for

$$A = A_0 + \xi^{(0)}(a_0). \quad (27.45)$$

Moreover, from the properties of the functions $\xi^{(s-1)}(a_0, \xi^{(s)})$ as functions of the variable a_0, it follows that $B^{(s)}(a_0)$, representing an analytic composite of the functions $\xi^{(s)}$, is uniformly differentiable on the set $\Omega_\varepsilon^{(s)}$. Differentiating the function $B^{(s)}(a_0)$ and making use of the inequalities (25.24) and (27.41) for $s \geqslant 1$, we find that

$$\left|\frac{\partial B^{(s)}(a_0)}{\partial a_{\alpha\beta}}\right| \leqslant \sum_{\alpha_1,\beta_1} \left|\frac{\partial \xi}{\partial \xi^{(1)}_{\alpha_1\beta_1}}\right| \left|\frac{\partial \xi^{(1)}_{\alpha_1\beta_1}}{\partial a_{\alpha_1\beta_1}}\right| + \sum_{\alpha_1,\beta_1,\alpha_2,\beta_2} \left|\frac{\partial \xi}{\partial \xi^{(1)}_{\alpha_1\beta_1}}\right| \left|\frac{\partial \xi^{(1)}_{\alpha_1\beta_1}}{\partial \xi^{(2)}_{\alpha_2\beta_2}}\right| \left|\frac{\partial \xi^{(2)}_{\alpha_2\beta_2}}{\partial a_{\alpha_2\beta_2}}\right| + \ldots + \sum_{\alpha_1\beta_1\ldots\alpha_s\beta_s} \left|\frac{\partial \xi}{\partial \xi^{(1)}_{\alpha_1\beta_1}}\right| \cdots \left|\frac{\partial \xi^{(s-1)}_{\alpha_{s-1}\beta_{s-1}}}{\partial \xi^{(s)}_{\alpha_s\beta_s}}\right| \left|\frac{\partial \xi^{(s)}_{\alpha_s\beta_s}}{\partial a_{\alpha\beta}}\right| \leqslant$$

$$\leqslant 2\left[M_1 + (1 + 4n^2 M_0^{\varkappa-1}) M_2 + \ldots + \prod_{j=0}^{s-1} (1 + 4n^2 M_j^{\varkappa-1}) M_s\right] <$$

$$< 2 \prod_{j=0}^{\infty} (1 + 4n^2 M_j^{\varkappa-1}) \sum_{j=1}^{\infty} M_j \leqslant c_4 M_1 < M_0, \qquad a_0 \in \Omega_\varepsilon^{(s)}. \quad (27.46)$$

Hence, we deduce that

$$| B^{(s)}(a_0') - B^{(s)}(a_0'') | \leqslant M_0 \, | a_0' - a_0'' |, \qquad a_0', a_0'' \in \mathfrak{M}_\varepsilon^{(s)}. \tag{27.47}$$

The stage is now set for the next theorem.

Theorem 18. *For sufficiently small M_0, the sequence of matrices $B^{(s)}(a_0)$ is defined on the set $\mathfrak{M}_\varepsilon^{(s)}$, is uniformly differentiable for $a_0 \in \Omega_\varepsilon^{(s)} = \mathfrak{M}_\varepsilon^{(s)} \setminus [\mathfrak{M}_\varepsilon^{(s)}]$ and satisfies the conditions*

$$\sum_{\alpha, \beta} \left| \frac{\partial B^{(s)}(a_0)}{\partial a^0_{\alpha\beta}} \right| \leqslant M_0 \qquad \textit{when } a_0 \in \Omega_\varepsilon^{(s)},$$

$$| B^{(s)}(a_0') - B^{(s)}(a_0'') | \leqslant M_0 \, | a_0' - a_0'' | \qquad \textit{when } a_0 \in \mathfrak{M}_\varepsilon^{(s)}, \tag{27.48}$$

$$\lim_{s \to \infty} B^{(s)}(a_0) = \xi^{(0)}(a_0)$$

uniformly for $a_0 \in \mathfrak{M}_\varepsilon$.

Proof. Recall the mapping $F : \mathfrak{M}_\varepsilon \to \mathfrak{M}(M_0(\varepsilon))$, defined by

$$Fa_0 = a_0 + \xi^{(0)}(a_0).$$

The relations (27.48) imply

$$Fa_0 = \lim_{s \to \infty} F^{(s)} a_0, \tag{27.49}$$

uniformly with respect to $a_0 \in \mathfrak{M}_\varepsilon$. Thus, since each $F^{(s)}$ is continuous on the set $\mathfrak{M}_\varepsilon$, so is the mapping F. Consequently, the set $F\{\mathfrak{M}_\varepsilon\} = \mathfrak{M}(M_0)$ is closed and measurable.

Applying to this set one of the properties of the Lebesgue measure of sets derived in § 26, we can set

$$\operatorname{mes} \mathfrak{M}(M_0) \geqslant \overline{\lim} \operatorname{mes} F^{(s)}\{\mathfrak{M}_\varepsilon\}. \tag{27.50}$$

Since, for every pair of points $a_0', a_0'' \in \mathfrak{M}_\varepsilon^{(s)}$ we have the inequality

$$| F^{(s)} a_0' - F^{(s)} a_0'' | \geqslant | a_0' - a_0'' | - | B^{(s)}(a_0') - B^{(s)}(a_0'') | \geqslant$$

$$\geqslant | a_0' - a_0'' | - M_0 \, | a_0' - a_0'' | \geqslant (1 - M_0) \, | a_0' - a_0'' | > 0$$

$$\text{for } a_0' \neq a_0'',$$

we see that the mapping $F^{(s)}$ is one-to-one on $\mathfrak{M}_\varepsilon^{(s)}$. Hence, the inequality (27.50) can be extended:

$$\operatorname{mes} \mathfrak{M}(M_0) \geqslant \overline{\lim} \lim_{l\to\infty} \operatorname{mes} F_s\{\mathfrak{M}_\varepsilon^{(s+l)}\} \geqslant \overline{\lim} \lim_{l\to\infty} \operatorname{mes} F^{(s)}\{\Omega_\varepsilon^{(s+l)}\}. \quad (27.51)$$

The function $F^{(s)}(a_0)$ is uniformly differentiable on $\Omega_\varepsilon^{(s)}$. Furthermore, the Jacobian of $F^{(s)}$ satisfies the inequality

$$0 < 1 - M_0^{\varkappa-1} \leqslant 1 - M_0 c_5 \leqslant \left| \det \frac{\partial F^{(s)}(a_0)}{\partial a_0} \right| < 1 + M_0 c_5 \leqslant 1 + M_0^{\varkappa-1},$$

which enables us to take advantage of Lemma 8 in measuring the set $F^{(s)}\{\Omega_\varepsilon^{(s+l)}\}$ and get the estimate

$$\operatorname{mes} F^{(s)}\{\Omega^{(s+l)}\} \geqslant \frac{\operatorname{mes} \Omega_\varepsilon^{(s+l)}}{1 + M_0^{\varkappa-1}}. \quad (27.52)$$

The relations (27.51) and (27.52) imply

$$\operatorname{mes} \mathfrak{M}(M_0) \geqslant \frac{1}{1 + M_0^{\varkappa-1}} \overline{\lim} \lim_{l\to\infty} \operatorname{mes} \Omega_\varepsilon^{(s+1)} \geqslant (1 - M_0^{\varkappa-1}),$$

$$\overline{\lim} \operatorname{mes} \Omega_\varepsilon^{(l)} = (1 - M_0^{\varkappa-1}) \overline{\lim} \mathfrak{M}_\varepsilon^{(l)} \setminus [\mathfrak{M}_\varepsilon^{(l)}].$$

Hence, by inspection of Lemma 9,

$$\operatorname{mes} \mathfrak{M}(M_0) \geqslant (1 - M_0^{\varkappa-1})(1 - \varepsilon). \quad (27.53)$$

The relation (27.53) establishes the basic metric result resolving the problem of the reducibility of linear systems, which can be formulated in the form of the next theorem [66].

Theorem 19. *Suppose that the right-hand sides of the following system of differential equations with quasi-periodic coefficients*

$$\frac{dx}{dt} = Ax + P(\varphi, \mu)x, \qquad \frac{d\varphi}{dt} = \omega \quad [\omega = (\omega_1, \omega_2, \ldots, \omega_m)] \quad (27.54)$$

satisfy the conditions:

(i) *The matrix* $P(\varphi, \mu)$ *is* 2π*-periodic in* $\varphi = (\varphi_1, \ldots, \varphi_m)$, *is analytic on the domain* $|\operatorname{Im} \varphi| = \sup_\alpha |\operatorname{Im} \varphi_\alpha| < \rho_0$ $(\rho_0 > 0)$, *is real for real* φ, *and satisfies the relation* $\lim_{\mu\to 0} P(\varphi, \mu) = 0$.

(ii) *For some positive* ε *and* d

$$|(k, \omega)| \geqslant \varepsilon |k|^{-d} \qquad (|k| \neq 0)$$

for every integral-valued vector $k = (k_1, \ldots, k_m)$.

Then, a positive number $\mu(\varepsilon)$, $\mu(\varepsilon) \to 0$ *as* $\varepsilon \to 0$ *can be found such that the set* $\mathfrak{M}(P)$ *of matrices* A *of the cube* $S_1 \{|a_{ij}| \leqslant 1\}$ *satisfies the relation*

$$\lim_{\varepsilon \to 0} \underline{\text{mes}}\, \mathfrak{M}(P) = 1,$$

if there exists for this set a change of variable

$$x = \Phi(\varphi, \mu)\, y,$$

which is non-singular, 2π*-periodic in* φ, *analytic on the domain* $|\operatorname{Im} \varphi| < \rho/2$, *real for* $\operatorname{Im} \varphi = 0$ *and which reduces the system* (27.54) *for* $\mu \leqslant \mu(\varepsilon)$ *to the system*

$$\frac{dy}{dt} = A_0(\mu)\, y.$$

This theorem proves the metric relation (24.8) derived in § 24 and shows that for small matrices $P(\varphi)$, analytic for $|\operatorname{Im} \varphi| \leqslant \rho_0$, the quasi-periodic system (27.54) with strongly non-commmensurable frequencies is reducible for the set of matrices A of almost full Lebesgue measure.

§ 28. Linear Systems with Smooth Right-Hand Sides

In the preceding sections linear systems, analytic in the arguments appearing on the right-hand sides, were considered. By applying the smoothing operators, we saw that the requirement of the analyticity of the right-hand side can be dispensed with and replaced by the condition of differentiability a finite number of times. As a smoothing operator, it is convenient to take here the operator S_N. This enables us to stick to the scheme of constructing a transformation matrix, outlined in the preceding section and to extend all the results deduced above to the system

$$\frac{dx}{dt} = Ax + P(\varphi)\, x, \qquad \frac{d\varphi}{dt} = \omega, \tag{28.1}$$

where $P(\varphi)$ is a matrix, l-times continuously differentiable and 2π-periodic in φ. Without loss of generality, it can be assumed that A is a real diagonal matrix : $A = \{\lambda_1, \ldots, \lambda_n\}$.

We introduce the notations

$$\bar{D} = \{\bar{p}_{11}, \bar{p}_{22}, \ldots, \bar{p}_{nn}\}, \quad r = \min_{\alpha \neq \beta} | \lambda_\alpha - \lambda_\beta |, \quad \bar{r} = \max\{r, \varepsilon\}. \quad (28.2)$$

Let $U(\varphi, N) = \sum_{|k|=N} Y_k e^{i(k,\varphi)}$, 2π-periodic in φ, be a solution of the equation

$$\left(\frac{\partial U}{\partial \varphi}, \omega\right) + UA = AU + S_N P(\varphi) - \bar{D}, \quad (28.3)$$

where, as before,

$$S_N P(\varphi) = \sum_{|k| \leqslant N} P_k e^{i(k,\varphi)}, \qquad P_k = \frac{1}{(2\pi)^m} \int_0^{2\pi} \ldots \int_0^{2\pi} P(\varphi) e^{-i(k,\varphi)} d\varphi_1 \ldots d\varphi_n.$$

Such a solution obviously exists and satisfies

$$| U(\varphi, N) |_\lambda = \max_{|\rho| = \rho_1 + \ldots + \rho_m \leqslant \lambda} | D_\varphi^{|\rho|} U(\varphi, N) | \leqslant C_1 N^{\lambda+m+1} | U(\varphi, N) |_0 \leqslant$$

$$\leqslant c N^{\lambda+m+1} \sum_{|k| \leqslant N} | Y_k | \leqslant c_1 N^{\lambda+m+1} \left[\frac{| P_0 |}{\min_{\alpha \neq \beta} | \lambda_\alpha - \lambda_\beta |} + \right.$$

$$\left. + \sum_{0 < |k| \leqslant N} \frac{| P_k | \, | k |^d}{\varepsilon} \right] \leqslant$$

$$\leqslant \frac{c_1 N^{\lambda+m+1}}{\max(r_0, \varepsilon)} \sum_{|k| \leqslant N} N^d | P_0 | \leqslant \frac{2^m c_1 N^{\lambda+d+2m+1}}{\bar{r}} | P_0 |. \quad (28.4)$$

In the system of equations (28.1), a change of variables is carried out by putting

$$x = (E + U(\varphi, N))\, y. \quad (28.5)$$

This yields a system of equations in the variable y,

$$\frac{dy}{dt} = (A + \bar{D})\, y + P_1(\varphi)\, y,$$

where

$$P_1(\varphi) = P_1(\varphi, N) = (E + U(\varphi, N))^{-1} (P(\varphi)\, U(\varphi, N) - U(\varphi, N)\, \bar{D} +$$

$$+ P(\varphi) - S_N P(\varphi)). \quad (28.6)$$

To estimate the matrix $P_1(\varphi)$, we make use of the estimate (28.4) and the properties of operator S_N given in § 14, to obtain

$$| P_1(\varphi) | \leqslant | (E + U(\varphi, N))^{-1} | \times$$
$$\times \left(\frac{2^{m+1} c_1 N^{d+2m+1}}{\bar{r}} | P(\varphi) |_0^2 + c N^{-l+m+1} | P(\varphi) |_l \right). \tag{28.7}$$

The differentiation of the function $P_1(\varphi)$ for its l-th derivatives, gives

$$| D_\varphi^l P_1(\varphi) | \leqslant c_2 [| (E + U(\varphi, N))^{-1} | (| P(\varphi) U(\varphi, N) |_l + | U(\varphi, N) |_l | \bar{D} | +$$
$$+ | P(\varphi) |_l + | S_N P(\varphi) |_l) + \max_{\alpha=1, \ldots, l} \{ | (E + U(\varphi, N))^{-1} |_\alpha \times$$
$$\times (| P(\varphi) U(\varphi, N) |_{l-\alpha} + | U(\varphi, N) |_{l-\alpha} | \bar{D} | +$$
$$+ | P(\varphi) |_{l-\alpha} + | (S_N P(\varphi) |_{l-\alpha}) \}] \leqslant$$
$$\leqslant c_3 \Bigg[| (E + U(\varphi, N))^{-1} | \Bigg(\max_{\alpha=1, \ldots, l} \left\{ | P(\varphi) |_\alpha \frac{N^{l-\alpha+d+2m+1}}{\bar{r}} \right\} \times$$
$$\times | P(\varphi) |_0 + \frac{N^{l+d+2m-1}}{\bar{r}} | P(\varphi) |_0^2 +$$
$$+ | P(\varphi) |_l + N^{l+m+1} | P(\varphi) |_0 \Bigg) +$$
$$+ | (E + U(\varphi, N))^{-1} |_0^{l+1} \frac{N^{d+2m+1}}{\bar{r}} | P(\varphi) |_0 \max_{\alpha=1, \ldots, l} \Bigg\{ N^\alpha \max_\beta \times$$
$$\times (| P(\varphi) |_\beta \cdot | U(\varphi, N) |_{l-\alpha-\beta}) + \frac{N^{l+d+2m+1}}{\bar{r}} | P(\varphi) |_0^2 +$$
$$+ N^{l+m+1} | P(\varphi) |_0 \Bigg\} \Bigg]. \tag{28.8}$$

The stage is now set for proving the following result.

Lemma 11. *Suppose that*

$$l > \frac{\varkappa}{\varkappa - 1} (\beta + m + 1), \tag{28.9}$$

where

$$1 < \varkappa < 2, \quad \beta > \frac{\varkappa(s_0 + d + 2m + 1)}{2 - \varkappa},$$

and s_0 *is a positive integer.*

Then a positive M^0 *can be found such that for* $M_0 < M^0$ *the inequalities*

$$| U(\varphi, N_1) |_{s_0} < \frac{M_0^{\varkappa - 1}}{4n}, \qquad | P^{(1)}(\varphi, N_1) | < M_0^\varkappa = M_1,$$
$$| D_\varphi^l P^{(1)}(\varphi, N_1) | \leqslant N_1^l, \tag{28.10}$$
$$N_1^{-\beta} = M_1 = N_0^{-\varkappa\beta}$$

hold.

Proof. For small M^0, inequality (28.4) in the notation (28.2) and subject to the conditions (28.9), implies the first inequality of (28.10):

$$| U(\varphi, N_1) |_{s_0} \leqslant \frac{2^m c_1}{\bar{r}} N_0^{\varkappa(s_0+d+2m+1)-(2-\varkappa)\beta} M_0^{\varkappa-1} \leqslant \frac{M_0^{\varkappa-1}}{4n}.$$

The second inequality of (28.10) also follows from (28.7), (28.2) and (28.9):

$$| P^{(1)}(\varphi, N_1) | \leqslant (n+1) \left[\frac{2^{m+1} c_1}{\bar{r}} N_0^{\varkappa(d+2m+1)-(2-\varkappa)\beta} M_1 + c N_1^{m+1} \left(\frac{N_0}{N_1} \right)^l \right] \leqslant$$

$$\leqslant \frac{M_1}{2} + (n+1) c N_0^{\varkappa(m+1)+(1-\varkappa)l+\beta\varkappa} M_1 \leqslant M_1.$$

To estimate the l-th derivatives of the matrix $P^{(1)}(\varphi, N)$, we appeal to a generalization of the Landau-Hadamard inequality in the form of the following lemma.

Lemma 12. *Suppose that* $f(x) \in C^r((I \times I \times \ldots \times I), I = [0, \infty))$. *Then*

$$|f|_\lambda \leqslant c |f|_0 \left(\frac{|f|_r}{|f|_0} \right)^{\lambda/r},$$

where $0 \leqslant \lambda \leqslant r$ *and* c *is a positive constant not depending on* f.

Proof. From this inequality we get

$$| P |_\alpha \leqslant c | P |_0 \left(\frac{| P |_l}{| P |_0} \right)^{\alpha/l} \leqslant c M_0 \left(\frac{N_0^l}{M_0} \right)^{\alpha/l} = c M_0^{\frac{l-\alpha}{l}} N_0^\alpha.$$

This inequality implies the estimates

$$\max_\alpha \left\{ | P |_\alpha \frac{N_1^{l-\alpha+d+2m+1}}{\bar{r}} \right\} \leqslant \frac{c}{\bar{r}} N_1^{l+d+2m+1},$$

$$\max_\alpha \left\{ N^\alpha \max_\beta (| P |_\beta | U |_{l-\alpha-\beta}) \right\} \leqslant \max \left\{ c M_0^{\frac{l-\beta}{l}} \frac{2^m c_1}{\bar{r}} N_1^{l+d+2m+1} M_0 \right\} \leqslant$$

$$\leqslant \frac{2^m c c_1}{\bar{r}} N_1^{l+d+2m+1} M_0. \qquad (28.11)$$

Combining inequalities (28.8) and (28.11), we obtain

$$| D_\varphi^l P^{(1)}(\varphi, N_1) | \leqslant c_4 \left[(n+1) \left(\frac{N_1^{l+d+2m+1} M_0 + N_1^{l+d+2m+1} M_0^2}{\bar{r}} + N_0^l + \right.\right.$$

$$\left. + N_1^{l+m+1} M_0 \right) + (n+1)^{l+1} \frac{N_1^{d+2m+1} M_0}{\bar{r}} \times$$

$$\left.\times \left(\frac{N_1^{l+d+2m+1} + N_1^{l+2m+d+1} M_0^2}{\bar{r}} + N_1^{l+m+1} M_0 \right) \right].$$

This implies, for small M^0, the last inequality of (28.10):

$$|D^l_\varphi P^{(1)}(\varphi, N_1)| \leqslant N_1^l.$$

This lemma proves that the process of transforming the original system (28.1) by means of the substitution (28.5) is capable of extension. In addition, if we set

$$M_j = M_{j-1}^{\varkappa} = N_j^{-\beta}, \qquad N_j = N_{j-1}^{\varkappa},$$

then the iteration of s substitutions of the form (28.5), i.e. the substitution

$$x = \prod_{j=1}^{s} (E + U^{(j)}(\varphi, N_j))\, x^{(s)},$$

reduces the original system (28.1) to the system

$$\frac{dx^{(s)}}{dt} = A_s x^{(s)} + P^{(s)}(\varphi)\, x^{(s)},$$

$$\frac{d\varphi}{dt} = \omega,$$

where the matrices $U^{(s)}(\varphi, N_s)$, $P^{(s)}(\varphi)$ and A_s are defined by

$$\left(\frac{\partial U^{(s)}}{\partial \varphi}, \omega\right) + U^{(s)}(\varphi, N_s)\, A_{s-1} = A_{s-1} U^{(s)}(\varphi, N_s) + S_{N_s} P^{(s-1)}(\varphi) - \overline{D^{(s-1)}},$$

$$P^{(s)}(\varphi) = (E + U^{(s)}(\varphi, N_s))^{-1} (P^{(s-1)}(\varphi)\, U^{(s)}(\varphi, N_s) - U^{(s)}(\varphi, N_s)\, \overline{D^{(s-1)}} + $$
$$+ P^{(s-1)}(\varphi) - S_{N_s} P^{(s-1)}(\varphi)), \tag{28.12}$$

$$A_s = A + \sum_{\alpha=0}^{s-1} \bar{D}^{(\alpha)}, \quad \bar{D}^{(0)} = \bar{D}, \quad \bar{D}^{(\alpha)} \{\bar{p}_{11}^{(\alpha)}, \bar{p}_{22}^{(\alpha)}, \ldots, \bar{p}_{nn}^{(\alpha)}\},$$

and for small M^0 satisfy the inequalities

$$|U^{(s)}(\varphi, N_s)| \leqslant \frac{M_{s-1}^{\varkappa-1}}{4n}, \quad |P^{(s)}(\varphi)| \leqslant M_s,$$

$$|A_s - A| \leqslant \sum_{\alpha=0}^{s-1} M_\alpha < \frac{r}{2}. \tag{28.13}$$

The inequalities (28.13) imply the estimates

$$\left| \prod_{j=1}^{s+1} (E + U^{(j)}(\varphi, N_j)) - \prod_{j=1}^{s} (E + U^{(j)}(\varphi, N_j)) \right| \leqslant$$

$$\leqslant \left| \prod_{j=1}^{s} (E + U^{(j)}(\varphi, N_j)) \right| \, | U^{(s+1)}(\varphi, N_{s+1}) | \leqslant$$

$$\leqslant \left| \prod_{j=1}^{s} \left(E + \frac{M_{j-1}^{\varkappa-1}}{4n} I \right) \right| \frac{M_s^{\varkappa-1}}{4n} \leqslant$$

$$\leqslant \frac{n}{4} \prod_{j=1}^{\infty} \left(1 + \frac{M_{j-1}^{\varkappa-1}}{4} \right) M_s^{\varkappa-1} \leqslant \frac{n}{2} M_s^{\varkappa-1},$$

which verifies the criterion of uniform convergence for the sequence $\prod\limits_{j=1}^{s} \times$ $\times\ (E + U^{(j)}(\varphi, N_j))$:

$$\left| \prod_{j=1}^{s+k_0} (E + U^{(j)}(\varphi, N_j)) - \prod_{j=1}^{s} (E + U^{(j)}(\varphi, N_j)) \right| \leqslant$$

$$\leqslant \frac{n}{2} \sum_{j=s}^{s+k_0-1} M_j^{\varkappa-1} \leqslant nM_s^{\varkappa-1}.$$

We introduce the notation

$$\Phi(\varphi) = \lim_{s \to \infty} \prod_{j=1}^{s} (E + U^{(j)}(\varphi, N_j)) = \prod_{j=1}^{\infty} (E + U^{(j)}(\varphi, N_j)),$$

$$A_0 = A + \overline{D} + \sum_{j=1}^{\infty} \overline{D}^{(j)},$$

and establish some properties of the matrix $\Phi(\varphi)$.

The matrix $\Phi(\varphi)$ is obviously 2π-periodic in φ. The inequality

$$\left| \prod_{j=1}^{s} (E + U^{(j)}(\varphi, N_j)) - E \right| \leqslant \left[\prod_{j=1}^{s} \left(1 + \frac{M_{j-1}^{\varkappa-1}}{4} \right) - 1 \right] | I | < 1$$

implies the convergence of the series $\sum\limits_{j=0}^{\infty} (E - \Phi(\varphi))^j$ and, consequently, the matrix $\Phi(\varphi)$ is non-singular. Furthermore,

$$\left| D_\varphi^{s_0} \left[\sum_{j=1}^{s+1} (E + U^{(j)}(\varphi, N_j)) - \prod_{j=1}^{s} (E + U^{(j)}(\varphi, N_j)) \right] \right| \leqslant$$

$$\leqslant \left| D_\varphi^{s_0} \left[\prod_{j=1}^{s} (E + U^{(j)}(\varphi, N_j)) \, U^{(s+1)}(\varphi, N_{s+1}) \right] \right| \leqslant$$

$$\leqslant c \max_{\alpha=0,1,\ldots,s_0} \left\{ \left| \prod_{j=1}^{s} (E + U^{(j)}(\varphi, N_j)) \right|_\alpha \, | U^{(s+1)}(\varphi, N_{s+1}) |_{s_0-\alpha} \right\} \leqslant$$

$$\leqslant c_5 \, | U^{(s+1)}(\varphi, N_{s+1}) |_{s_0} \leqslant c_5 \frac{M_s^{\varkappa-1}}{4n},$$

hence

$$\left| D_\varphi^{s_0} \prod_{j=1}^{s+k_0} (E + U^{(j)}(\varphi, N_j)) - D_\varphi^{s_0} \prod_{j=1}^{s} (E + U^{(j)}(\varphi, N_j)) \right| \leqslant$$

$$\leqslant \frac{c_5}{4n} \sum_{j=s}^{s+k_0-1} M_s^{\varkappa-1} \leqslant c_6 M_s^{\varkappa-1}.$$

The last inequality ensures that the matrix $\Phi(\varphi)$ is s_0-times differentiable.

The statements above crystallize into the following generalization of Theorem 17.†

Theorem 20. *Suppose that the right-hand sides of the system of equations*

$$\frac{dx}{dt} = Ax + P(\varphi)\,x,$$
$$\frac{d\varphi}{dt} = \omega \tag{28.14}$$

satisfy the following conditions:

(i) *The matrix $P(\varphi)$ is l-times continuously differentiable and 2π-periodic in φ.*

(ii) *For some positive ε and d, the inequality*

$$|(k, \omega)| \geqslant \varepsilon \, |k|^{-d} \tag{28.15}$$

holds for every integral-valued $k = (k_1, \ldots, k_m)$ ($|k| \neq 0$).

(iii) *The eigenvalues $\lambda = (\lambda_1, \ldots, \lambda_n)$ of the matrix A are real and distinct.*

Then a sufficiently small positive constant M_0 and an integer $l = l(s_0)$ can be

†See Appendix 8.

found such that for

$$| P (\varphi) | \leqslant M_0 < M^0, \quad | P (\varphi) |_l \leqslant c \tag{28.16}$$

the system of equations (28.14) *reduces to the system*

$$\frac{dy}{dt} = A_0 y,$$

$$\frac{d\varphi}{dt} = \omega,$$

where A_0 is a constant matrix, by means of the non-singular change of variable

$$x = \Phi (\varphi) y,$$

with $\Phi (\varphi)$ a matrix, 2π-periodic in φ and s_0-times continuously differentiable.

According to the above presentation, the scheme of constructing the transformation matrix $\Phi (\varphi)$ for a smooth right-hand side of the system of equations (28.14) differs from that for an analytic right-hand side only in that the functions $U^{(j)}(\varphi, N_j)$ are here constructed, not by the matrix $P^{(j-1)} (\varphi)$, but by its finite Fourier sum $S_N P^{(j-1)} (\varphi)$. The process of the construction of the matrix $\Phi (\varphi)$, therefore, remains rapidly convergent. All the propositions proved in § 25 for a system analytic in φ carry over to a smooth system without any significant change. Thus, in particular, the general solution of a second order system is defined in the first approximation by the formula (25.9), by restricting the finite sums there to the number of terms $N_1 \geqslant | P (\varphi) |_0^{-2\beta}$.

Now, to generalise the measure theorem for reducible systems, we consider the system of equations

$$\begin{aligned} \frac{dx}{dt} &= A_0 x + (P (\varphi) + \xi) x, \\ \frac{d\varphi}{dt} &= \omega \qquad [\omega = (\omega_1, \omega_2, \ldots, \omega_m)], \end{aligned} \tag{28.17}$$

where A_0, ξ are constants and $P (\varphi)$ is an $n \times n$ matrix, 2π-periodic in φ.

Suppose that the matrix $P (\varphi)$ is l-times continuously differentiable. The eigenvalues $\lambda = (\lambda_1, \ldots, \lambda_n)$ of the matrix A_0 satisfy the inequalities

$$| \lambda_\alpha - \lambda_\beta + i (k, \omega) | \geqslant \varepsilon | k |^{-d} \qquad (k \neq 0, \quad \alpha, \beta = 1, 2, \ldots, n)$$

for every integral-valued vector $k = (k_1, \ldots, k_n)$ and some constants ε and d.

Let $U^{(1)} (\varphi, \xi, N)$ be a periodic solution of the equation

$$\left(\frac{\partial U^{(1)}}{\partial \varphi}, \omega \right) + U^{(1)} A_0 = A_0 U^{(1)} + S_N P(\varphi) - \overline{P}(\varphi).$$

Then, for large l [it suffices that l should satisfy inequality (28.9)] and small M_0, the change of variables

$$x = (E + U^{(1)}(\varphi, \xi, N_1))\, x^{(1)},$$
$$\xi = \xi(\xi^{(1)}), \tag{28.18}$$

($N_1^{-\beta} = N_0^{-\varkappa\beta} = M_0$) reduces the system of equations (28.17) to the system

$$\frac{dx^{(1)}}{dt} = A_0 x^{(1)} + (P^{(1)}(\varphi, \xi^{(1)}) + \xi^{(1)})\, x^{(1)},$$
$$\frac{d\varphi}{dt} = \omega, \tag{28.19}$$

where

$$P^{(1)}(\varphi, \xi^{(1)}) = (E + U^{(1)}(\varphi, \xi(\xi^{(1)}), N_1))^{-1} [((P(\varphi) + \xi(\xi^{(1)})) \times$$
$$\times\, U^{(1)}(\varphi, \xi(\xi^{(1)}), N_1) - U^{(1)}(\varphi, \xi(\xi^{(1)}), N_1)\, \xi^{(1)} +$$
$$+ P(\varphi) - S_{N_1} P(\varphi)]. \tag{28.20}$$

It is clear from (28.18) and (28.20) that the matrices $U^{(1)}(\varphi, \xi(\xi^{(1)}), N_1)$, $\xi(\xi^{(1)})$ and $P^{(1)}(\varphi, \xi^{(1)})$ are 2π-periodic in φ, l-times continuously differentiable with respect to φ and analytic in $\xi^{(1)} = \{\xi^{(1)}_{\alpha\beta}\}$ for

$$|\xi^{(1)}| \leqslant M_0. \tag{28.21}$$

Moreover, for small M_0 in the domain (28.21), the matrices $U^{(1)}(\varphi, \xi(\xi^{(1)}), N_1)$, $\xi(\xi^{(1)})$ and $P^{(1)}(\varphi, \xi^{(1)})$ satisfy the inequalities

$$|U^{(1)}(\varphi, \xi(\xi^{(1)}), N_1)| < \frac{M_0^{\varkappa-1}}{4n}, \qquad |\xi(\xi^{(1)}) - \xi^{(1)}| < M_0,$$
$$|P^{(1)}(\varphi, \xi^{(1)})| < M_0^{\varkappa} = M_1, \quad |D_\varphi^l P^{(1)}(\varphi, \xi^{(1)})| < N_0^{l\varkappa} = N_1^l. \tag{28.22}$$

The properties of the matrix $P^{(1)}(\varphi, \xi^{(1)})$ permit us to effect the change of variables of the form (28.18) in the system (28.19) etc. Inequalities (28.22) and those analogous to them for $U^{(j)}(\varphi, \xi^{(j-1)}(\xi^{(j)}), N_j)$, $\xi^{(j-1)}(\xi^{(j)})$ and $P^{(j)}(\varphi, \xi^{(j)})$ ensure the convergence of the process indicated for the transformation of the original system (28.19), and the validity of the next generalized reducibility theorem.

Theorem 21. *Suppose that the right-hand sides of the system of equations*

$$\frac{dx}{dt} = A_0 x + (P(\varphi) + \xi)\, x,$$
$$\frac{d\varphi}{dt} = \omega \qquad [\omega = (\omega_1, \omega_2, \ldots, \omega_m)] \tag{28.23}$$

satisfy the conditions :

(i) *the matrix $P(\varphi)$ is l-times continuously differentiable and 2π-periodic in φ.*

(ii) *The eigenvalues $\lambda = (\lambda_1, \ldots, \lambda_n)$ of the matrix A_0 satisfy the inequalities*

$$|\lambda_\alpha - \lambda_\beta + i(k, \omega)| \geqslant \varepsilon |k|^{-d} \quad (|k| \neq 0,\ \alpha, \beta = 1, 2, \ldots, n) \tag{28.24}$$

for every integral-valued vector $k = (k_1, \ldots, k_n)$.

Then a sufficiently small positive constant M^0 and an integer $l = l(s_0)$ can be found such that for

$$|P(\varphi)| \leqslant M_0 < M^0, \quad |P(\varphi)|_l \leqslant c, \tag{28.25}$$

there exist a constant matrix ξ^0 satisfying the condition

$$|\xi^0| < 2M_0,$$

and a non-singular change of variable

$$x = \Phi(\varphi)\, y,$$

2π-periodic in φ and s_0-times continuously differentiable, which reduces the system (28.23) for $\xi = \xi^0$ to the system

$$\frac{dy}{dt} = A_0 y,$$
$$\frac{d\varphi}{dt} = \omega.$$

By Theorem 21, a mapping F is defined on the set $\mathfrak{M}_\varepsilon$ of matrices A_0, by

$$a_0 \to Fa_0 = a_0 + \xi^0(a_0).$$

The properties of this mapping coincide with those of the mapping F, established in the preceding section, where we have considered systems with analytic right-hand sides. This follows from the fact that the degree of smoothness of the matrix $P(\varphi)$ had no significant role to play in the proof of Theorem 19. Hence, for $\mathfrak{M}(M_0) = F\{\mathfrak{M}_\varepsilon\}$, the estimate

$$\text{mes}\, \mathfrak{M}(M_0) \geqslant (1 - M_0^{\varkappa - 1})(1 - \varepsilon)$$

remains valid and, what is more, the measure theorem for systems with non-analytic right-hand sides can be proved. This theorem will be stated here without proof.

Theorem 22. *For a system of equations*

$$\frac{dx}{dt} = Ax + P(\varphi, \mu)\, x, \qquad \frac{d\varphi}{dt} = \omega, \tag{28.26}$$

there can be found a positive $\mu(\varepsilon)$, $\mu(\varepsilon) \to 0$ *as* $\varepsilon \to 0$, *and an integer* $l = l(s_0)$ *such that whenever the matrix* $P(\varphi, \mu)$ *is* l*-times continuously differentiable,* 2π*-periodic in* φ, *and satisfies the relations*

$$\lim_{\varepsilon \to 0} | P(\varphi, \mu(\varepsilon)) + 0, \qquad | P(\varphi, \mu(\varepsilon)) |_l \leqslant c, \tag{28.27}$$

and the frequencies $\omega = (\omega_1, \ldots, \omega_m)$ *satisfy the inequality*

$$| (k, \omega) | \geqslant \varepsilon \, | k |^{-d} \qquad (| k | \neq 0) \tag{28.28}$$

for every integral-valued $k = (k_1, \ldots, k_m)$, *then the set* $\mathfrak{M}(P)$ *of matrices* A *of the cube* $S_1 \{ | a_{ij} | \leqslant 1 \}$, *for which there exists a change of variable*

$$x = \Phi(\varphi, \mu)\, y, \tag{28.29}$$

non-singular, 2π*-periodic* in φ *and* s_0*-times continuously differentiable, reducing the system* (28.26) *for* $\mu = \mu(\varepsilon)$ *to the system*

$$\frac{dy}{dt} = A_0(\mu)\, y, \qquad \frac{d\varphi}{dt} = \omega, \tag{28.30}$$

satisfies the relation

$$\lim_{\varepsilon \to 0} \operatorname{mes} \mathfrak{M}(P) = 1. \tag{28.31}$$

This theorem makes it explicit that the system (28.26) is reducible for a small and sufficiently smooth matrix $P(\varphi)$, as also is the case of an analytic matrix $P(\varphi)$ for a set of matrices A of almost full Lebesgue measure.†

†See Appendix 9.

Chapter 6

NEIGHBOURHOOD OF AN INVARIANT SMOOTH TOROIDAL MANIFOLD

§ 29. Behaviour of Integral Curves in the Neighbourhood of Toroidal Manifolds

Consider a system of non-linear differential equations

$$\frac{dy}{dt} = Y(y), \tag{29.1}$$

where $Y = \{Y_1, \ldots, Y_{m+n}\}$, $y = (y_1, \ldots, y_{m+n})$ are points of the $(m + n)$-dimensional Euclidean space E_{m+n}. Suppose that system (29.1) has an invariant manifold M, smoothly homeomorphic to the m-dimensional torus T. It is of interest to consider the positions of integral curves of the system (29.1) in a neighbourhood of the manifold M. With the classical work of Poincaré [58]† as the starting point, this problem has been investigated for simple manifolds, such as the equilibrium positions ($m = 0$) and the periodic trajectories ($m = 1$) of system (29.1).

Later Seigel [69] showed that in the general case the position of integral curves in the neighbourhood of simple manifolds is the same as that of a certain linear system. Suppose, for example, that $y = 0$ is the equilibrium position of system (29.1) and that $\mu = (\mu_1, \ldots, \mu_{m+n})$ is a vector of the eigenvalues of the matrix $A = \left.\frac{\partial Y(y)}{\partial y}\right|_{y=0}$. Then, we have the following theorem [69].

Theorem 23. *Assume that the right-hand side of the system of equations* (29.1) *is analytic in the neighbourhood of zero. If*

$$|\mu_i - (l, \mu)| > \varepsilon |l|^{-d} \qquad \left(|l| = \sum_{\alpha=1}^{n+m} l_\alpha > 1\right) \tag{29.2}$$

for some $\varepsilon > 0$, $d > 0$ *and every vector* $l = (l_1, \ldots, l_{m+n})$ *with non-negative integral-valued components* l_α, *then the change of variables*

$$x = y + g(y), \qquad g(0) = \left.\frac{\partial g(y)}{\partial y}\right|_{y=0} \tag{29.3}$$

†See Appendix 10.

exists, analytic in the neighbourhood of zero, and reduces the system of equations (29.1) *to the system*

$$\frac{dx}{dt} = Ax. \tag{29.4}$$

Belega [6] has investigated the position of integral curves of the system of equations (29.1) in the neighbourhood of a manifold M, analytically homeomorphic to the m-dimensional torus $\mathcal{T}$.

Assume that M is an invariant manifold such that in its neighbourhood we can introduce new variables $\varphi, x = (\varphi_1, \ldots, \varphi_m, x_1, \ldots, x_n)$, so that the system (29.1) takes the form

$$\begin{aligned} \frac{dx}{dt} &= Ax + F(\varphi, x), \\ \frac{d\varphi}{dt} &= \omega + f(\varphi, x), \end{aligned} \tag{29.5}$$

where A is a constant matrix, $F(\varphi, x)$ and $f(\varphi, x)$ are vector functions, analytic in the domain $|\operatorname{Im}\varphi| < \rho$, $|x| < \eta$ and 2π-periodic in the angular variables $\varphi = (\varphi_1, \ldots, \varphi_m)$, and

$$F(\varphi, 0) = \left.\frac{\partial F(\varphi, x)}{\partial x}\right|_{x=0} = f(\varphi, 0) = 0. \tag{29.6}$$

With these conventions, the following theorem has been proved in [6].

Theorem 24. *Let $\mu = (\mu_1, \ldots, \mu_n)$ be the eigenvalues of the matrix A. Suppose that the condition*

$$|(l, \mu) - \varepsilon_1 \mu_j + i(\omega, k)| > \varepsilon(|l| + |k|)^{-d} \tag{29.7}$$
$$(\varepsilon_1 = 0, 1; \;\; j = 1, \ldots, n; \;\; i^2 = -1)$$

is satisfied for some $\varepsilon > 0$, $d > 0$ and every integral-valued vector $l = (l_1, \ldots$ $\ldots, l_n)$, $k = (k_1, \ldots, k_m)$; here

$$|l| = \sum_{\alpha=1}^{n} l_\alpha > 1 + \varepsilon, \quad l_\alpha \geqslant 0,$$

$$|k| = \sum_{\beta=1}^{m} |k_\beta|.$$

Then there exists a change of variables

$$\begin{aligned} z &= x + X(\varphi, x), \\ \psi &= \varphi + \Psi(\varphi, x), \end{aligned} \tag{29.8}$$

analytic and analytically invertible in some neighbourhood $|x| < \eta_1$, $|\operatorname{Im} \varphi| < \rho_1$ *of the torus T with the functions*

$$X(\varphi, x), \quad \Psi(\varphi, x), \quad X(\varphi, 0) = \frac{\partial X(\varphi, x)}{\partial x}\bigg|_{x=0} = \Psi(\varphi, 0),$$

2π*-periodic in* φ*, which reduces the system of equations* (29.5) *to the system*

$$\frac{dz}{dt} = Az, \qquad \frac{d\psi}{dt} = \omega. \tag{29.9}$$

Since the condition (29.7) for $d > m + n$ is violated only on a set of vectors μ and ω of Lebesgue-measure zero, the position of the integral curves of the analytic system (29.5) in the neighbourhood of the torus $x = 0$ is, in the general case, the same as that of the linear system (29.9). The character of the position of the integral curves of system (29.1) with a smooth right-hand side essentially differs from the one with an analytic right-hand side. The passage from an analytic to a smooth system (under the homeomorphism of the integral curves of system (29.5) to the curves of system (29.9)) is connected with the transition from determining the inverse of linear operators acting on a coordinate space (the coefficient space of analytic functions) to the linear (differential) operators, acting on a function space (the space $C^r(D)$). This considerably complicates the problem and imparts a 'cruder' character to the reducibility conditions, restraining the reduction of the system of equations (29.5) to the system (29.9) where the values of μ_1 and ω span some special domain of the μ,ω-space. We have the following theorem

Theorem 25. *In the system of equations* (29.5) *suppose that the functions* $F(\varphi, x)$ *and* $f(\varphi, x)$ *are* 2π*-periodic in* $\varphi = (\varphi_1, \ldots, \varphi_m)$, l_0*-times continuously differentiable in the domain*

$$|x| < \eta \qquad (x = (x_1, \ldots, x_n)),$$

and satisfy the condition

$$F(\varphi, 0) = \frac{\partial F(\varphi, x)}{\partial x}\bigg|_{x=0} = f(\varphi, 0) = 0. \tag{29.10}$$

In addition, for given positive constants c_0, γ, $\bar{\eta}$ *and a given integer* s_0 $(s_0 \geqslant 2)$, *suppose that there exist in this domain a positive* $\delta_0 = \delta_0(c_0, \gamma, \bar{\eta}, s_0)$ *and an integer* $l_0 = l_0(s_0)$ *such that the functions* $F(\varphi, x)$ *and* $f(\varphi, x)$ *satisfy the inequalities*

$$|F(\varphi, x)| + |f(\varphi, x)| \leqslant \delta_0, \qquad |F(\varphi, x)|_{l_0} + |f(\varphi, x)|_{l_0} \leqslant c_0, \tag{29.11}$$

and the eigenvalues $\mu = (\mu_1, \ldots, \mu_n)$ *of the matrix A satisfy the inequality*

$$\operatorname{Re}(\mu_\alpha + \mu_\beta - \mu_j) \leqslant -\gamma \tag{29.12}$$

for any $\alpha, \beta, j = 1, \ldots, n$.

Then there exist the functions $Y(\theta, y)$ *and* $\Phi(\theta, y)$, 2π-*periodic in* $\theta = (\theta_1, \theta_2, \ldots, \theta_n)$, $(s_0 - 1)$-*times continuously differentiable with respect to* $y = (y_1, y_2, \ldots, y_n)$ *in the domain* $|y| \leqslant \eta - \bar{\eta}$, *satisfying the condition*

$$Y(\theta, 0) = \left.\frac{\partial Y(\theta, y)}{\partial y}\right|_{y=0} = \Phi(\theta, 0) = 0 \tag{29.13}$$

and the estimate

$$|Y(\theta, y)|_{s_0-1} + |\Phi(\theta, y)|_{s_0-1} \leqslant \bar{\eta}, \tag{29.14}$$

such that the system of equations

$$\begin{aligned} \frac{dx}{dt} &= Ax + F(\varphi, x), \\ \frac{d\varphi}{dt} &= \omega + f(\varphi, x) \end{aligned} \tag{29.15}$$

is reduced, by means of the change of variable

$$\begin{aligned} x &= y + Y(\theta, y), \\ \varphi &= \theta + \Phi(\theta, y), \end{aligned} \tag{29.16}$$

to the form

$$\begin{aligned} \frac{dy}{dt} &= Ay, \\ \frac{d\theta}{dt} &= \omega. \end{aligned} \tag{29.17}$$

The degree of the smoothness of the functions $F(\varphi, x)$ and $f(\varphi, x)$ plays a key role in the reducibility of system (29.15) with a smooth right-hand side. The elucidation of the influence of smoothness on the reducibility permits us to weaken the conditions (29.12) considerably and to replace them by the conditions on the functions $F(\varphi, x), f(\varphi, x)$, which are $l_0(s_0, \tau)$-times continuously differentiable:

$$\begin{aligned} |(\mu, l) - \mu_j + i(k, \omega)| &\geqslant \varepsilon(|l| + |k|)^{-d} \quad \text{for } 2 \leqslant |l| \leqslant \tau, \\ \operatorname{Re}[(\mu, l) - \mu_j] &\leqslant -\gamma < 0 \quad \text{for } \tau \leqslant |l| \leqslant l_0(s_0, \tau), \end{aligned} \tag{29.18}$$

where τ is any integer > 1 ; $l_0(s_0, \tau)$ is an integer-valued function of s_0, τ.

In the present chapter we aim to prove the theorem formulated above and to analyse the influence of small quasi-periodic perturbations on the manifold M and on the character of the position of integral curves in its neighbourhood.

§ 30. Auxiliary Propositions

Let A be a real $n \times n$ matrix, $\mu = (\mu_1, \ldots, \mu_n)$ a vector whose components are the eigenvalues of the matrix A and let $J = \{I_{\rho_1}(\mu_1), \ldots, I_{\rho_{k_0}}(\mu_{k_0})\}$ be its Jordan form.

Evidently, if μ_j is a real eigenvalue, then there corresponds to it in J the real $\rho_j \times \rho_j$ Jordan block

$$J_{\rho_j}(\mu_j) = \begin{pmatrix} \mu_j & 0 & 0 & \ldots & 0 & 0 \\ \varepsilon_1 & \mu_j & 0 & \ldots & 0 & 0 \\ 0 & \varepsilon_1 & \mu_j & \ldots & 0 & 0 \\ \vdots & \vdots & \vdots & & \vdots & \vdots \\ 0 & 0 & 0 & \ldots & \varepsilon_1 & \mu_j \end{pmatrix}, \tag{30.1}$$

where ε_1 is a non-zero real number. If $\mu_j = \alpha_j + i\beta_j$ is complex and $\bar{\mu}_j = \alpha_j - i\beta_j$ is its complex conjugate eigenvalue, then there corresponds to the pair μ_j, $\bar{\mu}_j$ in J, a pair $J_{\rho_j}(\mu_j)$, $J_{\rho_j}(\bar{\mu}_j)$ of the $\rho_j \times \rho_j$ Jordan cells of the form (30.1). Correponding to this pair we form a real $2\rho_j \times 2\rho_j$ matrix

$$J_{2\rho_j}\{\mu_j, \bar{\mu}_j\} = \begin{pmatrix} S_j & 0 & 0 & \ldots & 0 & 0 \\ \varepsilon_1 E_2 & S_j & 0 & \ldots & 0 & 0 \\ 0 & \varepsilon_1 E_2 & S_j & \ldots & 0 & 0 \\ \vdots & \vdots & \vdots & & \vdots & \vdots \\ 0 & 0 & 0 & \ldots & \varepsilon_1 E_2 & S_j \end{pmatrix}, \tag{30.2}$$

where E_2 is the 2×2 identity matrix, and

$$S_j = \begin{pmatrix} \alpha_j & -\beta_j \\ \beta_j & \alpha_j \end{pmatrix}.$$

The matrix form given by

$$B = \{D_1, \ldots, D_m\} = \begin{pmatrix} D_1 & 0 & \ldots & 0 \\ 0 & D_2 & \ldots & 0 \\ \vdots & \vdots & & \vdots \\ 0 & 0 & \ldots & D_m \end{pmatrix},$$

where

$$D_j = \begin{cases} J_{\rho_j}(\mu_j), & \text{if } \mu_j \text{ are real,} \\ J_{2\rho_j}(\mu_j, \bar{\mu}_j), & \text{if } \mu_j \text{ and } \bar{\mu}_j \text{ are complex conjugate,} \end{cases}$$

is called the real canonical form of a real matrix A. It it known [16] that for every real matrix A, there is a real non-singular matrix C reducing it to the canonical form B :

$$CAC^{-1} = B.$$

This motivates the next lemma.

Lemma 13. *Suppose that A is a matrix in real canonical form, and that $\mu_1, \ldots, \mu_n$ are its eigenvalues, and let $\varepsilon_1 > 0$, $\gamma = \max_j \operatorname{Re} \mu_j$. Then, for every $t \geqslant 0$,*

$$\| e^{At}x \| \leqslant e^{(\varepsilon_1+\gamma)t} \| x \|,$$

where

$$\| x \| = \left(\sum_{j=1}^{n} x_j^2 \right)^{1/2}.$$

Proof. Since

$$A = \{D_1, \ldots, D_m\},$$

it follows that

$$e^{At} = \{e^{D_1 t}, \ldots, e^{D_m t}\}.$$

For $D_j = J_{\rho_j}(\mu_j)$, the structure of the Jordan block

$$J_{\rho_j}(\mu_j) = \mu_j E_{\rho_j} + \varepsilon_1 Z_{\rho_j},$$

where

$$Z_{\rho_j} = \begin{pmatrix} 0 & 0 & \ldots & 0 & 0 \\ 1 & 0 & \ldots & 0 & 0 \\ \vdots & \vdots & & \vdots & \vdots \\ 0 & 0 & \ldots & 1 & 0 \end{pmatrix},$$

implies that

$$e^{D_j t} = e^{\mu_j E_{\rho_j} t + \varepsilon_1 Z_{\rho_j} t} = e^{\mu_j t} e^{\varepsilon_1 Z_{\rho_j} t} =$$

$$= e^{\mu_j t} \left(E + \frac{\varepsilon_1 Z_{\rho_j} t}{1!} + \ldots + \frac{(\varepsilon_1 Z_{\rho_j} t)^{\rho_j - 1}}{(\rho_j - 1)!} \right) =$$

$$= e^{\mu_j t} \begin{bmatrix} 1 & 0 & \ldots & 0 & 0 \\ \varepsilon_1 t & 1 & \ldots & 0 & 0 \\ \frac{(\varepsilon_1 t)^2}{2!} & \varepsilon_1 t & \ldots & 0 & 0 \\ \vdots & \vdots & & \vdots & \vdots \\ \frac{(\varepsilon_1 t)^{\rho_j - 1}}{(\rho_j - 1)!} & \frac{(\varepsilon_1 t)^{\rho_j - 2}}{(\rho_j - 2)!} & \ldots & \varepsilon_1 t & 1 \end{bmatrix}.$$

Hence, if $x_j = \{x_{j_1}, x_{j_2}, \ldots, x_{j_{\rho_j}}\}$ is a ρ_j-dimensional vector, the vector $e^{\rho_j t} x_j$ takes the form

$$e^{D_j t} x_j = \left\{ x_{j_1} e^{\mu_j t}, \left(\frac{\varepsilon_1 t}{1!} x_{j_1} + x_{j_2} \right) e^{\mu_j t}, \ldots, \sum_{\nu=1}^{\rho_j} \frac{(\varepsilon_1 t)^{\rho_j - \nu}}{(\rho_j - \nu)!} x_{j\nu} e^{\mu_j t} \right\}. \tag{30.3}$$

Furthermore, for $D_j = J_{2\rho_j}(\mu_j, \bar{\mu}_j)$, the structure of the matrix

$$J_{2\rho_j}(\mu_j, \bar{\mu}_j) = \{S_2, \ldots, S_2\} + \varepsilon_1 Z_{2\rho_j},$$

where

$$Z_{2\rho_j} = \begin{pmatrix} 0 & 0 & \ldots & 0 & 0 \\ E_2 & 0 & \ldots & 0 & 0 \\ 0 & E_2 & \ldots & 0 & 0 \\ \vdots & \vdots & & \vdots & \vdots \\ 0 & 0 & \ldots & E_2 & 0 \end{pmatrix},$$

implies that

$$e^{D_j t} = e^{\{S_2, \ldots, S_2\} t + \varepsilon_1 Z_{2\rho_j} t} = \{e^{S_2 t}, \ldots, e^{S_2 t}\} \times$$

$$\times \left(E + \frac{\varepsilon_1 Z_{2\rho_j} t}{1!} + \ldots + \frac{(\varepsilon_1 Z_{2\rho_j} t)^{\rho_j - 1}}{(\rho_j - 1)!} \right) =$$

$$= \begin{bmatrix} e^{S_2 t} & \ldots & 0 \\ \varepsilon_1 t e^{S_2 t} & \ldots & 0 \\ \vdots & \ldots & \vdots \\ \frac{(\varepsilon_1 t)^{\rho_j - 1}}{\rho_j - 1} e^{S_2 t} & \ldots & e^{S_2 t} \end{bmatrix}, \tag{30.4}$$

where

$$e^{S_2 t} = e^{\alpha_j t} \begin{pmatrix} \cos \beta_j t & -\sin \beta_j t \\ \sin \beta_j t & \cos \beta_j t \end{pmatrix}.$$

For $\varepsilon_1 t \geqslant 0$, (30.3) implies

$$\| e^{D_j t} x_j \|^2 = c^{2\mu_j t} \left[x_{j_1}^2 + \left(x_{j_2} + \frac{\varepsilon_1 t}{1!} x_{j_1} \right)^2 + \ldots \right.$$

$$\left. \ldots + \left(x_{\rho_j} + \frac{\varepsilon_1 t}{1!} x_{\rho_j - 1} + \ldots + \frac{(\varepsilon_1 t)^{\rho_j - 1}}{(\rho_j - 1)!} x_{j_1} \right)^2 \right].$$

However,

$$\left(x_{j_2} + \frac{\varepsilon_1 t}{1!} x_{j_1} \right)^2 \leqslant x_{j_2}^2 \left(1 + \frac{\varepsilon_1 t}{1!} \right) + x_{j_1}^2 \frac{\varepsilon_1 t}{1!} \left(1 + \frac{\varepsilon_1 t}{1!} \right) \leqslant$$

$$\leqslant \left(x_{j_2}^2 + \frac{\varepsilon_1 t}{1!} x_{j_1}^2 \right) e^{\varepsilon_1 t}, \qquad (30.5)$$

$$\left(x_{j_3} + \frac{\varepsilon_1 t}{1!} x_{j_2} + \frac{(\varepsilon_1 t)^2}{2!} x_{j_1} \right)^2$$

$$\leqslant x_{j_3}^2 \left(1 + \frac{\varepsilon_1 t}{1!} + \frac{(\varepsilon_1 t)^2}{2!} \right) +$$

$$+ x_{j_2} \frac{\varepsilon_1 t}{1!} \left(1 + \frac{\varepsilon_1 t}{1!} + \frac{(\varepsilon_1 t)^2}{2!} \right) +$$

$$+ x_{j_1}^2 \frac{(\varepsilon_1 t)^2}{2!} \left(1 + \frac{\varepsilon_1 t}{1!} + \frac{(\varepsilon_1 t)^2}{2!} \right) \leqslant$$

$$\leqslant \left(x_{j_3}^2 + \frac{\varepsilon_1 t}{1!} x_{j_2}^2 + \frac{(\varepsilon_1 t)^2}{2!} x_{j_1}^2 \right) e^{\varepsilon_1 t},$$

etc.

Inspection of the inequalities (30.5) yields the estimate

$$\| e^{D_j t} x_j \|^2 \leqslant e^{2\mu_j t} e^{\varepsilon_1 t} \left[x_{j_1}^2 \left(1 + \frac{\varepsilon_1 t}{1!} + \frac{(\varepsilon_1 t)^2}{2!} + \ldots + \frac{(\varepsilon_1 t)^{\rho_j - 1}}{(\rho_j - 1)!} \right) + \right.$$

$$\left. + x_{j_2}^2 \left(1 + \frac{\varepsilon_1 t}{1!} + \ldots + \frac{(\varepsilon_1 t)^{\rho_i - 2}}{(\rho_j - 2)!} \right) + \ldots + x_{\rho_j}^2 \right] \leqslant$$

$$\leqslant e^{2(\mu_j + \varepsilon_1)t} \| x_j \|^2. \qquad (30.6)$$

Since $x_j \{ x_{j_1}, y_{j_1}, \ldots, x_{j\rho_j}, y_{j\rho_j} \}$ is a $2\rho_j$-dimensional vector, (30.4) implies

$$\| e^{D_j t} x_j \|^2 = e^{2\alpha_j t} \left\{ (\cos \beta_j t x_{j_1} - \sin \beta_j t y_{j_1})^2 + (\sin \beta_j t x_{j_1} + \cos \beta_j t y_{j_1})^2 + \right.$$

$$+ [\varepsilon_1 t (\cos \beta_j t x_{j_1} - \sin \beta_j t y_{j_1}) + (\cos \beta_j t x_{j_2} - \sin \beta_j t y_{j_2})]^2 +$$

$$+ [\varepsilon_1 t (\sin \beta_j t x_{j_1} + \cos \beta_j t y_{j_1}) + (\sin \beta_j t x_{j_2} + \cos \beta_j t y_{j_2})]^2 + \ldots$$

$$\ldots + \left[\sum_{\nu=1}^{\rho_j} \frac{(\varepsilon_1 t)^{\rho_j - \nu}}{(\rho_j - \nu)!} (\cos \beta_j t x_{j_\nu} - \sin \beta_j t y_{j_\nu}) \right]^2 +$$

$$\left. + \left[\sum_{\nu=1}^{\rho_j} \frac{(\varepsilon_1 t)^{\rho_j - \nu}}{(\rho_j - \nu)!} (\sin \beta_j t x_{j_\nu} + \cos \beta_j t y_{j_\nu}) \right]^2 \right\}.$$

However,

$$(\cos\beta_j t x_{j_1} - \sin\beta_j t y_{j_1})^2 + (\sin\beta_j t x_{j_1} + \cos\beta_j t y_{j_1})^2 = a_{j_1}^2 + b_{j_1}^2 = x_{j_1}^2 + y_{j_1}^2,$$

$$[\varepsilon_1 t(\cos\beta_j t x_{j_1} - \sin\beta_j t y_{j_1}) + (\cos\beta_j t x_{j_2} - \sin\beta_j t y_{j_2})]^2 +$$

$$+ [\varepsilon_1 t(\sin\beta_j t x_{j_1} + \cos\beta_j t y_{j_1}) + (\sin\beta_j t x_{j_2} + \cos\beta_j t y_{j_2})]^2 =$$

$$= [\varepsilon_1 t\, a_{j_1} + a_{j_2}]^2 + [\varepsilon_1 t b_{j_1} + b_{j_2}]^2 \leqslant$$

$$\leqslant \left(a_{j_2}^2 + \frac{\varepsilon_1 t}{1!} a_{j_1}^2\right) e^{\varepsilon_1 t} + \left(b_{j_2}^2 + \frac{\varepsilon_1 t}{1!} b_{j_1}^2\right) e^{\varepsilon_1 t} =$$

$$= e^{\varepsilon_1 t}\left[(x_{j_2}^2 + y_{j_2}^2) + \frac{\varepsilon_1 t}{1!}(x_{j_1}^2 + y_{j_1}^2)\right],$$

etc. Thus, it is possible to set

$$\| e^{D_j t} x_j \|^2 \leqslant e^{2\alpha_j t} e^{\varepsilon_1 t}\left[(x_{j_1}^2 + y_{j_1}^2)\left(1 + \frac{\varepsilon_1 t}{1!} + \ldots + \frac{(\varepsilon_1 t)^{\rho_j - 1}}{(\rho_j - 1)!}\right) + \right.$$

$$+ (x_{j_2}^2 + y_{j_2}^2)\left(1 + \frac{\varepsilon_1 t}{1!} + \ldots + \frac{(\varepsilon_1 t)^{\rho_j - 2}}{(\rho_j - 2)!}\right) + \ldots$$

$$\left. \ldots + \left((x_{j\rho_j}^2 + y_{j\rho_j}^2)\right)\right] \leqslant e^{2(\alpha_j + \varepsilon_1) t} \| x_j \|^2. \qquad (30.7)$$

The inequalities (30.6) and (30.7) yield the estimate

$$\| e^{At} x \|^2 = \| \{e^{D_1 t} x_1, e^{D_2 t} x_2, \ldots, e^{D_m t} x_m\} \|^2 =$$

$$= \| e^{D_1 t} x_1 \|^2 + \| e^{D_2 t} x_2 \|^2 + \ldots + \| e^{D_m t} x_m \|^2 \leqslant$$

$$\leqslant e^{2(\mathrm{Re}\,\mu_1 + \varepsilon_1)t} \| x_1 \|^2 + e^{2(\mathrm{Re}\,\mu_2 + \varepsilon_1)t} \| x_2 \|^2 + \ldots$$

$$\ldots + e^{2(\mathrm{Re}\,\mu_m + \varepsilon_1)t} \| x_m \|^2 \leqslant$$

$$\leqslant e^{2(\varepsilon_1 + \max_j \mathrm{Re}\,\mu_j)t} (\| x_1 \|^2 + \ldots + \| x_m \|^2) = e^{2(\varepsilon_1 + \gamma)t} \| x \|^2,$$

which completes the proof of the lemma.

Further, consider the function $f(x_1, \ldots, x_n) = f(x)$, defined for x, belonging to a bounded closed domain D. Let $r_1, \ldots, r_n$ be non-negative integers.

The function $f(x)$ is said to belong to the space $C_0^{r_1, \ldots, r_n}(D)$, if in the domain D it has all derivatives whose order in every argument x_i does not exceed r_i, and if those derivatives are continuous in D. Evidently, if $C^r(D)$ is the space of functions of variables $x_1, \ldots, x_n$, r-times continuously differentiable in D, then

$$C^r(D) = \bigcap_{r_1 + \ldots + r_n = r} \overset{0}{C}{}^{r_1, \ldots, r_n}(D). \tag{30.8}$$

The following lemma is now proved as an auxiliary proposition.

Lemma 14. *Let $F(\varphi_1, \ldots, \varphi_m, x_1, \ldots, x_n)$ be a function, 2π-periodic in $\varphi_1, \ldots, \varphi_m$ and belonging to the space $\overset{0}{C}{}^{l_1, \ldots, l_m, r_1, \ldots, r_n}(D)$; let $F^{(k)}(x) = F^{(k_1, \ldots, k_m)}(x_1, \ldots, x_n)$ be its Fourier coefficients and let $k_{\beta_1}, \ldots, k_{\beta_j}$ be non-zero coordinates of the vector $k = (k_1, \ldots, k_m)$.*

Then, $F^{(k)}(x) \in C_0^{r_1, \ldots, r_n}(D)$, and

$$\left| \frac{\partial^\alpha F^{(k)}(x)}{\partial x_1^{\alpha_1} \ldots \partial x_n^{\alpha_n}} \right| \leqslant \min_{\substack{0 \leqslant l_1' \leqslant l_{\beta_1} \\ \vdots \\ 0 \leqslant l_j' \leqslant l_{\beta_j}}} \left[|k_{\beta_1}|^{-l_1'} \ldots |k_{\beta_j}|^{-l_j'} \times \times \max_{\varphi, x} \left| \frac{\partial^{l_1' + \ldots + l_j' + \alpha} F(\varphi, x)}{\partial \varphi_{\beta_1}^{l_1'} \ldots \partial \varphi_{\beta_j}^{l_j'} \partial x_1^{\alpha_1} \ldots \partial x_n^{\alpha_n}} \right| \right], \tag{30.9}$$

where

$$\alpha = \alpha_1 + \ldots + \alpha_n, \quad 0 \leqslant \alpha_\nu \leqslant r_j, \qquad \nu = 1, 2, \ldots, n.$$

This lemma characterizes the rate of decrease with respect to x of the derivatives of the Fourier coefficients of the function $F(\varphi, x)$ as $|k| \to \infty$. For $l_1 = l_2 = \ldots = l_m = l$ the relation (30.9), in particular, yields

$$\left| \frac{\partial^\alpha F^{(k)}(x)}{\partial x_1^{\alpha_1} \ldots \partial x_n^{\alpha_n}} \right| \leqslant (\max_i |k_i|)^{-1} \max_{i, \varphi, x} \left| \frac{\partial^{l+\alpha} F(\varphi, x)}{\partial \varphi_i^l \partial x_1^{\alpha_1} \ldots \partial x_n^{\alpha_n}} \right|,$$

whenever $|k| \neq 0$.

Proof. Since the function $[\partial^\alpha F(\varphi, x)]/(\partial x_1^{\alpha_1} \ldots \partial x_n^{\alpha_n})$ is continuous in the variables φ, $x \in D$, we get the identity

$$\frac{1}{(2\pi)^m} \int_0^{2\pi} \ldots \int_0^{2\pi} \frac{\partial^\alpha F(\varphi, x)}{\partial x_1^{\alpha_1} \ldots \partial x_n^{\alpha_n}} e^{-i(k, \varphi)} d\varphi_1 \ldots d\varphi_m =$$

$$= \frac{1}{(2\pi)^m} \frac{\partial^\alpha}{\partial x_1^{\alpha_1} \ldots \partial x_n^{\alpha_n}} \int_0^{2\pi} \ldots \int_0^{2\pi} F(\varphi, x) e^{-i(k, \varphi)} d\varphi_1 \ldots d\varphi_m =$$

$$= \frac{\partial^\alpha F^{(k)}(x)}{\partial x_1^{\alpha_1} \ldots \partial x_n^{\alpha_n}}, \tag{30.10}$$

with $F^{(k)}(x) \in C_0^{r_1, \ldots, r_n}(D)$, and

$$F^{(k)}(x) = \frac{1}{(2\pi)^m} \int_0^{2\pi} \ldots \int_0^{2\pi} F(\varphi, x)\, e^{-i(k,\varphi)}\, d\varphi_1 \ldots d\varphi_m.$$

Integrating the left-hand side of (30.10) by parts, we have

$$\frac{\partial^\alpha F^{(k)}(x)}{\partial x_1^{\alpha_1} \ldots \partial x_n^{\alpha_n}} = \frac{i^{-(l_1' + \ldots + l_j')}}{(2\pi)^m k_{\beta_1}^{l_1'} k_{\beta_2}^{l_2'} \ldots k_{\beta_j}^{l_j'}} \int_0^{2\pi} \ldots \int_0^{2\pi} \frac{\partial^{l_1' + \ldots + l_j' + \alpha} F(\varphi, x)}{\partial \varphi_{\beta_1}^{l_1'} \ldots \partial \varphi_{\beta_j}^{l_j'} \partial x_1^{\alpha_1} \ldots \partial x_n^{\alpha_n}} \times$$

$$\times\, e^{-i(k,\varphi)}\, d\varphi_1 \ldots d\varphi_m,$$

implying

$$\left| \frac{\partial^\alpha F^{(k)}(x)}{\partial x_1^{\alpha_1} \ldots \partial x_n^{\alpha_n}} \right| \leqslant \min_{\substack{0 \leqslant l_1' \leqslant l_{\beta_1} \\ \vdots \\ 0 \leqslant l_j' \leqslant l_{\beta_j}}} \left[|k_{\beta_1}|^{-l_1'} \ldots |k_{\beta_j}|^{-l_j'} \right] \times$$

$$\times \max_{\varphi, x} \left| \frac{\partial^{l_1' + \ldots + l_j' + \alpha} F(\varphi, x)}{\partial \varphi_{\beta_1}^{l_1'} \ldots \partial \varphi_{\beta_j}^{l_j'} \partial x_1^{\alpha_1} \ldots \partial x_n^{\alpha_n}} \right|.$$

This inequality completes the proof of Lemma 14.

Further, let the functions $F(\varphi, x) = (F_1, \ldots, F_n)$ and $f(\varphi, x) = (f_1, \ldots, f_m)$ be 2π-periodic in $\varphi = (\varphi_1, \ldots, \varphi_m)$ and defined for $\| x \| \leqslant \eta$; let A be an $n \times n$ matrix in real canonical form; let $\mu = (\mu_1, \ldots, \mu_n)$ be the eigenvalues of A and let $\omega = (\omega_1, \ldots, \omega_m)$.

Suppose that the functions $F(\varphi, x)$ and $f(\varphi, x)$ belong to the space

$$C_0^{l,\tau}(\|x\| \leqslant \eta) = C_0^{l, \ldots, l, \tau, \ldots, \tau}(\| x \| \leqslant \eta) \qquad (l \geqslant m + 2, \quad \tau \geqslant 2),$$

and satisfy the condition

$$F(\varphi, 0) = \left. \frac{\partial F(\varphi, x)}{\partial x} \right|_{x=0} = f(\varphi, 0) = 0. \tag{30.11}$$

Consider the solutions $[u = (u_1, u_2, \ldots, u_n];\ w = (w_1, w_2, \ldots, w_m)]$, 2π-periodic in φ, of the systems of equations

$$\frac{\partial u}{\partial x} Ax + \left(\frac{\partial u}{\partial \varphi}, \omega \right) = Au + F(\varphi, x), \tag{30.12}$$

$$\frac{\partial w}{\partial x} Ax + \left(\frac{\partial w}{\partial \varphi}, \omega \right) = f(\varphi, x), \tag{30.13}$$

which satisfy the condition

$$u(\varphi, 0) = \left.\frac{\partial u(\varphi, x)}{\partial x}\right|_{x=0} = w(\varphi, 0) = 0. \tag{30.14}$$

For the case of functions $F(\varphi, x)$ and $f(\varphi, x)$, analytic in the domain $|\operatorname{Im} \varphi| < \rho$, $\| x \| < \eta$, the existence proof for and the determination of such solutions are considered in [6]. It is shown there that the systems (30.12) and (30.13) have periodic solutions in the class of analytic functions when the eigenvalues of the matrix A and the frequencies ω satisfy the inequality (29.7).

We shall now investigate the existence problem for periodic solutions of the systems (30.12) and (30.13) for smooth functions, but we first consider two simple examples.

Example 1. For the system of equations

$$\frac{\partial w}{\partial x_1} \mu_1 x_1 + \frac{\partial w}{\partial x_2} \mu_2 x_2 + \left(\frac{\partial w}{\partial \varphi}, \omega\right) = x_1^\alpha x_2^\beta \sum_{(k)} f_k e^{i(k, \varphi)}, \tag{30.15}$$

the expression

$$w = x_1^\alpha x_2^\beta \sum_{(k)} \frac{f_k}{\alpha\mu_1 + \beta\mu_2 + i(k, \omega)} e^{i(k, \varphi)} \tag{30.16}$$

represents a formal periodic solution. For real μ_1 and μ_2, satisfying the condition

$$\alpha\mu_1 + \beta\mu_2 \neq 0, \tag{30.17}$$

(30.16) is the solution sought for any non-negative α and β ($\alpha \geqslant 2$, $\beta \geqslant 2$, when α and β are non-integers). The relation (30.17) is plausible for arbitrary α, β provided that $\mu_1 \mu_2 > 0$. Consequently, it is to be expected that for the real parts of the eigenvalues of matrix A to be negative (positive) the requirement is just the condition that ensures the existence of periodic solutions of system (30.13) with the requisite properties, for any function $f(\varphi, x)$.

Example 2. Consider the system of equations

$$\sum_{\alpha=1}^{n} \frac{\partial u^{(j)}}{\partial x_\alpha} \mu_\alpha x_\alpha + \left(\frac{\partial u^{(j)}}{\partial \varphi}, \omega\right) = \mu_j u^{(j)} + x_1^{\alpha_1} \dots x_n^{\alpha_n} \sum_k F_k e^{i(k,\varphi)}$$

$$(j = 1, \dots, n; \quad \alpha_1 + \dots + \alpha_n \geqslant 2, \quad \alpha_j \geqslant 2 \text{ for } \alpha_j \neq [\alpha_j]), \tag{30.18}$$

with its formal periodic solution

$$u^{(j)} = x_1^{\alpha_1} \dots x_n^{\alpha_n} \sum_k \frac{F_k^{(j)}}{(\alpha, \mu) - \mu_j + i(k, \omega)} e^{i(k,\varphi)}. \tag{30.18'}$$

For real $\mu_1, \ldots, \mu_n$ the expression (30.18′) is the desired periodic solution of system (30.18), whenever

$$(\alpha, \mu) - \mu_j \neq 0 \quad (j = 1, \ldots, n). \tag{30.19}$$

The inequalities (30.19) remain valid without any additional restrictions on the magnitude of α, provided that

$$\mu_i + \mu_j - \mu_\beta < 0, \quad \text{or } \mu_i + \mu_j - \mu_\beta > 0 \quad (i, j, \beta = 1, \ldots, n). \tag{30.20}$$

With the notation $\mu^0 = \max_i |\mu_i|$, $\mu_0 = \min_i |\mu_i|$, the inequalities (30.20) are equivalent to

$$\mu_i - \mu_j > 0, \qquad \frac{\mu^0}{\mu_0} < 2 \qquad (i, j = 1, \ldots, n). \tag{30.21}$$

The relations (30.21) imply that the condition (30.19) is satisfied for every α ($\alpha_1 + \ldots + \alpha_n \geqslant 2$, $\alpha_j > 2$ with $\alpha_j \neq [\alpha_j]$), whenever the eigenvalues of the matrix A lie in either of the half-open intervals $-2\mu_0 < \mu_i \leqslant \mu_0$, or $\mu_0 \leqslant \mu_i < < 2\mu_0$.

Obviously, the inequalities

$$\operatorname{Re}[\mu_i + \mu_j - \mu_\beta] < 0, \quad \text{or} \quad \operatorname{Re}[\mu_i + \mu_j - \mu_\beta] > 0 \qquad (i, j, \beta = 1, \ldots, n) \tag{30.22}$$

also ensure the existence of periodic solutions of the system (30.12).

The use of the notation

$$\mu^0 = \max_i |\operatorname{Re} \mu_i|, \quad \mu_0 = \min_i |\operatorname{Re} \mu_i|$$

gives the inequalities

$$\operatorname{Re} \mu_i \operatorname{Re} \mu_j > 0, \quad \frac{\mu^0}{\mu_0} < 2 \qquad (i, j = 1, \ldots, n), \tag{30.23}$$

equivalent to the inequalities (30.22). From this it follows that the conditions (30.22) are satisfied for every α, whenever the eigenvalues of the matrix A lie either in the strip $-2\mu_0 < \operatorname{Re} \mu_i \leqslant -\mu_0$, or $\mu_0 \leqslant \operatorname{Re} \mu_i < 2\mu_0$ in the complex μ-plane.

The assertions derived below highlight the role played by the conditions (30.19) and (30.23) vis-a-vis the periodic solutions of the equations under consideration. The starting point is the following lemma.

Lemma 15. *For the system* (30.13) *to have a unique periodic solution, satisfying the condition* (30.14), *it is necessary and sufficient that*

$$\operatorname{Re} \mu_i \operatorname{Re} \mu_j > 0 \tag{30.24}$$

for all $i, j = 1, \ldots, n$.

Proof. We first show the necessity of the condition (30.24).

Suppose that a solution of the system of equations (30.13), which satisfies the condition (30.14), exists and is unique. Then, obviously, $w = 0$ is a unique solution of the equations

$$\frac{\partial w}{\partial x} Ax = 0. \tag{30.25}$$

satisfying the condition

$$w(0) = 0. \tag{30.26}$$

In proving the necessity of conditions (30.24) it is, consequently, enough to show that their violation implies the violation of the uniqueness of the solution of system (30.25), satisfying $w(0) = 0$. Hence, we assume that $\operatorname{Re} \mu_1 = 0$, or $\operatorname{Re} \mu_1 > 0$, $\operatorname{Re} \mu_2 < 0$, and show that the trivial solution of system (30.25) is not unique among the solutions satisfying the condition (30.26).

Let μ_1 and μ_2 be real numbers, then system (30.25) has the form

$$\frac{\partial w}{\partial x_1} \mu_1 x_1 + \ldots + \frac{\partial w}{\partial x_\rho} \mu_2 x_\rho + \ldots = 0. \tag{30.27}$$

For $\mu_1 = 0$, the system (30.27) is satisfied by the functions $w = w^0 = x_1^2$, and for $\mu_1 > 0$, $\mu_2 < 0$ by the functions

$$w = w^0 = \begin{cases} x_1^{-\mu_2 d} \quad x_\rho^{\mu_1 d}, & \text{if } x_1 \geqslant 0, \quad x_\rho \geqslant 0, \\ 0, & \text{if } x_1 < 0, \quad x_\rho < 0, \end{cases}$$

where d is an arbitrary positive number.

Let μ_1 be real and $\mu_2 = \alpha + i\beta$ complex, then the system of equations (30.25) assumes the form

$$\frac{\partial w}{\partial x_1} \mu_1 x_1 + \ldots + \frac{\partial w}{\partial x_\rho} (\alpha x_\rho - \beta y_\rho) + \frac{\partial w}{\partial y_\rho} (\beta x_\rho + \alpha y_\rho) + \ldots = 0,$$

and its solution, for $\mu_1 > 0$ and $\alpha \leqslant 0$, is the function

$$w = w^0 = x_1^{2|\alpha| d} \, (x_\rho^2 + y_\rho^2)^{\mu_1 d}.$$

If $\mu_1 = \alpha_1 + i\beta_1$ and $\mu_2 = \alpha_2 + i\beta_2$ are both complex, then the system is of the form

$$\frac{\partial w}{\partial x_1} (\alpha_1 x_1 - \beta_1 y_1) + \frac{\partial w}{\partial y_1} (\beta_1 x_1 + \alpha_1 y_1) + \ldots + \frac{\partial w}{\partial x_\rho} (\alpha_2 x_\rho - \beta_2 y_\rho) +$$

$$+ \frac{\partial w}{\partial y_\rho} (\beta_2 x_\rho + \alpha_2 y_\rho) + \ldots = 0,$$

and its solution is given by the functions

$$w = w^0 = x_1^2 + y_1^2\,,$$

and
$$w = w^0 = (x_1^2 + y_1^2)^{-\alpha_2 d}(x_\rho^2 + y_\rho^2)^{\alpha_1 d},$$

respectively for $\alpha_1 = 0$ and $\alpha_1 > 0$, $\alpha_2 < 0$.

For large d the function w^0 belongs to the space $C^{\tau_1}(\| x \| \leqslant \eta)$ with every $\tau_1 < \infty$ and satisfies (30.26). Consequently, $w = w^0$ is a non-trivial solution of system (30.25), satisfying the condition (30.26), a contradiction proving the necessity of the conditions (30.24) for the existence of a unique periodic solution of the system (30.13) with the properties indicated in the lemma.

The sufficiency of the conditions (30.24) for the existence of such a solution is proved in the next lemma.

Lemma 16. *Suppose that the eigenvalues of the matrix A are real and distinct. Then, for the system of equations* (30.12) *to have a unique periodic solution in the space* $C^{\bar{l},2}$ $(\| x \| \leqslant \eta)$, $\bar{l} \geqslant 1$, *satisfying the condition* (30.14), *it is necessary and sufficient that*

$$\mu_\alpha + \mu_\beta - \mu_j < 0, \quad \textit{or} \quad \mu_\alpha + \mu_\beta - \mu_j > 0$$

for all $\alpha, \beta, j = 1, \ldots, n$.

This lemma is less general than its predecessor. It shows that the violation of the conditions (30.23) leads to the fact that the system (30.12) either has no periodic solution or has more than one solution (for a definite form of the matrix A). From the proof of Lemma 16 it will become clear that larger the 'dispersion' of the eigenvalues of matrix A, i.e. the larger the ratio μ^0/μ_0, the higher is the order τ_1 of the space $C^{\bar{l},\tau_1}(\| x \| \leqslant \eta)$ in which the desired solution of the system (30.12) is not unique.

Proof of Lemma 16. Suppose that the system (30.12) has a unique periodic solution, satisfying the condition (30.14). Then, obviously, the trivial solution $u = 0$ is a unique solution of equations

$$\frac{\partial u}{\partial x} Ax = Au, \tag{30.28}$$

satisfying

$$u(0) = \left.\frac{\partial u(x)}{\partial x}\right|_{x=0} = 0. \tag{30.29}$$

Set $|\mu_1| = \max\limits_i |\mu_i|$, $|\mu_2| = \min\limits_i |\mu_i|$. Then, by inspection of the structure of the matrix A, the system (30.28) can be expressed in the form

$$\frac{\partial u_i}{\partial x_i}\mu_1 x_1 + \ldots + \frac{\partial u_i}{\partial x_\rho}\mu_2 x_\rho + \ldots = \mu_i u_i \quad (i = 1, \ldots, n). \tag{30.30}$$

Suppose that the conditions of this lemma are not satisfied. This is plausible if the eigenvalues of the matrix A have different signs, say, $\mu_1 > 0$, $\mu_2 \leqslant 0$ or $\mu_i\mu_j > 0$, $\mu^0/\mu_0 \geqslant 2$ for $i, j = 1, \ldots, n$.

For $\mu_2 = 0$ the solution of the system (30.30) is $u_2^0 = x_\rho^2$, $u_1^0 = u_3^0 = \ldots = u_n^0 = 0$; for $\mu_1 > 0$, $\mu_2 < 0$ it is

$$u_1^0 = \begin{cases} x_1^d x_\rho^{(d-1)\frac{\mu_1}{-\mu_2}}, & \text{if } x_1 \geqslant 0, \quad x_\rho \geqslant 0, \\ 0 \quad, & \text{if } x_1 < 0, \ x_\rho < 0, \ u_2^0 = \ldots = u_n^0 = 0, \end{cases} \tag{30.31}$$

and for $\mu_i\mu_j > 0$, $\mu^0/\mu_0 \geqslant 2$ $(i, j = 1, \ldots, n)$ it is

$$u_1^0 = \begin{cases} x_\rho^{\frac{\mu_1}{\mu_2}}, & \text{if } x_\rho \geqslant 0, \\ 0, & \text{if } x_\rho < 0, \ u_2^0 = \ldots = u_n^0 = 0. \end{cases} \tag{30.32}$$

The function $u = u^0$, obviously, satisfies the condition (30.29). The system (30.28) has a non-trivial solution satisfying (30.30) at least in the space $C^{[\mu_1/\mu_2]}$ $(\| x \| \leqslant \eta)$. This contradicts the uniqueness and hence establishes the necessity of the condition (30.11) for the existence in $C^{\bar{l},2}$ $(\| x \| \leqslant \eta)$ of a unique periodic solution, satisfying the condition (30.14).

The sufficiency of the condition remains to be proved. For this we first recall some notations. We denote by $\rho = (\rho_1, \ldots, \rho_m)$ and $r = (r_1, \ldots, r_n)$ respectively, m- and n-dimensional vectors with non-negative integral-valued coordinates

$$| \rho | = \sum_{\alpha=1}^{m} \rho_\alpha, \quad | r | = \sum_{\beta=1}^{n} r_\beta,$$

$$D_\varphi^\rho D_x^r = D_{\varphi_1}^{\rho_1} \ldots D_{\varphi_m}^{\rho_m} D_{x_1}^{r_1} \ldots D_{x_n}^{r_n} = \frac{\partial^{|\rho|+|r|}}{\partial\varphi_1^{\rho_1} \ldots \partial\varphi_m^{\rho_m} \partial x_1^{r_1} \ldots \partial x_n^{r_n}},$$

and introduce the notation

$$| u(x) | = \max_i | u_i(x) |, \quad | D_x^r u(x) | = \max_i | D_x^r u_i(x) |.$$

Lemma 17. *Suppose that the functions $F(\varphi, x)$ and $f(\varphi, x)$ are 2π-periodic in φ, belong to the space $C_0^{l,\tau}$ $(\| x \| \leqslant \eta)$, where $\tau \geqslant 2$, $l = m + 1 + \sigma$, and satisfy the relations* (30.11). *Suppose also that the eigenvalues of the matrix A satisfy the inequalities*

$$\operatorname{Re}[\mu_\alpha + \delta(\mu_\beta - \mu_j)] \leqslant -\gamma \quad (\delta = 0, 1) \tag{30.33}$$

for any $\alpha, \beta, j = 1, \ldots, n$ and some constant $\gamma > 0$.

Then, the system of equations

$$\frac{\partial u}{\partial x} Ax + \left(\frac{\partial u}{\partial \varphi}, \omega\right) = \delta\,[Au + F(\varphi, x)] + (1 - \delta) f(\varphi, x) \tag{30.34}$$

has a solution

$$u^{(\delta)}(\varphi, x) = \overline{\{\delta F(\varphi, x) + (1 - \delta) f(\varphi, x)\}} = \sum_{|k|=-\infty}^{\infty} u_k^{(\delta)}(x)\, e^{i(k,\varphi)},$$

2π-periodic in φ, belonging to the space $C_0^{\sigma,\tau}$ ($\| x \| \leqslant \eta$) and satisfying the conditions

$$u^{(1)}(\varphi, 0) = \left.\frac{\partial u^{(1)}(\varphi, x)}{\partial x}\right|_{x=0} = u^{(0)}(\varphi, 0) = 0\,; \tag{30.35}$$

$$| u^{(\delta)}(\varphi, x) | \leqslant a_0 \Big[\max_{\substack{i, |r_1|=1+\delta \\ \|x\|\leqslant\eta}} | D^l_{\varphi_i} D^{r_1}_x F^{(\delta)}(\varphi, x) | + \max_{\substack{|r_1|=1+\delta \\ \|x\|\leqslant\eta}} | D^{r_1}_x F^{(\delta)}(\varphi, x) | \Big] \tag{30.36}$$

$$| D^\rho_\varphi D^r_x u^{(\delta)}(\varphi, x) | \leqslant a_1 \Big[\max_{\substack{i, |r_1|=1+\delta \\ \|x\|\leqslant\eta}} | D^l_{\varphi_i} D^{r_1}_x F^{(\delta)}(\varphi, x) | + \max_{\substack{|r_1|=1+\delta \\ \|x\|\leqslant\eta}} | D^{r_1}_x F^{(\delta)}(\varphi, x) | \Big]$$

for $| r | = 1$, $| \rho | \leqslant \sigma$, and

$$| D^\rho_\varphi D^r_x u^{(\delta)}(\varphi, x) | \leqslant a_2 \Big[\max_{\substack{i, |r_1|=|r| \\ \|x\|\leqslant\eta}} | D^l_{\varphi_i} D^{r_1}_x F^{(\delta)}(\varphi, x) | + \max_{\substack{|r_1|=|r| \\ \|x\|\leqslant\eta}} | D^{r_1}_x F^{(\delta)}(\varphi, x) | \Big] \tag{30.37}$$

for $2 \leqslant | r | \leqslant \tau$, $| \rho | \leqslant \sigma$, where a_0, a_1, a_2 are positive constants, not depending on $F^{(\delta)}(\varphi, x)$.

It is quite obvious that for $\delta = 0$, $F^{(0)}(\varphi, x) = f(\varphi, x)$ and the system of equations (30.34) is identical to the system (30.13); for $\delta = 1$, $F^{(1)}(\varphi, x) = Au + F(\varphi, x)$, and the system (30.34) coincides with (30.12).

Proof of Lemma 17. For $\delta = 0$ or $\delta = 1$, the lemma is proved in an identical manner. Its proof is, therefore, derived for the case $\delta = 1$, i.e. for the system

$$\frac{\partial u}{\partial x} Ax + \left(\frac{\partial u}{\partial \varphi}, \omega\right) = Au + F(\varphi, x). \tag{30.38}$$

The matrix A has been taken to be in real canonical form. Let ε_1 be one of its parameters, ranging over numbers different from zero. Suppose that ε_1 is positive, fixed and satisfies the inequality

$$\varepsilon_1 < \gamma. \tag{30.39}$$

With these restrictions on the matrix A, the estimate of Lemma 13 for the function $e^{At} x$, namely

$$\| e^{At} x \| \leqslant e^{(\varepsilon_1 + \max_j \operatorname{Re} \mu_j) t} \| x \| \qquad (t \geqslant 0),$$

holds, implying that

$$\| e^{At} x \| \leqslant e^{(\varepsilon_1 - \gamma) t} \| x \| \leqslant \| x \| \tag{30.40}$$

for every $t \geqslant 0$.

Taking note of (30.40), we see that x can be replaced by $e^{At} x$ and φ by $\varphi_t = \omega t + \varphi$ in system (30.38). Then, equations (30.38) are replaced by

$$\frac{\partial u (\varphi_t, e^{At} x)}{dt} = Au (\varphi_t, e^{At} x) + F (\varphi_t, e^{At} x), \qquad t \geqslant 0, \tag{30.41}$$

whose solution, taken at the point $t = 0$, satisfies the original equations (30.38).

If, for $t \geqslant 0$, the integral

$$\int_t^\infty e^{A(t-\tau)} F (\varphi_\tau, e^{A\tau} x) \, d\tau$$

exists, then the expression

$$u(\varphi_t, e^{At} x) = - \int_t^\infty e^{A(t-\tau)} F (\varphi_\tau, e^{A\tau} x) \, d\tau, \qquad t \geqslant 0 \tag{30.42}$$

can be taken as a solution of the system (30.41). Putting $t = 0$,

$$u (\varphi, x) = - \int_0^\infty e^{-A\tau} F (\varphi_t, e^{A\tau} x) \, d\tau. \tag{30.43}$$

It will be shown that the function $u (\varphi, x)$, as defined by (30.43), exists in the space $C_0^{l, \tau}$ ($\| x \| \leqslant \eta$), satisfies the condition (30.35) as well as the inequalities (30.36), (30.37) and is a solution of the system of equations (30.38).

Since $F (\varphi, x) \in C_0^{l, 2}$ ($\| x \| \leqslant \eta$) and $F (\varphi, 0) = \left. \dfrac{\partial F (\varphi, x)}{\partial x} \right|_{x=0} = 0,$

by Taylor's formula, we can write

$$F(\varphi_\tau, e^{A\tau}x) = \frac{1}{2}\sum_{\alpha, \beta} \frac{\partial^2 F(\varphi_\tau, e^{A\tau}x)}{\partial x_\alpha \partial x_\beta} \{e^{A\tau}x\}_\alpha \{e^{A\tau}x\}_\beta, \tag{30.44}$$

where $\{e^{A\tau}x\}_j$ is the j-th coordinate of the vector $e^{A\tau}x$.

Substituting (30.44) in (30.43), we obtain

$$u^{(j)}(\varphi, x) = -\frac{1}{2}\int_0^\infty \sum_{\nu, \alpha, \beta} \{e^{-A\tau}\}_{j\nu} \frac{\partial^2 F^{(\nu)}(\varphi_\tau, e^{A\tau}x)}{\partial x_\alpha \partial x_\beta} \{e^{A\tau}x\}_\alpha \{e^{A\tau}x\}_\beta \, d\tau \tag{30.45}$$

$$(j = 1, \ldots, n).$$

However, for the $j\nu$-th element of the matrix $e^{A\tau}$, we have the estimate

$$|\{e^{A\tau}\}_{j\nu}| \leqslant P_\nu^{(j)}(\tau)\, e^{\mathrm{Re}\,\mu_j \tau},$$

where $P_\nu^{(j)}(\tau)$ is a polynomial in τ of degree less than n.

Taking note of this estimate, we see that (30.45) implies the inequalities

$$|u^{(j)}(\varphi, x)| \leqslant \frac{1}{2} \max_{\substack{\alpha, \beta, \nu \\ \|x\| \leqslant \eta}} \left| \frac{\partial^2 F^{(\nu)}(\varphi, x)}{\partial x_\alpha \partial x_\beta} \right| \int_0^\infty \sum_{\alpha, \beta, \nu} |\{e^{-A\tau}\}_{j\nu} \times$$

$$\times \{e^{A\tau}x\}_\alpha \{e^{A\tau}x\}_\beta | \, d\tau \leqslant \frac{1}{2} \max_{\substack{\alpha, \beta \\ \|x\| \leqslant \eta}} \left| \frac{\partial^2 F(\varphi, x)}{\partial x_\alpha \partial x_\beta} \right| +$$

$$+ \int_0^\infty \sum_{\alpha, \beta} \left[\sum_{\nu=1}^n P_\nu^{(j)}(\tau) \sum_{\nu=1}^n P_\nu^{(\alpha)}(\tau) \sum_{\nu=1}^m P_\nu^{(\beta)}(\tau) \right] \times$$

$$\times e^{\mathrm{Re}[-\mu_{\rho_j} + \mu_{\rho_\alpha} + \mu_{\rho_\beta}]\tau} \eta^2 \tau \leqslant$$

$$\leqslant \frac{a\eta^2}{2} \max_{\substack{|r|=2 \\ \|x\| \leqslant \eta}} |D_x^r F(\varphi, x)| \int_0^\infty P(\tau)\, e^{-\gamma\tau} d\tau \leqslant$$

$$\leqslant a_0 \max_{\substack{|r|=2 \\ \|x\| \leqslant \eta}} |D_x^r F(\varphi, x)|. \tag{30.46}$$

The inequalities (30.46) signify that the integral (30.43) converges uniformly; consequently, for $\|x\| \leqslant \eta$, there is a function $u(\varphi, x)$ continuous in x and φ and satisfying the inequalities (30.36) for $|\rho| + |r| = 0$. Moreover, by (30.43), the function $u(\varphi, x)$ is 2π-periodic in φ and satisfies the condition $u(\varphi, 0) = 0$.

We now recall that

$$\frac{\partial^{|\rho|}F(\varphi, 0)}{\partial\varphi_1^{\rho_1}\dots\partial\varphi_m^{\rho_m}} = \left.\frac{\partial^{|\rho|+1}F(\varphi, x)}{\partial\varphi_1^{\rho_1}\dots\partial\varphi_m^{\rho_m}\partial x_\alpha}\right|_{x=0} = 0$$

for $|\rho| \leqslant l$, $\alpha = 1, \dots, n$ and make use of expansion (30.44) for the functions

$$\frac{\partial^{|\rho|}F(\varphi, x)}{\partial\varphi_1^{\rho_1}\dots\partial\varphi_m^{\rho_m}}, \quad \frac{\partial^{|\rho|+1}F(\varphi, x)}{\partial\varphi_1^{\rho_1}\dots\partial\varphi_m^{\rho_m}\partial x_\alpha},$$

to arrive at the estimate

$$|D_\varphi^\rho D_x^r F(\varphi_\tau, e^{A\tau}x)| \leqslant a \max_{\substack{|r_1|=2\\ \|x\|\leqslant\eta}} |D_\varphi^\rho D_x^{r_1} F(\varphi, x)| \times \times P(\tau)\, e^{\max\limits_{\alpha,\beta} \mathrm{Re}[\mu_\alpha + \mu_\beta]\tau}, \tag{30.47}$$

where $P(\tau)$ is some polynomial in τ, $|r| \leqslant 1$.

The estimate (30.47) demonstrates the convergence of the integral

$$\int_0^\infty e^{-A\tau} D_\varphi^\rho D_x^r \{F(\varphi_\tau, e^{A\tau}x)\}\, d\tau. \tag{30.48}$$

Consequently, the function $u(\varphi, x)$ belongs to the space $C_0^{l,1}$ ($\|x\| \leqslant \eta$) and the inequality (30.36) holds for its derivatives. The expression

$$\frac{\partial u(\varphi, x)}{\partial x} = -\int_0^\infty e^{-A\tau} \frac{\partial F(\varphi_\tau, e^{A\tau}x)}{\partial y} e^{A\tau} d\tau$$

implies the relation $(\partial u(\varphi, x)/\partial x)|_{x=0} = 0$, and the convergence of the integral (30.48) yields the inequality

$$|D_\varphi^\rho D_x^r u(\varphi, x)| = \left| \int_0^\infty e^{-A\tau} D_\varphi^\rho D_x^r \{F(\varphi_\tau, e^{A\tau}x)\}\, d\tau \right| \leqslant$$

$$\leqslant a \max_{\substack{|r_1|=|r|\\ \|x\|\leqslant\eta}} |D_\varphi^\rho D_x^{r_1} F(\varphi, x)| \int_0^\infty P(\tau) \times$$

$$\times e^{\max\limits_{\alpha,\beta,j} \mathrm{Re}[\mu_\alpha + \mu_\beta - \mu_j]\tau} d\tau \leqslant$$

$$\leqslant a_0 \max_{\substack{|r_1|=|r|\\ \|x\|\leqslant\eta}} |D_\varphi^\rho D_x^{r_1} F(\varphi, x)|.$$

For $|\rho| \leqslant l$ and $2 \leqslant |r| \leqslant \tau$, it is obvious that the function $u(\varphi, x)$ belongs to the space $C_0^{l,\tau} \|x\| \leqslant \eta)$ and satisfies the inequality (30.37).

It is not difficult to deduce that the function $u(\varphi, x)$ defined by (30.43) is, indeed, a solution of equations (30.38).

Thus, Lemma 17 is proved.

It is to be shown that, subject to the fulfilment of condition (30.33), the system (30.34) has, in the space $C_0^{l,1+\delta}$ $(\|x\| \leqslant \eta)$, a unique periodic solution, satisfying the condition (30.35).

In fact, suppose that for $\delta = 1$ the system of equations (30.34) has, in the space $C_0^{l,2}$ $(\|x\| \leqslant \eta)$, two solutions with the properties indicated. Then, their difference will also be a solution of the system (30.34), satisfying the condition 30.35), i.e. of the system (30.12) obeying the condition (30.14). However, every solution of the system (30.12) satisfies the relation

$$u(\varphi_t, e^{At} x) = e^{At} u(\varphi, x) \qquad \text{for } t \geqslant 0. \tag{30.49}$$

We differentiate (30.49) to obtain

$$e^{-At} \frac{\partial u(\varphi_t, e^{At}x)}{\partial y} e^{At} = \frac{\partial u(\varphi, x)}{\partial x}, \qquad t \geqslant 0,$$

whence, by applying Taylor's formula, we have

$$e^{-At} \sum_{\alpha=1}^{n} \frac{\partial u'_y(\varphi_t, e^{At}\tilde{x})}{\partial y_\alpha} \{e^{At}x\}_\alpha e^{At} = \frac{\partial u(\varphi, x)}{\partial x}.$$

Passage to the limit $t \to \infty$ in the last expression gives $\partial u(\varphi, x)/\partial x = 0$, i.e.

$$u(\varphi, x) = u(\varphi, 0) = 0,$$

contradicting the assumption of the existence of two solutions.

Similarly, it can be established that the desired solution of (30.34) for $\delta = 0$, i.e. of system (30.13), is unique.

Consider now the system (30.34) for $\delta = 1$, assuming $F(\varphi, x)$ to be a sufficiently smooth function, expressed by

$$F(\varphi, x) = \sum_{i=2}^{\tau_0} \frac{1}{i!} \left(x_1 \frac{\partial}{\partial x_1} + \ldots + x_n \frac{\partial}{\partial x_n} \right)^i F(\varphi, 0) + F_1(\varphi, x) =$$

$$= \sum_{|r| \geqslant 2}^{\tau_0} a_r(\varphi) x_1^{r_1} \ldots x_n^{r_n} + F_1(\varphi, x), \tag{30.50}$$

where $2 \leqslant \tau_0 < \tau$, $F_1(\varphi, x) = F(\varphi, x) - \Sigma\, a_r x_1^{r_1} \ldots x_n^{r_n}$.

The solution $u(\varphi, x)$ of equations (30.34) is representable in the form

$$u(\varphi, x) = u_1(\varphi, x) + \sum_{|r| \geqslant 2}^{\tau_0} u_r(\varphi, x), \tag{30.51}$$

where $u_1(\varphi, x)$ is a solution of the system of equations (30.12) for $F(\varphi, x) = F_1(\varphi, x)$ and $u_r(\varphi, x)$ is a solution of the system

$$\frac{\partial u_r}{\partial x} Ax + \left(\frac{\partial u_r}{\partial \varphi}, \omega\right) = Au_r + a_r(\varphi) x_1^{r_1} \dots x_n^{r_n}. \tag{30.52}$$

Assume, as before, that

$$\operatorname{Re} \mu_\alpha \leqslant -\gamma < 0$$

for every $\alpha = 1, \dots, n$; $\gamma > 0$.

The function $u_1(\varphi, x)$ can be defined by the expression

$$u_1(\varphi, x) = -\int_0^\infty e^{-A\tau} F_1(\varphi_\tau, e^{A\tau} x)\, d\tau. \tag{30.53}$$

Since $F_1(\varphi, x)$, as well as $F(\varphi, x)$, belong to the space $C_0^{l,\tau}(\| x \| \leqslant \eta)$ and satisfy the condition

$$F_1(\varphi, 0) = \left.\frac{\partial F_1(\varphi, x)}{\partial x_{\alpha_1}}\right|_{x=0} = \dots = \left.\frac{\partial^{\tau_0} F_1(\varphi, x)}{\partial x_{\alpha_1} \dots \partial x_{\alpha_{\tau_0}}}\right|_{x=0} = 0 \tag{30.54}$$

for all $\alpha_1, \dots, \alpha_{\tau_0} = 1, \dots, n$, the integral (30.53) converges uniformly, provided that

$$\operatorname{Re}[(r, \mu) - \mu_j] \leqslant = -\gamma \qquad (\gamma > 0) \tag{30.55}$$

for all $j = 1, \dots, n$ and $|r| = \tau_0 + 1$. What is more, the arguments deduced above make it easy to establish that the function $u_1(\varphi, x)$ is 2π-periodic in φ, belongs to the space $C_0^{l,\tau}(\| x \| \leqslant \eta)$ and satisfies the conditions

$$u_1(\varphi, 0) = \left.\frac{\partial u_1(\varphi, x)}{\partial x_{\alpha_1}}\right|_{x=0} = \dots = \left.\frac{\partial^{\tau_0} u_1(\varphi, x)}{\partial x_{\alpha_1} \dots \partial x_{\alpha_{\tau_0}}}\right|_{x=0} = 0, \tag{30.56}$$

and the estimate

$$|D_\varphi^\rho D_x^r u_1(\varphi, x)| \leqslant a_0 \max_{\substack{|r_1| = |\bar{r}| \\ \| x \| \leqslant \eta}} |D_\varphi^\rho D_x^{r_1} F(\varphi, x)| \tag{30.57}$$

for $|\rho| \leqslant l$, $|r| \leqslant \tau$, where

$$\bar{r} = \begin{cases} \tau_0 + 1 & \text{for } |r| \leqslant \tau_0, \\ |r| & \text{for } \tau_0 < |r| \leqslant \tau. \end{cases}$$

We now consider the system of equations defining the functions $u_r(\varphi, x)$. The solution of system (30.52) is sought in the form

$$u_r(\varphi, x) = \sum_{|k|} u_r^{(k)}(x)\, e^{i(k,\varphi)}. \tag{30.58}$$

Expanding the coefficients $a_r(\varphi)$ that appear on the right-hand side of (30.50) in the Fourier series, i.e. writing

$$a_r(\varphi) = \sum_{|k|} a_r^{(k)} e^{i(k,\varphi)}, \tag{30.59}$$

and substituting (30.58) and (30.59) in (30.52), we obtain a system of equations for $u_r^{(k)}(x)$:

$$\frac{\partial u}{\partial x} Ax + i(k, \omega)\, u = Au + a_r^{(k)} x_1^{r_1} \dots x_n^{r_n}. \tag{30.60}$$

Since $A = \{D_1, \dots, D_\nu\}$, the system of equations (30.60) decomposes into the systems

$$\frac{\partial u}{\partial x} Ax = [D_1 - i(k, \omega)]\, u + a_r^{(k)} x_1^{r_1} \dots x_n^{r_n}, \tag{30.61}$$

where $u = (u_1, \dots, u_{\rho_1})$ is a ρ_1-dimensional vector, ρ_1 being the dimension of the matrix D_1.

If the eigenvalue μ_1 of the matrix A is real, then the system (30.52) assumes the form

$$\begin{aligned} \frac{\partial u_1}{\partial x} Ax &= [\mu_1 - i(k, \omega)]\, u_1 + a_{r_1}^{(k)} x_1^{r_1} \dots x_n^{r_n}, \\ \frac{\partial u_j}{\partial x} Ax &= \varepsilon_1 u_{j-1} + [\mu_1 - i(k, \omega)]\, u_j + a_{r_j}^{(k)} x_1^{r_1} \dots x_n^{r_n}. \end{aligned} \tag{30.62}$$

The solution of the first equation of the system (30.62) is sought in terms of V, a homogeneous form of order $|r|$. Let $b_1, \dots, b_N$ be the coefficients of V. Substituting V in the first equation of (30.62) and equating the coefficients, we obtain for b_j ($j = 1, 2, \ldots, N$), the system of linear non-homogeneous equations

$$[B - \mu_1 + i(k, \omega)]\, b = b_0; \tag{30.63}$$

here B is a matrix whose entries are linear combinations of the coefficients of the matrix A:

$$b = (b_1, \ldots, b_N), \quad b_0 = \{0, \ldots, 0, a_{r_1}^{(k)}, 0, \ldots, 0\}.$$

The roots of the equation [40]

$$D^0_{|r|}(\lambda) = 0$$

are defined by the formula

$$\lambda = m_1\mu_1 + \ldots + m_n\mu_n = (m, \mu),$$

where $m_1, \ldots, m_n$ are any non-negative integers, related by the dependence

$$|m| = m_1 + \ldots + m_n = |r|.$$

Assuming that

$$|(m, \mu) - \mu_1 + i(k, \omega)| \neq 0 \qquad \text{for } |m| = |r|,$$

and solving the system of equations (30.63) for b, we find that

$$b = [B - \mu_1 + i(k, \omega)]^{-1} b_0. \tag{30.64}$$

Hence, we find a unique form of $|r|$-th order, which satisfies the first equation of (30.62).

Having obtained the form V, we consider the second equation of the system (30.62) and define u_2 from it as an $|r|$-th order form, and so on.

If the eigenvalue μ_1 of the matrix A is complex: $\mu_1 = \alpha_1 + i\beta_1$, then the system (30.62) takes the form

$$\begin{aligned} \frac{\partial u_1}{\partial x} Ax &= [\alpha_1 - i(k, \omega)]\, u_1 - \beta_1 u_j + a_{r_1}^{(k)} x_1^{r_1} \ldots x_n^{r_n}, \\ \frac{\partial u_j}{\partial x} Ax &= \beta_1 u_{j-1} + [\alpha_1 - i(k, \omega)]\, u_j + a_{r_j}^{(k)} x_1^{r_1} \ldots x_n^{r_n}, \\ & j = 2, 3, \ldots, \rho_1, \end{aligned} \tag{30.65}$$

and the change of variables

$$v_1 = u_1 + iu_2, \quad v_2 = iu_1 + u_2$$

reduces the first two equations of this system to the equations

$$\begin{aligned} \frac{\partial v_1}{\partial x} Ax &= [\alpha_1 + i\beta_1 - i(k, \omega)]\, v_1 + (a_{r_1}^{(k)} + ia_{r_2}^{(k)})\, x_1^{r_1} \ldots x_n^{r_n}, \\ \frac{\partial v_2}{\partial x} Ax &= [\alpha_1 - i\beta_1 - i(k, \omega)]\, v_2 + (ia_{r_1}^{(k)} + a_{r_2}^{(k)})\, x_1^{r_1} \ldots x_n^{r_n}, \end{aligned} \tag{30.66}$$

each of which has the form of the first equation of (30.62) and is solved independently.

Assuming that

$$|(m, \mu) - \alpha_1 \pm i\beta_1 + i(k, \mu)| \neq 0 \qquad \text{for } |m| = |r|,$$

we can find $|r|$-th order forms, satisfying the system (30.66) and, consequently, the first two equations of the system (30.65). Having found u_1 and u_2, the next two equations of system (30.65) can be considered and u_3 and u_4 determined as $|r|$-th order forms, etc.

Suppose that

$$|(m, \mu) - \mu_j + i(k, \omega)| \geqslant \varepsilon(|m| + |k|)^{-d} \tag{30.67}$$

for $j = 1, \ldots, n$, $2 \leqslant |m| \leqslant \tau$ and some $\varepsilon > 0$, $d > 0$. Then, the system (30.60) has a unique solution

$$u_r^{(k)}(x) = \sum_{|l|=|r|} b_l^{(k)} x_1^{l_1} \ldots x_n^{l_n}, \tag{30.68}$$

where $b_l^{(k)}$ are the coefficients of terms of $|r|$-th order forms, satisfying system (30.61).

It is to be shown now that the condition (30.67) guarantees not only the existence of solutions $u_r^{(k)}(x)$ but also the convergence of the series (30.58).

Considering the solution of the system (30.62) we observe that the coefficients of the form V, satisfying the first equation of (30.62), are defined by the formula (30.64) and the coefficients of the form $b^{(j)}$, satisfying the j-th equation of system (30.62) are expressible in terms of $a_{r_j}^{(k)}, a_{r_{j-1}}^{(k)}, \ldots, a_{r_1}^{(k)}$, by the formula

$$b^{(j)} = [B - \mu_1 + i(k, \omega)]^{-1} (\varepsilon_1 b^{(j-1)} + b_0^{(j)}) =$$

$$= [B - \mu_1 + i(k, \omega)]^{-1} \sum_{\alpha=0}^{j-1} \varepsilon_1^{\alpha} [B - \mu_1 + i(k, \omega)]^{-\alpha} b_0^{(j-\alpha)}, \tag{30.69}$$

$$j = 2, \ldots, \rho_1,$$

where $b_0^{(j)} = \{0, 0, \ldots, 0, a_{r_j}^{(k)}, 0, \ldots, 0\}$.

To evaluate the matrix $[B - \mu_1 + i(k, \omega)]^{-1}$, we have

$$|[B - \mu_1 + i(k, \omega)]^{-1} \leqslant \frac{|B_1((k, \omega))|}{|D^0(\mu_1 - i(k, \omega))|} \leqslant$$

$$\leqslant \frac{c(|(k, \omega)|^{N-1} + 1)}{\prod_{|m|=|r|} [(m, \mu) - \mu_1 + i(k, \omega)]} \leqslant$$

$$\leqslant \frac{c_1(|k| + |r|)^{N-1}}{\varepsilon^N(|r| + |k|)^{-Nd}} = c_1(\varepsilon)(|r| + |k|)^{N(d+1)-1},$$

where $B_1((k, \omega))$ is a matrix made up of the cofactors of the matrix $[B - \mu_1 + i(k, \omega)]$. This inequality yields the estimate

$$|b^{(j)}| \leqslant c_2(\varepsilon)(|r| + |k|)^{\rho_1[N(d+1)-1]}|a_r^{(k)}| \tag{30.70}$$

for $j = 1, 2, \ldots, \rho_1$.

Considering the system (30.65) we are soon convinced of the validity of an estimate, similar to (30.70), for the coefficients of the $|r|$-th order form satisfying the system (30.65). The above assertions imply that if ρ_0 is the highest order of the Jordan block components of the matrix A, then

$$|b_i^{(k)}| \leqslant c_2(\varepsilon)(|r| + |k|)^{\rho_0[N_{|r|}(d+1)-1]}|a_r^{(k)}|. \tag{30.71}$$

The constants $a_r^{(k)}$, the Fourier coefficients of the function $a_r(\varphi)$, are defined to within an integral factor by the function $\left.\dfrac{\partial^{|r|}F(\varphi, x)}{\partial x_1^{r_1}\ldots\partial x_n^{r_n}}\right|_{x=0}$.

Since $F(\varphi, x) \in C_0^{l,\tau}(\|x\| \leqslant \eta)$, then by the assertions of the lemma

$$|a_r^{(k)}| \leqslant c_3\left(\max_i |k_i|\right)^{-l} \max_{\substack{i,\,|r_1|=|r|\\ \|x\|\leqslant\eta}} |D_{\varphi_i}^l D_x^{r_1} F(\varphi, x)| \quad (|k| \neq 0).$$

This inequality implies the estimate

$$|b_i^{(k)}| \leqslant c_2(\varepsilon)\, c_3(|r| + |k|)^{d(|r|)}\left(\max_i |k_i|\right)^{-l} \max_{\substack{i,\,|r_1|=|r|\\ \|x\|\leqslant\eta}} |D_{\varphi_i}^l D_x^{r_1} F(\varphi, x)| \leqslant$$

$$\leqslant c_4(\varepsilon, |r|)\left(\max_i |k_i|\right)^{-l+d(|r|)} \max_{\substack{i,\,|r_1|=|r|\\ \|x\|\leqslant\eta}} |D_{\varphi_i}^l D_x^{r_1} F(\varphi, x)|,$$

$$(|k| \neq 0) \tag{30.72}$$

with the notation

$$d(|r|) = \rho_0[N_{|r|}(d + 1) - 1]. \tag{30.73}$$

Suppose that

$$l = m + d(|r|) + \sigma + 1,$$

then

$$
\begin{aligned}
|u_r(\varphi, x)| &\leqslant \sum_{(|k|)} |u_r^{(k)}(x)| \leqslant \sum_{(|k|)} \sum_{|l|=|r|} |b_l^{(k)}|\,|x_1^{l_1} \ldots x_n^{l_n}| \leqslant \\
&\leqslant c_5 \sum_{(|k|)} |b_l^{(k)}| \leqslant c_5 \Bigg[|b_l^{(0)}| + c_4 \max_{\substack{i, |r_1|=|r| \\ \|x\| \leqslant \eta}} |D_{\varphi_i}^l D_x^{r_1} F(\varphi, x)| \times \\
&\times \sum_{|k| \neq 0} \left(\max_i |k_i|^{-l+d} \right) \Bigg] \leqslant c_6 \Bigg[\max_{\substack{|r_1|=|r| \\ \|x\| \leqslant \eta}} |D_x^{r_1} F(\varphi, x)| + \\
&+ \sum_{\nu=1}^{\infty} \nu^{-l+d+m-1} \max_{\substack{i, |r_1|=|r| \\ \|x\| \leqslant \eta}} |D_{\varphi_i}^l D_x^{r_1} F(\varphi, x)| \Bigg] \leqslant \\
&\leqslant c_7 \Bigg[\max_{\substack{|r_1|=|r| \\ \|x\| \leqslant \eta}} |D_x^{r_1} F(\varphi, x)| + \max_{\substack{i, |r_1|=|r| \\ \|x\| \leqslant \eta}} |D_{\varphi_i}^l D_x^{r} F(\varphi, x)| \Bigg].
\end{aligned} \tag{30.74}
$$

Inequality (30.74) implies that the series (30.58) is uniformly convergent. We now examine the series

$$
D_\varphi^\rho u_r(\varphi, x) = \sum_{|k| \neq 0} (ik_1)^{l_1} \ldots (ik_n)^{l_n} u_r^{(k)}(x)\, e^{i(k, \varphi)},
$$

obtained by term-by-term differentiation of series (30.58), and confirm that the function $u_r(\varphi, x)$ has σ continuous derivatives in φ and satisfies the inequality

$$
\begin{aligned}
|D_\varphi^\rho u_r(\varphi, x)| &\leqslant c_5 \sum_{|k| \neq 0} \left(\max_i |k_i| \right)^{|\rho|} |b_l^{(k)}| \leqslant \\
&\leqslant c_6 \sum_{|k| \neq 0} \left(\max_i |k_i| \right)^{|\rho|} \left(\max_i |k_i| \right)^{-l+d} \times \\
&\times \max_{\substack{i, |r_1|=|r| \\ \|x\| \leqslant \eta}} |D_{\varphi_i}^l D_x^{r_1} F(\varphi, x)| \leqslant \\
&\leqslant c_6 \sum_{\nu=1}^{\infty} \nu^{-\sigma+|\rho|-2} \max_{\substack{i, |r_1|=|r| \\ \|x\| \leqslant \eta}} |D_{\varphi_i}^l D_x^{r_1} F(\varphi, x)| \leqslant \\
&\leqslant c_7 \max_{\substack{i, |r_1|=|r| \\ \|x\| \leqslant \eta}} |D_{\varphi_i}^l D_x^{r_1} F(\varphi, x)| \qquad \text{for } |\rho| \leqslant \sigma.
\end{aligned} \tag{30.75}
$$

The above assertions are summed up in the the next theorem.

Theorem 26. *Suppose that*

$$\begin{aligned} &\operatorname{Re}[(r, \mu) - \mu_j] \leqslant -\gamma \qquad \text{when } |r| = \tau_0 + 1, \\ &|(r, \mu) - \mu_j + i(k, \omega)| \geqslant \varepsilon(|r| + |k|)^{-d} \quad \text{when } 2 \leqslant |r| \leqslant \tau_0 \end{aligned} \tag{30.76}$$

for all $j = 1, \ldots, n$, *some integer* τ_0 *and constants* $\varepsilon > 0$, $\gamma > 0$.

If $F(\varphi, x) \in C_0^{l,\tau}(\|x\| \leqslant \eta)$, *and if*

$$\tau > \tau_0, \qquad l = m + d(\tau_0) + \sigma + 1, \qquad \sigma \geqslant 1, \tag{30.77}$$

then the system (30.12) *has a solution* $u(\varphi, x) = L^{(1)} F(\varphi, x)$, 2π-*periodic in* φ *belonging to the space* $C_0^{\sigma,\tau}(\|x\| \leqslant \eta)$ *and satisfying the condition* (30.14) *as well as the inequality*

$$|D_\varphi^\rho D_x^r u(\varphi, x)| \leqslant a_0 \Bigg[\max_{\substack{i,\, 2 \leqslant |r_1| \leqslant \tau_0 \\ \|x\| \leqslant \eta}} |D_{\varphi_i}^l D_x^{r_1} F(\varphi, x)| + \max_{\substack{|r_1| = |\bar{r}| \\ \|x\| \leqslant \eta}} |D_\varphi^\rho D_x^{r_1} F(\varphi, x)| \Bigg] \tag{30.78}$$

for $|\rho| \leqslant \sigma$, $|r| \leqslant \tau$, *where*

$$\bar{r} = \begin{cases} \tau_0 + 1, & \text{when } |r| \leqslant \tau_0, \\ |r| & \text{when } |r| > \tau_0, \end{cases}$$

and a_0 *is a positive constant depending on* ε *and* τ_0.

As before, we write $\mu_0 = \min_j |\operatorname{Re} \mu_j|$, $\mu^0 = \max_j |\operatorname{Re} \mu_j|$. The first condition in (30.76) is equivalent to the inequalities

$$\operatorname{Re} \mu_j < 0, \qquad \frac{\mu^0}{\mu_0} < \tau_0 + 1 \qquad (j = 1, \ldots, n).$$

The eigenvalues of the matrix A satisfying the first condition in (30.76) must have negative real parts, the distance between which does not exceed $(\tau_0+1)\,\mu_0$ units.

The second condition in (30.76) demands that the eigenvalues of the matrix lie either outside the hyperplanes

$$\operatorname{Re}[(r, \mu) - \mu_j] = 0, \qquad 2 \leqslant |r| \leqslant \tau_0, \qquad j = 1, \ldots, n$$

of the eigenvalue space, or on these hyperplanes but exterior to the point

$$\operatorname{Im}[(r, \mu) - \mu_j + (k, \omega)] = 0, \qquad 2 \leqslant |r| \leqslant \tau_0, \qquad j = 1, \ldots, n.$$

§ 31. Iteration Theorem

Consider the s-th step of an iteration process, first for a 'general' and then for the particular iteration process based predominantly on the degree of smoothness of the right-hand side of system (29.1) and the manifold M.

Assume that for the system of equations

$$\frac{dx}{dt} = Ax + F(\varphi, x), \qquad \frac{d\varphi}{dt} = \omega + f(\varphi, x) \tag{31.1}$$

the matrix A and the functions $F(\varphi, x), f(\varphi, x)$ have the properties identified in § 29 and are first order infinitesimals in the domain

$$\| x \| \leqslant \eta. \tag{31.2}$$

It is required to find a change of variable, 2π-periodic in φ, which reduces the system (31.1) to a system of the same form but with the functions $F^{(1)}(\varphi, x)$, $f^{(1)}(\varphi, x)$ of higher order of smallness, for definiteness say, of n-th order, where

$$1 < \varkappa < 2.$$

To solve the problem posed, we set

$$N_s = N_{s-1}^{\varkappa}, \quad \delta_s = \delta_{s-1}^{\varkappa}, \quad \delta_s = N_s^{-\beta} \qquad (s = 1, 2, 3, \ldots). \tag{31.3}$$

Further, we suppose that the constants $\varkappa_1$, β, s_0 and l_0 satisfy the conditions

$$\varkappa = \frac{3}{2}, \quad \beta = \frac{10}{9} + \frac{3}{2}(s_0 + 3), \quad l_0 \geqslant \frac{13}{2} + \frac{9}{2}(s_0 + 3), \quad s_0 \geqslant 7. \tag{31.4}$$

Subject to the choice of the parameters indicated, we have the following theorem.

Inductive Theorem 27. *For the system* (31.1), *suppose that the eigenvalues of the matrix A satisfy the inequality*

$$\operatorname{Re}\,[\mu_\alpha + \mu_\beta - \mu_j] \leqslant -\gamma < 0 \tag{31.5}$$

for all $\alpha, \beta, j = 1, 2, \ldots, n$ and that the functions $F(\varphi, x)$ and $f(\varphi, x)$ are l_0-times continuously differentiable with respect to φ and x for $\| x \| \leqslant \eta$ and satisfy the condition

$$F(\varphi, 0) = \left.\frac{\partial F(\varphi, x)}{\partial x}\right|_{x=0} = f(\varphi, 0) = 0 \tag{31.6}$$

and the inequalities

$$| F(\varphi, x) | + | f(\varphi, x) | \leqslant \delta_{s-1} \qquad (\| x \| \leqslant \eta), \tag{31.7}$$

$$|D_\varphi^\rho D_x^r F(\varphi, x) | + | D_\varphi^\rho D_x^r f(\varphi, x) | \leqslant N_{s-1}^{l_0} \tag{31.8}$$

for $| \rho | + | r | = l_0$.

Then, for sufficiently small δ_0, *there exists a change of variables*

$$\begin{aligned} x &= y + Y(\theta, y), \\ \varphi &= \theta + \Phi(\theta, y) \end{aligned} \tag{31.9}$$

with the functions $Y(\theta, y)$ *and* $\Phi(\theta, y)$, 2π*-periodic in* θ, *defined in the domain*

$$\| y \| \leqslant \eta - N_{s-1}^{-1}, \tag{31.10}$$

and satisfying the condition

$$Y(\theta, 0) = \left. \frac{\partial Y(\theta, y)}{\partial y} \right|_{y=0} = \Phi(\theta, 0) = 0 \tag{31.11}$$

and the inequality

$$| Y(\theta, y) |_{s_0} + | \Phi(\theta, y) |_{s_0} \leqslant N_{s-1}^{-1}, \tag{31.12}$$

such that in the variables y, θ *the system* (31.1) *assumes the form*

$$\begin{aligned} \frac{dy}{dt} &= Ay + F^{(1)}(\theta, y), \\ \frac{d\theta}{dt} &= \omega + f^{(1)}(\theta, y), \end{aligned} \tag{31.13}$$

where the functions $F^{(1)}(\theta, y)$ *and* $f^{(1)}(\theta, y)$, 2π*-periodic in* θ, *are* l_0*-times continuously differentiable with respect to* θ *and* y *in the domain* (31.10) *and satisfy the condition*

$$F^{(1)}(\theta, 0) = \left. \frac{\partial F^{(1)}(\theta, y)}{\partial y} \right|_{y=0} = f^{(1)}(\theta, 0) = 0, \tag{31.14}$$

and the inequalities

$$| F^{(1)}(\theta, y) | + | f^{(1)}(\theta, y) | \leqslant \delta_s \quad (\| x \| \leqslant \eta - 2N_{s-1}^{-1}), \tag{31.15}$$

$$| D_\theta^\rho D_y^r F^{(1)}(\theta, y) | + | D_\theta^\rho D_y^r f^{(1)}(\theta, y) | \leqslant N_s^{l_0} \quad \text{for } | \rho | + | r | = l_0. \tag{31.16}$$

Proof. We set

$$Y(\theta, y) = L^{(1)} T^{1}_{N_s N_s} F(\theta, y),$$
$$\Phi(\theta, y) = L^{(0)} T^{0}_{N_s N_s} f(\theta, y), \tag{31.17}$$

where $\overset{1}{T}_{N_s N_s}$, $\overset{0}{T}_{N_s N_s}$ are the smoothing operators $\overset{1}{T}_{NM}$ and $\overset{0}{T}_{NM}$ $(N = M = N_s)$, examined in § 13, and $L^{(1)}$, $L^{(0)}$ are the operators introduced in Lemma 17.

Taking account of the properties of smoothing operators $\overset{1}{T}_{NM}$ and $\overset{0}{T}_{NM}$ established in Lemma 3 and the assertions of Lemma 17, we infer that the functions $Y(\theta, y)$ and $\Phi(\theta, y)$, defined in the domain

$$\| y \| \leqslant \eta - N_s^{-1} \tag{31.18}$$

and 2π-periodic in θ, are continuously differentiable any number of times with respect to θ and y, and satisfy the condition (31.11). What is more, by manipulating the inequalities (30.57) and (31.8) we find that the estimates for $Y(\theta, y)$ and its derivatives are given by

$$| D_\theta^\rho D_y^r Y(\theta, y) | \leqslant a_0 \max_{\substack{|r_1| = |\bar{r}| \\ \|y\| \leqslant \eta}} | D_\theta^\rho D_y^{r_1} T^{(1)}_{N_s N_s} F(\theta, y) | \leqslant a_0 C N^{|\rho| + |\bar{r}| + 1} \delta_{s-1} =$$
$$= a_0 C N_{s-1}^{\varkappa(|\rho| + |\bar{r}| + 1) - \beta}, \tag{31.19}$$
$$|\bar{r}| = \begin{cases} 2 & \text{for } |r| \leqslant 1, \\ |r| & \text{for } |r| \geqslant 2, \end{cases}$$

and for $\Phi(\theta, y)$ and its derivatives by

$$| D_\theta^\rho D_y^r \Phi(\theta, y) | \leqslant a_0 \max_{|r_1| = |\bar{r}|} | D_\theta^\rho D_y^{r_1} T^{(0)}_{N_s N_s} f(\theta, y) | \leqslant$$
$$\leqslant a_0 C N_s^{|\rho| + |\bar{r}|} \delta_{s-1} = a_0 C N_{s-1}^{\varkappa(|\rho| + |\bar{r}|) - \beta}, \tag{31.20}$$
$$|\bar{r}| = \begin{cases} 1 & \text{for } |r| = 0 \\ |r| & \text{for } |r| \geqslant 1. \end{cases}$$

For sufficiently small δ_0, inequalities (31.19), (31.20) and the relations (31.3), (31.4) imply the estimate (31.12), given by

$$| Y(\theta, y) |_{s_0} + | \Phi(\theta, y) |_{s_0} \leqslant 2 a_0 C N_{s-1}^{\varkappa(s_0 + 3) - \beta} =$$
$$= 2 a_0 C N_{s-1}^{-\frac{1}{9}} N_{s-1}^{-1} \leqslant N_{s-1}^{-1}. \tag{31.21}$$

Differentiating (31.9) and using the formulas (31.17), we find that

$$\left(E+\frac{\partial Y}{\partial y}\right)\left(\frac{dy}{dt}-Ay\right)+\frac{\partial Y}{\partial \theta}\left(\frac{d\theta}{dt}-\omega\right)=$$
$$=F(\theta+\Phi, y+Y)-T^1F(\theta, y),$$
$$(31.22)$$
$$\frac{\partial \Phi}{\partial y}\left(\frac{dy}{dt}-Ay\right)+\left(E+\frac{\partial \Phi}{\partial t}\right)\left(\frac{d\theta}{dt}-\omega\right)=$$
$$=f(\theta+\Phi, y+Y)-T^0f(\theta, y).$$

Solving the system (31.22) in $\left(\frac{dy}{dt}-Ay\right)$, $\frac{d\theta}{dt}-\omega$, which is always possible because of (31.21), we obtain the system of equations

$$\frac{dy}{dt}=Ay+F^{(1)}(\theta, y),$$

$$\frac{d\theta}{dt}=\omega+f^{(1)}(\theta, y),$$

with the notation

$$\begin{pmatrix} F^{(1)}(\theta, y) \\ f^{(1)}(\theta, y)\end{pmatrix}=\begin{pmatrix} E+\frac{\partial Y}{\partial y}, & \frac{\partial Y}{\partial \theta} \\ \frac{\partial \Phi}{\partial y}, & E+\frac{\partial \Phi}{\partial \theta}\end{pmatrix}^{-1}\begin{pmatrix} F(\theta+\Phi, y+Y)-T^1F(\theta, y) \\ f(\theta+\Phi, y+Y)-T^0f(\theta, y)\end{pmatrix}. \tag{31.23}$$

The functions $F^{(1)}(\theta, y)$ and $f^{(1)}(\theta, y)$, defined in the domain (31.18), are 2π-periodic in θ and l_0-times continuously differentiable and, as is easily verified, satisfy the condition (31.14).

To establish the inequalities (31.15) and (31.16), we use equation (31.23) to obtain

$$|F^{(1)}(\theta, y)|+|f^{(1)}(\theta, y)|\leqslant$$
$$\leqslant c_1[|F(\theta+\Phi, y+Y)-T^1F(\theta, y)|+|f(\theta+\Phi, y+Y)-T^0f(\theta, y)|]\leqslant$$
$$\leqslant c_1[|F(\theta+\Phi, y+Y)-T^1F(\theta+\Phi, y+Y)|+|f(\theta+\Phi, y+Y)$$
$$-T^0f(\theta+\Phi, y+Y)|+|T^1F(\theta+\Phi, y+Y)-T^1F(\theta, y)|+$$
$$+|T^0f(\theta+\Phi, y+Y)-T^0f(\theta, y)|]. \tag{31.24}$$

By the properties of the smoothing operators and the inequalities (31.7), (31.8),

(31.19) and (31.20), we get

$$\begin{aligned}
&| F(\varphi, x) - T'F(\varphi, x) | + | f(\varphi, x) - T^0 f(\varphi, x) | \leqslant \\
&\leqslant CN_s^{-l_0} \Bigg[\sup_{\substack{|\rho|+|r|=l_0 \\ \|x\| \leqslant \eta}} (N_s | D_\varphi^\rho D_x^r F(\varphi, x) | + | D_\varphi^\rho D_x^r f(\varphi, x) | \Bigg] \leqslant \\
&\leqslant CN_s \left(\frac{N_{s-1}}{N_s} \right)^{-l_0} = CN_{s-1}^{\varkappa + (1-\varkappa) l_0} = CN_{s-1}^{\varkappa(1+\beta) + (1-\varkappa) l_0} \delta_s = \\
&= CN_{s-1}^{-\frac{1}{12}} \delta_s \leqslant \frac{1}{2} \delta_s,
\end{aligned} \tag{31.25}$$

$$\begin{aligned}
&| T^1 F(\varphi, x) - T^1 F(\varphi, y) | + | T^0 f(\varphi, x) - T^0 f(\theta, y) | \leqslant \\
&\leqslant \left| \frac{\partial T^1 F}{\partial \theta} \Phi + \frac{\partial T^1 F}{\partial y} Y \right| + \left| \frac{\partial T^0 f}{\partial \theta} \Phi + \frac{\partial T^0 f}{\partial y} Y \right| \leqslant \\
&\leqslant c_2 [2N_s | \Phi | + N_s (N_s + 1) | Y |] \delta_{s-1} \leqslant \\
&\leqslant c_3 N_s^2 N_{s-1}^2 = c_3 N^{-5\varkappa - (2-\varkappa)\beta} \delta_s < c_3 N_{s-1}^{-1} \delta_s < \frac{1}{2} \delta_s.
\end{aligned}$$

The expression (31.23) together with the inequalities (31.24) and (31.25) implies the estimate (31.15), since

$$| F^{(1)}(\theta, y) | + | f^{(1)}(\theta, y) | \leqslant \delta_s.$$

To obtain the estimate (31.16), we proceed as follows. We consider derivatives of the functions

$$\begin{aligned}
&Y_1(\theta_1, y_1) = N_s Y\left(\frac{\theta_1}{N_s}, \frac{y_1}{N_s} \right), \quad \Phi_1(\theta_1, y_1) = N_s \Phi\left(\frac{\theta_1}{N_s}, \frac{y_1}{N_s} \right), \\
&F_1(\theta_1, y_1) = N_s F\left(\frac{\theta_1}{N_s}, \frac{y_1}{N_s} \right), \quad f_1(\theta_1, y_1) = N_s f\left(\frac{\theta_1}{N_s}, \frac{y_1}{N_s} \right),
\end{aligned} \tag{31.26}$$

defined in the domain

$$| y_1 | \leqslant N_s \eta - 1. \tag{31.27}$$

The structure of the functions $Y_1(\theta_1, y_1)$, $\Phi_1(\theta_1, y_1)$ and inequalities (31.19), (31.20) and (31.7), (31.8) yield

$$\begin{aligned}
&| D_{\theta_1}^\rho D_{y_1}^r Y_1(\theta_1, y_1) | + | D_{\theta_1}^\rho D_{y_1}^r \Phi_1(\theta_1, y_1) | \leqslant \\
&\leqslant N_s^{1-|\rho|-|r|} [| D_\theta^\rho D_y^r Y(\theta, y) | + | D_\theta^\rho D_y^r \Phi(\theta, y) |] \leqslant \\
&\leqslant a_0 C N_s [N_s^{|\bar{r}|+1-|r|} + N_s^{|\bar{r}_1|-|r|}] \delta_{s-1} \leqslant 2a_0 C N_s^4 \delta_{s-1} \leqslant 1
\end{aligned} \tag{31.28}$$

for $|\rho| + |r| \leqslant l_0 + 1$; and

$$| F_1(\theta_1, y_1) | + | f_1(\theta_1, y_1) | \leqslant N_s \delta_{s-1} \leqslant 1, \tag{31.29}$$

$$| D^{\rho}_{\theta_1} D^{r}_{y_1} F_1(\theta_1, y_1) | + | D^{\rho}_{\theta_1} D^{r}_{y_1} f_1(\theta_1, y_1) | \leqslant N_s \left(\frac{N_{s-1}}{N_s} \right)^{l_0} \leqslant 1$$

for $|\rho| + |r| = l_0$.

From (31.29) it follows that the functions $F_1(\theta_1, y_1)$ and $f_1(\theta_1, y_1)$ and their l_0-th order derivatives are bounded by unity. This enables us to infer that the derivatives of any order lower than l_0 satisfy, in the domain (31.27), the inequality

$$| D^{\rho}_{\theta_1} D^{r}_{y_1} F_1(\theta_1, y_1) | + | D^{\rho}_{\theta_1} D^{r}_{y_1} f(\theta_1, y_1) | \leqslant c_4 \tag{31.30}$$

for $|\rho| + |r| \leqslant l_0$, c_4 being a constant not depending on N_s.

On the basis of inequalities (31.28) — (31.30), the functions

$$F_1^{(1)}(\theta_1, y_1) = N_s F^{(1)}\left(\frac{\theta_1}{N_s}, \frac{y_1}{N_s} \right), \quad f_1^{(1)}(\theta_1, y_1) = N_s f^{(1)}\left(\frac{\theta_1}{N_s}, \frac{y_1}{N_s} \right)$$

can be expressed in terms of functions whose derivatives with respect to θ_1, y_1 up to order l_0 inclusive are bounded by a constant not depending on N_s. Hence

$$| D^{\rho}_{\theta_1} D^{r}_{y_1} F_1^{(1)}(\theta_1, y_1) | + | D^{\rho}_{\theta_1} D^{r}_{y_1} f_1^{(1)}(\theta_1, y_1) | \leqslant c_5$$

for $|\rho| + |r| = l_0$. However,

$$D^{\rho}_{\theta_1} D^{r}_{y_1} [F_1^{(1)}(\theta_1, y_1) + f_1^{(1)}(\theta_1, y_1)] = N_s^{1-|\rho|-|r|} D^{\rho}_{\theta} D^{r}_{y} [F^{(1)}(\theta, y) + f^{(1)}(\theta, y)];$$

and the last inequality gives the estimate

$$| D^{\rho}_{\theta} D^{r}_{y} F^{(1)}(\theta, y) | + | D^{\rho}_{\theta} D^{r}_{y} f^{(1)}(\theta, y) | \leqslant N_s^{l_0}$$

for $|\rho| + |r| = l_0$, completing the proof of Theorem 27.

Theorem 27 has been established with the hypothesis that the eigenvalues $\mu_1, \ldots, \mu_n$ of matrix A satisfy the inequality (31.5). This restriction can be considerably weakened by raising the degree of smoothness of the functions $F(\varphi, x)$ and $f(\varphi, x)$. In fact, if we choose the integers l_0 and s_0 satisfying the condition

$$l_0 > \frac{\varkappa}{\varkappa - 1}(\beta + 1), \qquad s_0 \leqslant \frac{\varkappa - 1}{2 - \varkappa} d_0 + \frac{3\varkappa - 2}{\varkappa(2 - \varkappa)}, \tag{31.31}$$

where the constants $\varkappa$, β, d_0 are connected by the relations

$$1 < \varkappa < 2, \quad \beta > 1 + \varkappa (d_0 + s_0), \quad d_0 = m + d(\tau_0) + \tau_0 + 2, \tag{31.32}$$

with $d(\tau_0)$ a constant defined by (30.73) and $\tau_0 > 1$ an integer, it is then trivial to prove the following statement.

Suppose that the system (31.1) *satisfies the hypotheses of Theorem* 27 *with the difference that the eigenvalues of the matrix A satisfy the inequalities* (29.2), *and that l_0, s_0 are defined by* (31.31). *Then the assertions of Theorem* 27 *remain valid in this case also for the system* (31.1).

§ 32. Reducibility Theorem in the Neighbourhood of a Toroidal Manifold

We recall the system of equations

$$\frac{dy}{dt} = Y(y). \tag{32.1}$$

Suppose that in the space $E_{n+m}(y)$ a neighbourhood of the manifold M is expressible as a direct product of the n-dimensional cube $K\{| x_i | \leqslant \eta, i = 1, \ldots, n\}$ and the m-dimensional torus $T\{\varphi_i, i = 1, \ldots, m\}$.

Let $x = 0$ be the equation of manifold M in the coordinate system x, φ. In the neighbourhood of the manifold M, the system (32.1) assumes the form

$$\frac{dx}{dt} = P(\varphi)\, x + G(\varphi, x), \qquad \frac{d\varphi}{dt} = \omega + R(\varphi, x)\,; \tag{32.2}$$

here the matrix $P(\varphi)$ and the vector functions $G(\varphi, x)$ and $R(\varphi, x)$, defined in the domain $\| x \| \leqslant \eta$, are 2π-periodic in φ and sufficiently smooth, and satisfy the condition

$$G(\varphi, 0) = \left.\frac{\partial G(\varphi, x)}{\partial x}\right|_{x=0} = R(\varphi, 0) = 0. \tag{32.3}$$

Suppose that the linear system

$$\frac{dx}{dt} = P(\varphi)\, x, \qquad \frac{d\varphi}{dt} = \omega \tag{32.4}$$

is reducible in the sense that there exists a change of variable

$$x \to C(\varphi)\, x, \tag{32.5}$$

which is non-singular, real, 2π-periodic and sufficiently smooth in φ and reduces (32.4) to the system

$$\frac{dx}{dt} = Ax, \qquad \frac{d\varphi}{dt} = \omega, \tag{32.6}$$

where A is a constant real matrix†.

The change of variable (32.5) reduces (32.2) to the system

$$\frac{dx}{dt} = Ax + F(\varphi, x), \qquad \frac{d\varphi}{dt} = \omega + f(\varphi, x), \tag{32.6}$$

where the functions $F(\varphi, x)$ and $f(\varphi, x)$ have properties similar to those indicated above for the functions $G(\varphi, x)$ and $R(\varphi, x)$.

With the above set-up, we shall investigate the positions of the trajectories of system (32.6) in the neighbourhood of the torus $x = 0$ and establish a series of results for them.

Theorem 28. *Suppose that the vector functions $F(\varphi, x)$ and $f(\varphi, x)$ are 2π-periodic in φ; for given positive $c_0, \gamma, \varepsilon, d, \eta$ and integers $s_0, \tau_0 (s_0, \tau_0 \geqslant 2)$, suppose that there exist a positive $\delta_0 = \delta_0(c_0, \gamma, \varepsilon, \eta, s_0, \tau_0)$ and an integer $l_0 = l_0(s_0, \tau_0)$, such that the functions $F(\varphi, x)$ and $f(\varphi, x)$ are l_0-times continuously differentiable in the domain $\| x \| \leqslant \eta$, and satisfy the condition*

$$F(\varphi, 0) = \left.\frac{\partial F(\varphi, x)}{\partial x}\right|_{x=0} = f(\varphi, 0) = 0 \tag{32.7}$$

as well as the inequalities

$$| F(\varphi, x) | + | f(\varphi, x) | \leqslant \delta_0, \qquad | F(\varphi, x) |_{l_0} + | f(\varphi, x) |_{l_0} \leqslant c_0; \tag{32.8}$$

moreover, let the eigenvalues $\mu = (\mu_1, \ldots, \mu_n)$ of matrix A satisfy the conditions

$$\operatorname{Re} [(r, \mu) - \mu_j)] \leqslant -\gamma \quad \text{when} \quad | r | = \tau_0 + 1,$$
$$| (r, \mu) - \mu_j + i(k, \omega) | \geqslant \varepsilon(| r | + | k |)^{-d} \quad \text{when} \quad 2 \leqslant | r | \leqslant \tau_0 \tag{32.9}$$

†See Appendix 11.

for every $j = 1, 2, \ldots, n$, *all integral-valued* $k = (k_1, \ldots, k_m)$ *and integral-valued non-negative* $r = (r_1, \ldots, r_n)$.

Then, there exist functions $Y(\theta, y)$ *and* $\Phi(\theta, y)$, 2π*-periodic in* θ, $(s_0 - 1)$*-times continuously differentiable in the domain*

$$\| y \| \leqslant \eta - \bar{\eta},$$

satisfying the condition

$$Y(\theta, 0) = \left. \frac{\partial Y(\theta, y)}{\partial y} \right|_{y=0} = \Phi(\theta, 0) = 0, \tag{32.10}$$

and the estimate

$$| Y(\theta, y) |_{s_0-1} + | \Phi(\theta, y) |_{s_0-1} \leqslant \bar{\eta}, \tag{32.11}$$

so that through the change of variables

$$\begin{aligned} x &= y + Y(\theta, y), \\ \varphi &= \theta + \Phi(\theta, y), \end{aligned} \tag{32.12}$$

the system (32.6) *is reducible to the form*

$$\begin{aligned} \frac{dy}{dt} &= Ay, \\ \frac{d\theta}{dt} &= \omega. \end{aligned} \tag{32.13}$$

This theorem implies that a neighbourhood of the manifold $x = 0$ for system (32.6) has the same structure as a neighbourhood of the torus $y = 0$ for the linear system (32.13). However, a neighbourhood of the manifold $y = 0$ for (32.13) is filled by the trajectories

$$\begin{aligned} y &= e^{At} y_0 \\ \theta &= \omega t + \theta_0. \end{aligned} \tag{32.14}$$

Consequently, a neighbourhood of the invariant manifold $x = 0$ for (32.6) is filled by the semi-trajectories

$$\begin{aligned} x &= e^{At} y_0 + Y(\omega t + \theta_0, e^{At} y_0), \\ \varphi &= \omega t + \theta_0 + \Phi(\omega t + \theta_0, e^{At} y_0) \quad (t \geqslant 0, \| y_0 \| \leqslant \eta - \bar{\eta}). \end{aligned} \tag{32.15}$$

It will now be shown that violation of conditions (32.9) destroys reducibility of system (32.6) to the form (32.13), even by means of a twice continuously differentiable substitution. To this end, we consider the system

$$\frac{dx_1}{dt} = 2\mu_1 x_1 + x_2^2 F_1,$$

$$\frac{dx_2}{dt} = \mu_1 x_2, \tag{32.16}$$

$$\frac{d\varphi}{dt} = \omega,$$

whose solutions are

$$x_1 = e^{2\mu t} x_0 + F_1 t e^{2\mu_1 t} y_0^2,$$

$$x_2 = e^{\mu_1 t} y_0, \tag{32.17}$$

$$\varphi = \omega t + \varphi_0.$$

The representation of functions (32.17) in the form (32.15) is possible by means of the vector functions

$$Y(x, y) = \{Y_1(x, y), 0\},$$

where $Y_1(x, y)$ is a solution of the equation

$$\frac{\partial Y_1}{\partial x} 2\mu_1 x + \frac{\partial Y_1}{\partial y} \mu_1 y = 2\mu_1 Y_1 + y^2 F_1, \tag{32.18}$$

and satisfies the condition

$$Y_1(0, 0) = \left.\frac{\partial Y_1(0, x)}{\partial x}\right|_{x=0} = \left.\frac{\partial Y_1(0, y)}{\partial y}\right|_{y=0} = 0.$$

From (32.18) it follows that the function $Y_1(0, y)$ satisfies the equation

$$\frac{dY_1(0, y)}{dy} \mu_1 y = 2\mu_1 Y(0, y) + y^2 F_1,$$

so that

$$Y_1(0, y) = Cy^2 + \frac{F_1}{\mu_1} y^2 \ln | y |.$$

The function $Y_1(0, y)$ has only one continuous derivative, hence for the system (32.16) the substitution (32.12) cannot be twice continuously differentiable.

The proof of Theorem 28, as also of Theorem 25, rests on the construction of functions $Y(\theta, y)$ and $\Phi(\theta, y)$. The inductive theorem plays a key role in this, and so the following proof applies equally to both the theorems.

Proof of Theorem 28. Since the matrix A is real, the system (32.6) can be reduced by the real linear transformation $x \to C(\varphi)$ to a system of the same form with the matrix A in real canonical form and with a parameter ε_1 satisfying the condition (30.39). On effecting such a reduction, we choose δ_0 so small that the inductive Theorem 27 is applicable to the system (32.6) with $s = 1$. This admits the change of variables

$$\begin{aligned} x &= x^{(1)} + Y^{(1)}(\varphi^{(1)}, x^{(1)}), \\ \varphi &= \varphi^{(1)} + \Phi^{(1)}(\varphi^{(1)}, x^{(1)}), \end{aligned} \tag{32.19}$$

reducing the system (32.6) to the system

$$\begin{aligned} \frac{dx^{(1)}}{dt} &= Ax^{(1)} + F^{(1)}(\varphi^{(1)}, x^{(1)}), \\ \frac{d\varphi^{(1)}}{dt} &= \omega + f^{(1)}(\varphi^{(1)}, x^{(1)}), \end{aligned} \tag{32.20}$$

etc.

At the s-th step, the change of variables

$$\begin{aligned} x^{(s-1)} &= x^{(s)} + Y^{(s)}(\varphi^{(s)}, x^{(s)}), \\ \varphi^{(s-1)} &= \varphi^{(s)} + \Phi^{(s)}(\varphi^{(s)}, x^{(s)}) \end{aligned} \tag{32.21}$$

reduces the system of equations in $x^{(s-1)}$ and $\varphi^{(s-1)}$ to the system

$$\begin{aligned} \frac{dx^{(s)}}{dt} &= Ax^{(s)} + F^{(s)}(\varphi^{(s)}, x^{(s)}), \\ \frac{d\varphi^{(s)}}{dt} &= \omega + f^{(s)}(\varphi^{(s)}, x^{(s)}). \end{aligned} \tag{32.22}$$

In addition, by the preceding theorem, the functions $Y^{(s)}$, $\Phi^{(s)}$, $F^{(s)}$ and $f^{(s)}$ are 2π-periodic in $\varphi^{(s)}$ and l_0-times continuously differentiable for

$$\| x^{(s)} \| \leqslant \eta - (N_0^{-1} + N_1^{-1} + \dots + N_{s-1}^{-1}), \tag{32.23}$$

and satisfy the relations

$$\begin{aligned} Y^{(s)}(\varphi, 0) &= \left.\frac{\partial Y^{(s)}(\varphi, x)}{\partial x}\right|_{x=0} = \Phi^{(s)}(\varphi, 0) = F^{(s)}(\varphi, 0) = \\ &= \left.\frac{\partial F^{(s)}(\varphi, x)}{\partial x}\right|_{x=0} = f^{(s)}(\varphi, 0) = 0, \end{aligned} \tag{32.24}$$

$$\begin{gathered} | Y^{(s)}(\varphi, x) |_{s_0} + | \Phi^{(s)}(\varphi, x) |_{s0} \leqslant N_{s-1}^{-1}, \\ | F^{(s)}(\varphi^{(s)}, x^{(s)}) | + | f^{(s)}(\varphi^{(s)}, x^{(s)}) | \leqslant \delta_s, \\ | D_\varphi^\rho D_x^r F^{(s)}(\varphi^{(s)}, x^{(s)}) | + | D_\varphi^\rho D_x^r f^{(s)}(\varphi^{(s)}, x^{(s)}) | \leqslant N_s^{l_0} \quad \text{for } | \rho | + | r | = l_0. \end{gathered} \tag{32.25}$$

From what has been said it follows that the combination of substitutions (32.19)—(32.21), i.e. the substitution

$$x = x^{(s)} + X^{(s)}(\varphi^{(s)}, x^{(s)}),$$
$$\varphi = \varphi^{(s)} + \Psi^{(s)}(\varphi^{(s)}. x^{(s)}), \tag{32.26}$$

reduces the system (32.6) to the system (32.22).

To determine some properties of the functions $X^{(s)}(\varphi^{(s)}, x^{(s)})$ and $\Psi^{(s)}(\varphi^{(s)}, x^{(s)})$, we start from the relations

$$X^{(s+1)}(\theta, y) = Y^{(s+1)}(\theta, y) + X^{(s)}(\theta + \Phi^{(s+1)}(\theta, y), y + Y^{(s+1)}(\theta, y)),$$
$$\Psi^{(s+1)}(\theta, y) = \Phi^{(s+1)}(\theta, y) + \Psi^{(s)}(\theta + \Phi^{(s+1)}(\theta, y), y + Y^{(s+1)}(\theta, y)). \tag{32.27}$$

Let $\| y \| \leqslant \eta - \sum_{v=0}^{s} N_v^{-1}$. We differentiate (32.27) to obtain

$$\sum_{|\rho|+|r|=1} | D_\theta^\rho D_y^r X^{(s+1)}(\theta, y) |_0 \leqslant \sum_{|\rho|+|r|=1} | D_\theta^\rho D_y^r Y^{(s+1)}(\theta, y) |_0 +$$
$$+ \sum_{\beta=1}^{m} \left| \frac{\partial X^{(s)}(\theta + \Phi^{(s+1)}(\theta, y), y + Y^{(s+1)}(\theta, y))}{\partial \theta_\beta} \right|_0 \times$$
$$\times \left(1 + \sum_{\alpha=1}^{m} \left| \frac{\partial \Phi_\beta^{(s+1)}(\theta, y)}{\partial \theta_\alpha} \right|_0 + \sum_{\alpha=1}^{n} \left| \frac{\partial \Phi^{(s+1)}(\theta, y)}{\partial y_\alpha} \right|_0 \right) +$$
$$+ \sum_{\beta=1}^{n} \left| \frac{\partial X^{(s)}(\theta + \Phi^{(s+1)}(\theta, y), y + Y^{(s+1)}(\theta, y))}{\partial y_\beta} \right|_0 \times$$
$$\times \left(1 + \sum_{\alpha=1}^{m} \left| \frac{\partial Y_\beta^{(s+1)}(\theta, y)}{\partial \theta_\alpha} \right|_0 + \sum_{\alpha=1}^{m} \left| \frac{\partial Y_\beta^{(s+1)}(\theta, y)}{\partial y_\alpha} \right|_0 \right).$$

Using (32.25), we see that this implies the estimate

$$\sum_{|\rho|+|r|=1} | D_\theta^\rho D_y^r X^{(s+1)}(\theta, y) |_0 \leqslant$$
$$\leqslant (m + n) N_s^{-1} + \sum_{|\rho|+|r|=1} | D_\theta^\rho D_y^r X^{(s)}(\theta, y) |_0 (1 + (m + n) N_s^{-1}). \tag{32.28}$$

Setting

$$\sum_{\rho|+|r|=1} | D_\theta^\rho D_y^r X^{(s+1)}(\theta, y) |_0 = z_{s+1},$$

we see that the inequality (32.28) assumes the form

$$z_{s+1} + 1 \leqslant (z_s + 1)(1 + (m + n) N_s^{-1}), \quad z_0 = 0, \tag{32.29}$$

which implies, for $\| y \| \leqslant \eta - \sum_{\nu=0}^{s} N_\nu^{-1}$, that

$$\sum_{|\rho|+|r|=1} | D_\theta^\rho D_y^r X^{(s)}(\theta, y) |_0 \leqslant \prod_{0 \leqslant \nu < \infty} (1 + (m + n) N_\nu^{-1}) - 1 = b_1, \tag{32.30}$$

where $b_1 \to 0$, as $N_0^{-1} \to 0$. Similarly, the estimate for $\Psi^{(s)}(\theta, y)$ is given by

$$\sum_{|\rho|+|r|=1} | D_\theta^\rho D_y^r \Psi^{(s)}(\theta, y) |_0 \leqslant b_1. \tag{32.31}$$

We now choose δ_0 so small that the earlier arguments hold and, in addition, the inequality

$$\sum_{\nu=0}^{\infty} N_\nu^{-1} \leqslant \bar{\eta}$$

is satisfied. Then, with the variables θ, y in the domain $\| y \| \leqslant \eta - \tilde{\eta}$ we obtain the estimate

$$\begin{aligned} &| X^{(s+1)}(\theta, y) - X^{(s)}(\theta, y) | + | \Psi^{(s+1)}(\theta, y) - \Psi^{(s)}(\theta, y) | \\ &\quad \leqslant | Y^{(s+1)}(\theta, y) | + | \Phi^{(s+1)}(\theta, y) | + | X^{(s)}(\theta + \Phi^{(s+1)}(\theta, y), y + \\ &\quad + Y^{(s+1)}(\theta, y)) - X^{(s)}(\theta, y) | + \\ &\quad + | \Psi^{(s)}(\theta + \Phi^{(s+1)}(\theta, y), y + Y^{(s+1)}(\theta, y)) - \Psi^{(s)}(\theta, y) | \leqslant \\ &\leqslant 2N_s^{-1} + 2b_1 N_s^{-1} = 2(1 + b_1) N_s^{-1}, \end{aligned}$$

which implies that the following criterion for the uniform convergence of the sequences of functions $X^{(s)}(\theta, y)$ and $\Psi^{(s)}(\theta, y)$ is satisfied :

$$| X^{(s+k_0)}(\theta, y) - X^{(s)}(\theta, y) | + | \Psi^{(s+k_0)}(\theta, y) - \Psi^{(s+k_0)}(\theta, y) | \leqslant$$

$$\leqslant 2(1 + b_1) \sum_{\nu=0}^{k_0-1} N_{s+\nu}^{-1}, \tag{32.32}$$

s and k_0 being any natural numbers.

We now write

$$X^{(\infty)}(\theta, y) = \lim_{s \to \infty} X^{(s)}(\theta, y), \qquad \Psi^{(\infty)}(\theta, y) = \lim_{s \to \infty} \Psi^{(s)}(\theta, y). \tag{32.33}$$

The periodicity of the functions $X^{(s)}(\theta, y)$ and $\Psi^{(s)}(\theta, y)$ in θ implies that the functions $X^{(\infty)}(\theta, y)$ and $\Psi^{(\infty)}(\theta, y)$ are periodic; moreover, (32.24) implies the equality

$$X^{(\infty)}(\theta, 0) = \left.\frac{\partial X^{(\infty)}(\theta, y)}{\partial y}\right|_{y=0} = \Psi^{(\infty)}(\theta, 0) = 0,$$

and (30.32) the estimate

$$|X^{(\infty)}(\theta, y)| + |\Psi^{(\infty)}(\theta, y)| \leqslant 2(1 + b_1)\sum_{\nu=0}^{\infty} N_\nu^{-1} \leqslant \bar{\eta}.$$

Since $|F^{(s)}(\theta, y)| + |f^{(s)}(\theta, y)| \to 0$ uniformly in θ and y as $s \to \infty$, hence in order to complete the proof of Theorem 28 we have only to prove, by setting

$$Y(\theta, y) = X^{(\infty)}(\theta, y), \quad \Phi(\theta, y) = \Psi^{(\infty)}(\theta, y), \tag{32.34}$$

that the functions $X^{(\infty)}(\theta, y)$ and $\Psi^{(\infty)}(\theta, y)$ are $(s - 1)$-times contiuously differentiable and obtain the estimate (32.11) for the derivatives.

Starting with the expressions (32.27) and (32.33), if we have to show that the functions $X^{(\infty)}(\theta, y)$ and $\Psi^{(\infty)}(\theta, y)$ are $(s_0 - 1)$-times continuously differtiable, we need only establish the boundedness of the first s_0 derivatives of the functions $X^{(s)}(\theta, y)$ and $\Psi^{(s)}(\theta, y)$. This we can establish by induction. The inequalities (32.30) and (32.31) imply the boundedness of the first partial derivatives of $X^{(s)}(\theta, y)$ and $\Psi^{(s)}(\theta, y)$. We assume that the derivatives of the functions $X^{(s)}(\theta, y)$ and $\Psi^{(s)}(\theta, y)$ up to order s_1 $(1 \leqslant s_1 < s_0)$ are bounded by a constant $b_{s_1} \to 0$ as $N_0^{-1} \to 0$; it is to be shown that the $(s_1 + 1)$-th derivatives of these functions are also bounded by the constant $b_{s_1+1} \to 0$ as $N_0^{-1} \to 0$.

If we differentiate (32.27) and make use of the estimate (32.25), we obtain

$$\begin{aligned}
&\sum_{|\rho|+|r|=s_1+1} |D_\theta^\rho D_y^r X^{(s+1)}(\theta, y)|_0 \leqslant \\
&\leqslant \sum_{|\rho|+|r|=s_1+1} |D_\theta^\rho D_y^r Y^{(s+1)}(\theta, y)|_0 + \\
&\quad + \sum_{|\rho|+|r|=s_1+1} |D_\theta^\rho D_y^r \{X^{(s)}(\theta + \Phi^{(s+1)}(\theta, y), y + Y^{(s+1)}(\theta, y))\}| \leqslant \\
&\leqslant N_s^{-1} \sum_{|\rho|+|r|=s_1+1} 1 + \sum_{|\rho|+|r|=s_1+1} |D_\varphi^\rho D_x^r X^{(s)}(\varphi, x)|_0 \times \\
&\quad \times \Big(1 + N_s^{-1} \sum_{|\rho|+|r|=s_1+1} 1\Big)^{s_1+1} + \bar{b} N_s^{-1},
\end{aligned} \tag{32.35}$$

where $\bar{b}$ is a constant depending on $b_1, \ldots, b_{s_1}$.

Writing

$$\sum_{|\rho|+|r|=s_1+1} |D_\theta^\rho D_y^r X^{(s+1)}(\theta, y)|_0 = z_{s+1},$$

we see that the inequality (32.35) assumes the form

$$z_{s+1} \leqslant dN_s^{-1} + z_s(1 + dN_s^{-1})^{s_1+1} + \bar{b}N_s^{-1}, \qquad z_0 = 0, \qquad (32.36)$$

where $d = \sum_{|\rho|+|r|=s_1+1} 1.$

Solving inequality (32.36), we find that

$$\bar{z}_s \leqslant (d + \bar{b}) \sum_{\nu=0}^{s-1} N_\nu^{-1}, \qquad (32.37)$$

implying the estimate

$$\sum_{|\rho|+|r|=s_1+1} |D_\theta^\rho D_y^r X^{(s)}(\theta, y)| \leqslant (d + \bar{b}) \sum_{\nu=0}^{\infty} N_\nu^{-1} \prod_{0 \leqslant \nu < \infty} (1 + dN_\nu^{-1})^{s_1+1} =$$

$$= b_{s_1+1}, \qquad (32.38)$$

which, taken together with the estimate

$$\sum_{|\rho|+|r|=s_1+1} |D_\theta^\rho D_y^r \Psi^{(s)}(\theta, y)| \leqslant b_{s_1+1}, \qquad (32.39)$$

obtainable by a similar method, establishes the boundedness of the $(s_1 + 1)$-th partial derivatives of $X^{(s)}(\theta, y)$ and $\Psi^{(s)}(\theta, y)$ by the constant $b_{s_1+1} \to 0$ as $N_0^{-1} \to 0$.

It has been verified that the functions $X^{(\infty)}(\theta, y)$ and $\Psi^{(\infty)}(\theta, y)$ are $(s_0 - 1)$-times differentiable. The inequality (32.11) for the derivatives of $X^{(\infty)}(\theta, y)$ and $\Psi^{(\infty)}(\theta, y)$ follows from the estimates (32.38) and (32.39) by virtue of the relations

$$b_{s_1} \to 0 \quad \text{as } N_0^{-1} \to 0, \qquad s_1 = 1, \ldots, s_0.$$

Thus, Theorems 25 and 28 are completely proved.

By conditions (31.4), the relationship between l_0 and s_0 disscussed in Theorem 25, is defined by the inequality†

$$l_0 \geqslant \frac{9(s_0 + 3) + 13}{2}, \qquad s_0 \geqslant 7. \qquad (32.40)$$

Now, recalling that in a neighbourhood of the equilibrium position $y = 0$

†See Appendix 12.

the system (32.1) reduces to the system (32.6) without loss of smoothness, we see that (32.40) implies, in particular, that for transforming equation (32.1) to the form (32.13) in the neighbourhood of an equilibrium position with the aid of a six-times continuously-differentiable change of variable $y \to y + Y(y)$, it is sufficient that the function $Y(y)$ be 52 times continuously differentiable. The smoothness necessary on the right-hand side of the system (32.1) is the same for its reduction to system (32.13) in a neighbourhood of the cycle $y = y_0(\varphi)$.

The relationship of l_0 to s_0 and τ_0, figuring in Theorem 28, is given by (31.31) and (31.32). For $\varkappa = \frac{3}{2}$, the inequalities (31.31) are satisfied by

$$l_0 \geqslant \frac{9(d_0 + s_0) + 13}{2}, \qquad s_0 \geqslant d_0 + 4, \tag{32.41}$$

where

$$d_0 = m + \rho_0 \left[\frac{n(n+1)\dots(n+\tau_0-1)}{\tau_0!} (d+1) - 1 \right],$$

ρ_0 being the highest order of the Jordan block components of the matrix A.

§ 33. Behaviour Under Perturbation of Integral Curves in the Neighbourhood of an Invariant Manifold

Suppose that the original system of equations (29.1), with a smooth toroidal invariant manifold M, is subjected to a small time-independent perturbation. In this section, we shall discuss the problem of the behaviour of the manifold M and the integral curves in its neighbourhood under such a perturbation. We assume, as before, that in a neighbourhood of the manifold M the unperturbed system (29.1) admits the representation

$$\begin{aligned} \frac{dx}{dt} &= Ax + F_1(\varphi, x), \\ \frac{d\varphi}{dt} &= \lambda + f_1(\varphi, x). \end{aligned} \tag{33.1}$$

This assumption implies that, in a neighbourhood of the manifold M, the perturbed system is representable in the form

$$\begin{aligned} \frac{dx}{dt} &= Ax + F_0(\varphi) + P(\varphi)x + F(\varphi, x), \\ \frac{d\varphi}{dt} &= \lambda + f_0(\varphi) + f(\varphi, x), \end{aligned} \tag{33.2}$$

where $F_0(\varphi)$ and $f_0(\varphi)$ are small functions, 2π-periodic in φ; $P(\varphi)$ is a small matrix, 2π-periodic in φ; the functions $F(\varphi, x)$ and $f(\varphi, x)$ are 2π-periodic in φ and satisfy the condition (32.7).

Since the right-hand side of the system (33.2) is smooth, but not analytic, this system is similar to the system studied in Chaps. 1 and 2. In the preceding Chaps. 3-5 we have discussed, in sufficient detail, the special forms of such a system, defined by the conditions

$$\begin{gathered} f(\varphi, x) = F_0(\varphi) = F(\varphi, x) = A = P(\varphi) \equiv 0, \\ f_0(\varphi) = f(\varphi, x) = F_0(\varphi) = F(\varphi, x) \equiv 0, \quad f_0(\varphi) = F_0(\varphi) \equiv 0. \end{gathered} \tag{33.3}$$

Based on the results of foregoing chapters, the discussions here shall be concerned with one more special form of the system (33.2), namely the form defined by the condition

$$f(\varphi, x) = F_0(\varphi) \equiv 0. \tag{33.4}$$

For this, the disturbance

$$\Delta = \lambda - \omega \tag{33.5}$$

can be chosen such that by the transformation

$$\varphi \to \psi + \Phi(\psi) \tag{33.6}$$

of the angular coordinates, the original system is transformed into the system

$$\begin{aligned} \frac{dx}{dt} &= Ax + P_1(\psi)\, x + F_1(\psi, x), \\ \frac{d\psi}{dt} &= \omega, \end{aligned} \tag{33.7}$$

with the frequencies $\omega = (\omega_1, \ldots, \omega_m)$, satisfying the inequality

$$|(k, \omega)| \geqslant \varepsilon |k|^{-d} \qquad (|k| \neq 0, \ \varepsilon > 0, \ d > 0) \tag{33.8}$$

for every integral-valued vector $k = (k_1, \ldots, k_m)$. For the linear part of system (33.7) we can construct a non-singular change of variable

$$x = \Phi(\psi)\, y, \tag{33.9}$$

which reduces this system, for almost every A (in the sense of Lebesgue measure), to the system

$$\begin{aligned} \frac{dy}{dt} &= A_0 y + F(\psi, y), \\ \frac{d\psi}{dt} &= \omega, \end{aligned} \tag{33.10}$$

where A_0 is a matrix, whose eigenvalues satisfy the inequality

$$|\mu_\alpha^0 - \mu_\beta^0 + i(k, \omega)| \geqslant \varepsilon |k|^{-d} \qquad (|k| \neq 0) \tag{33.11}$$

for all $\alpha, \beta = 1, \ldots, n$ and every integral-valued vector k.

Carrying out the transformations as in § 32, we see that the system (33.10) finally reduces to the form

$$\begin{aligned} \frac{dx}{dt} &= A_0 x, \\ \frac{d\psi}{dt} &= \omega. \end{aligned} \tag{33.12}$$

In all the cases considered above, the problem of the behaviour of the invariant torus $x = 0$ under perturbation was trivial. The character of the perturbations was such that the torus $x = 0$ remained fixed; it is only the trajectories on it or in its neighbourhood that were somewhat affected.

We now proceed to consider the system (33.2). As in Chap. 1, we introduce the disurbance Δ by formula (33.5) and set up system (33.2) in the form

$$\begin{aligned} \frac{dx}{dt} &= Ax + F_0(\varphi) + P(\varphi)x + F(\varphi, x), \\ \frac{d\varphi}{dt} &= \omega + \Delta + f_0(\varphi) + f(\varphi, x). \end{aligned} \tag{33.13}$$

We suppose that the eigenvalues $\mu_1, \ldots, \mu_n$ of the matrix A are real and distinct and satisfy the inequality

$$\mu_\alpha + \mu_\beta - \mu_j \leqslant -\gamma < 0 \qquad (\alpha, \beta, j = 1, 2, \ldots, n). \tag{33.14}$$

In the system (33.13), we effect the change of variables

$$\begin{aligned} x &= y + v(\psi, y, \Delta_1), \\ \varphi &= \psi + w(\psi, y, \Delta_1), \\ \Delta &= \Delta(\Delta_1), \end{aligned} \tag{33.15}$$

where $v(\psi, y, \Delta_1)$ is a solution of the the equation

$$\frac{\partial v}{\partial y} Ay + \left(\frac{\partial v}{\partial \psi}, \omega\right) = Av + TF_0 + [TP - \bar{D}]y + T^1 F, \tag{33.16}$$

and $w(\psi, y, \Delta_1)$ is a solution of the equation

$$\frac{\partial w}{\partial y} Ay + \left(\frac{\partial w}{\partial \psi}, \omega\right) = Tf_0 - \bar{f}_0 + T^0 f, \tag{33.17}$$

both periodic in ψ ; also Δ (Δ_1) is a solution of the equation

$$\Delta_1 = \Delta + \bar{f_0}. \tag{33.18}$$

Here T, T^0, T^1 are smoothing operators, and $\bar{D} = \{\bar{p}_{11}, \ldots, \bar{p}_{nn}\}$ is a constant diagonal matrix, whose elements are mean values of the diagonal elements of the matrix $P(\varphi)$.

The periodic solution of equation (33.16) is defined by

$$v(\psi, y, \Delta_1) = \int_{-\infty}^{0} e^{-A\tau} T F_0 (\omega\tau + \psi) \, d\tau + v_1(\psi) \, y + + \int_{0}^{\infty} e^{-A\tau} T^1 F(\omega\tau + \psi, \, e^{A\tau} y) \, d\tau, \tag{33.19}$$

where $v_1(\psi)$, periodic in ψ, is a solution of the matrix equation

$$\left(\frac{\partial v_1}{\partial \psi}, \omega\right) + v_1 A = A v_1 + P(\varphi) - \bar{D}. \tag{33.20}$$

The function $v(\psi, y, \Delta_1)$ differs from the solution $u(\varphi, x)$ of equation (30.38), studied in § 30, by its first two terms; the functions $w(\psi, y, \Delta_1)$ and $\Delta(\Delta_1)$, however, consist of those terms whose properties were derived in earlier discussions. Noticing the structure of the first two terms of the function $v(\psi, y, \Delta_1)$, it can be established that, subject to an appropriate choice of the parameters M_s, N_s determining the character of smoothing operators T, T^0 and T^1, the reduction process for system (33.13) by substitutions of the form (33.15) can be continued ad infinitum and is convergent. Thus, for the system (33.16) we can indicate the disturbance

$$\Delta = D^{(\infty)} \tag{33.21}$$

such that by the change of variables

$$\begin{aligned} x &= y + Y(\theta, y), \\ \varphi &= \theta + \Phi(\theta, y) \end{aligned} \tag{33.22}$$

the system (33.13) corresponding to it, reduces to the system

$$\begin{aligned} \frac{dy}{dt} &= A_0 y, \\ \frac{d\theta}{dt} &= \omega. \end{aligned} \tag{33.23}$$

The above discussion yields the following theorem.

Theorem 29. *Suppose that the functions* $F_0(\varphi)$, $F(\varphi, x)$, $f_0(\varphi)$ *and* $f(\varphi)$, x *and the matrix* $P(\varphi)$ *are* 2π*-periodic in* φ; *for given positive* c_0, γ, ε, $\underline{d}$, η *and an integer* s_0 ($s_0 \geqslant 2$), *suppose that there exist a positive* $\delta_0 = \delta_0(c_0, \gamma, \bar{\eta}, r, s_0)$ *and an integer* $l = l(s_0)$ *such that the functions and the matrix referred to are l-times continuously differentiable for*

$$\| x \| \leqslant \eta, \tag{33.24}$$

satisfy the condition

$$F(\varphi, 0) = \frac{\partial F(\varphi, 0)}{\partial x} = f(\varphi, 0) = 0, \tag{33.25}$$

and obey the inequalities

$$\begin{aligned} &| F_0(\varphi) | + | F(\varphi, x) | + | f_0(\varphi) | + | f(\varphi, x) | + | P(\varphi) x | \leqslant \delta_0 \\ &| F_0(\varphi) |_l + | F(\varphi, x) |_l + | f_0(\varphi) |_l + | f(\varphi, x) |_l + | P(\varphi) x |_l \leqslant c_0 ; \end{aligned} \tag{33.26}$$

further, suppose that the eigenvalues $\mu_1, \ldots, \mu_n$ *of the matrix A are real and distinct, and satisfy the inequalities* (33.14) *and that the frequencies* $\omega = (\omega_1, \ldots, \omega_m)$ *satisfy the inequalities* (33.8).

Then there exist a constant $\Delta = (\Delta_1, \ldots, \Delta_m)$, *satisfying the inequality*

$$| \Delta | \leqslant \bar{\eta} \qquad (\Delta = 0 \quad \text{for } | f_0(\varphi) | + | f(\varphi, x) | = 0), \tag{33.27}$$

and the functions $Y(\theta, y)$, *and* $\Phi(\theta, y)$, 2π*-periodic in* θ, $(s - 1)$*-times continuously differentiable for*

$$\| y \| \leqslant \eta - \bar{\eta}, \tag{33.28}$$

and satisfying the inequality

$$| Y(\theta, y) |_{s_0 - 1} + | \Phi(\theta, y) |_{s_0 - 1} \leqslant \bar{\eta}, \tag{33.29}$$

such that the system of equations

$$\begin{aligned} \frac{dx}{dt} &= Ax + F_0(\varphi) + P(\varphi) x + F(\varphi, x), \\ \frac{d\varphi}{dt} &= \omega + \Delta + f_0(\varphi) + f(\varphi, x), \end{aligned} \tag{33.30}$$

by means of the change of variables

$$\begin{aligned} x &= y + Y(\theta, y), \\ y &= \theta + \Phi(\theta, y), \end{aligned} \tag{33.31}$$

is reducible to the form

$$\frac{dy}{dt} = A_0 y,$$
$$\frac{d\theta}{dt} = \omega. \tag{33.32}$$

Proof. This theorem implies that the general solution of system (33.30) has the form

$$x_t = e^{A_0 t} y_0 + Y(\omega t + \theta_0, e^{A_0 t} y_0),$$
$$\varphi_t = \omega t + \theta_0 + \Phi(\omega t + \theta_0, e^{A_0 t} y_0), \tag{33.33}$$

where y_0 and θ_0 are arbitrary constants, and $\| y_0 \| \leqslant \eta - \bar{\eta}$. In particular, the system (33.30) has the asymptotically stable invariant toroidal manifold

$$x = Y(\theta, 0),$$
$$\varphi = \theta + \Phi(\theta, 0), \tag{33.34}$$

filled by the conditionally periodic motions

$$\theta_t = \omega t + \theta_0. \tag{33.35}$$

Moreover, by the change-of-variable formula (33.15), the functions $Y(\theta, y)$ and $\Phi(\theta, y)$ are determined in first approximation by

$$Y(\theta, y) \approx Y_1(\theta, y) = \int_{-\infty}^{0} e^{-A\tau} F_0(\omega\tau + \theta)\, d\tau + v_1(\theta)\, y + \\ + \int_{0}^{\infty} e^{-A\tau} F(\omega\tau + \theta, e^{A\tau} y)\, d\tau, \tag{33.36}$$

$$\Phi(\theta, y) \approx \Phi_1(\theta, y) = \sum_{|k| \neq 0} \frac{f_k^0 e^{i(k,\theta)}}{i(k, \omega)} - \int_{0}^{\infty} f_0(\omega\tau + \theta, e^{A\tau} y)\, d\tau,$$

where f_k^0 are the Fourier coefficients of the function $f_\theta(\theta)$.

Thus, the expression

$$x = \int_{-\infty}^{0} e^{-A\tau} F_0(\omega\tau + \theta)\, d\tau,$$
$$\varphi = \theta + \sum_{|k| \neq 0} \frac{f_k^0 e^{i(k,\theta)}}{i(k, \omega)} \tag{33.37}$$

is a first approximation of the invariant manifold (33.34), and

$$x_t = e^{(A+\bar{D})t}y_0 + Y_1\,(\omega t + \theta_0,\, e^{(A+\bar{D})t}y_0),$$
$$\varphi_t = \omega t + \theta_0 + \Phi_1\,(\omega t + \theta_0,\, e^{(A+\bar{D})t}y_0), \tag{33.38}$$

is a first approximation of the general solution.

Remark. Since the process of successive substitutions converges rapidly, a few successive approximations give a good approximation of the functions $Y(\theta, y)$ and $\Phi(\theta, y)$ and, consequently, also a good approximation of the invariant manifold (33.34).

The restrictions imposed on the eigenvalues of the matrix A can be considerably weakened, as was done in Chap. 2, by introducing in the system (33.2) apart from the disturbance Δ, also a correction to the eigenvalues of matrix A, given by

$$\xi x = \left\{ \sum_{j=1}^{n} \xi_{ij} x_j \right\} = (A - A_0)\, x, \tag{33.39}$$

where A_0 is a matrix, whose eigenvalues $\mu_0 = (\mu_1^0, \ldots, \mu_n^0)$ satisfy the inequalities

$$\operatorname{Re}\,[(r, \mu^0) - \mu_j^0\,] \leqslant -\gamma \qquad \text{for} \quad |\,r\,| = r_0 + 1,$$
$$|\,(r, \mu^0) - \mu_j + i(k, \omega)\,| \geqslant K\,(\,|\,r\,| + |\,k\,|)^{-d} \qquad \text{for} \quad 1 \leqslant |\,r\,| \leqslant r_0 \tag{33.40}$$

for every integral-valued k, non-negative r, $r_0 \geqslant 1$ an integer and positive K, γ and d.

Then, it is possible to effect in the system (33.2) the change of variables

$$x = y + v\,(\psi, y, \Delta_1, \xi_1), \qquad \varphi = \psi + w\,(\psi, y, \Delta_1, \xi_1),$$
$$\Delta = \Delta\,(\Delta_1, \xi_1), \qquad \xi = \xi\,(\xi_1, \Delta_1), \tag{33.41}$$

where v and ω satisfy equations (33.16) and (33.17) with A replaced by A_0 and $\bar{D}$ by

$$\bar{P} = \frac{1}{(2\pi)^m} \int_0^{2\pi} P\,(\varphi)\, d\varphi,$$

and $\Delta\,(\Delta_1, \xi_1)$, $\xi\,(\xi_1, \Delta_1)$ taken as a solution of the system

$$\Delta_1 = \Delta + F_0,$$
$$\xi_1 = \xi + \bar{P}. \tag{33.42}$$

Recalling the arguments employed in proving the preceding theorem, we see that we can extend the process of transforming the system (33.2) by successive substitutions (33.41) and also that it converges. These conclusions motivate the formulation of the next theorem.

Theorem 30. *Suppose that the functions $F_0(\varphi)$, $F(\varphi, x)$, $f_0(\varphi)$ and $f(\varphi, x)$ and the matrix $P(\varphi)$ are 2π-periodic in φ; for given positive $c_0, \gamma, \varepsilon, d, \tau$ and integers τ_0 and $s_0(\tau_0 \geqslant 1, s_0 \geqslant 2)$ suppose that there exist a positive $\delta_0 = \delta(c_0, \gamma, \varepsilon, \tau, \tau_0, s_0)$ and an integer $l = l(\tau_0, s_0)$ such that the functions and the matrix referred to are l-times continuously differentiable in the domain* (33.24) *and satisfy the conditions* (33.25) *and* (33.26); *further, suppose that the eigenvalues $\mu^0 = (\mu_1^0, \ldots, \mu_n^0)$ of the matrix A_0 and the frequencies $\omega = (\omega_1, \ldots, \omega_m)$ satisfy respectively the inequalities* (33.14) *and* (33.8).

Then, there exist a constant vector $\Delta = (\Delta_1, \ldots, \Delta_m)$, a constant matrix $\xi = \{\xi_{\alpha\beta}\}$, satisfying the inequalities

$$|\Delta| \leqslant \bar{\eta}, \quad |\xi| \leqslant \bar{\eta} \qquad (\Delta = 0 \quad \text{for } |f_0(\varphi)| + |f(\varphi, x)| = 0), \tag{33.43}$$

and also the functions $Y(\theta, y)$ and $\Phi(\theta, y)$, 2π-periodic in θ, $(s_0 - 1)$-times continuously differentiable in the domain (33.28), *satisfying the inequality* (33.29) *and such that the system of equations*

$$\begin{aligned} \frac{dx}{dt} &= (A_0 + \xi)\, x + F_0(\varphi) + P(\varphi)\, x + F(\varphi, x), \\ \frac{d\varphi}{dt} &= \omega + \Delta + f_0(\varphi) + f(\varphi, x) \end{aligned} \tag{33.44}$$

by the change of variables

$$\begin{aligned} x &= y + Y(\theta, y), \\ \varphi &= \theta + \Phi(\theta, y) \end{aligned} \tag{33.45}$$

reduces to the system

$$\frac{dy}{dt} = A_0 y, \qquad \frac{d\varphi}{dt} = \omega.$$

Without dilating further on the problem of the behaviour of an invariant toroidal manifold under perturbation, we conclude the present chapter with the remark that the results of this section can be interpreted measure-theoretically for the purpose of ascertaining the scope of generality of the reduction of the system (33.1) to the canonical form (33.32).†

†See Appendix 13.

Chapter 7

NEIGHBOURHOOD OF A COMPACT INVARIANT MANIFOLD OF A NON-AUTONOMOUS SYSTEM

§ 34. Statement of the Problem and Basic Postulates

Consider a system of non-linear differential equations

$$\frac{dy}{dt} = Y(t, y), \tag{34.1}$$

($Y = (Y_1, \ldots, Y_{n+m})$ and $y = (y_1, \ldots, y_{n+m})$ being points of an $(n + m)$-dimensional Euclidean space $E_{n+m}(y)$), whose right-hand sides are defined for all t, y in the domain

$$(t, y) \in (-\infty, \infty) \times D, \tag{34.2}$$

and are sufficiently smooth and, together with their partial derivatives, bounded on that domain.

Suppose that the system (34.1) has an $(m + 1)$-dimensional ($m \geqslant 0$) smooth invariant manifold M_t belonging to some compact set D_2,

$$M_t \subset D_2 \subset D \quad \text{for } t \in (-\infty, \infty).$$

To investigate the solutions of the system (34.1) in a neighbourhood of M_t, we suppose that for every fixed $t \in (-\infty, \infty)$ the manifold M_t is smoothly homeomorphic to the manifold D_1, $c = (c_1, \ldots, c_m)$ is a coordinate system over D_1, and

$$y = f(t, c) \quad ((t, c) \in (-\infty, \infty) \times D_1) \tag{34.3}$$

is a function effecting that homeomorphism. In the space $E_{n+m}(y)$ a neighbourhood of the manifold D_1 must be representable as a direct product of D and an n-dimensional cube $K\{|z_i| \leqslant \eta, i = 1, \ldots, n\}$ which, from the viewpoint of the topology of manifolds, is equivalent to the requirement that D_1 is a normalizable submanifold of that space.

The restriction introduced ensures that the system of equations (34.1) reduces, in the neighbourhood of M_t, to the system

$$\begin{aligned} \frac{dc}{dt} &= Q_1(t, c)\, z + G_1(t, c, z), \\ \frac{dz}{dt} &= Q_2(t, c)\, z + G_2(t, c, z), \end{aligned} \tag{34.4}$$

with the matrices $Q_1(t, c)$ and $Q_2(t, c)$ and the functions $G_1(t, c, z)$, $G_2(t, c, z)$, continuously differentiable a finite number of times (here and everywhere in the sequel continuity implies uniform continuity) for $(t, c) \in (-\infty, \infty) D_1$, $\| z \| \leqslant \varepsilon$, and together with their partial derivatives bounded and satisfying the condition

$$G_i(t, c, 0) = \frac{\partial G_i(t, c, 0)}{\partial z} = 0 \qquad (i, = 1, 2).$$

In order to make such a reduction the following procedure can be adopted. Denote by $y(t, y_0)$ a solution of the system (34.1), passing at $t = 0$ through the point y_0. By the definition of an invariant manifold, every solution $y = y(t, f(0, c))$ of the system (34.1) lies on the manifold M_t for $t \in (-\infty, \infty)$ and is bounded since M_t is compact. The system of equations (34.1) has, consequently, an m-parameter family of bounded solutions. A neighbourhood of the manifold M_t is a neighbourhood of the family $y(t, f(0, c)) = Y_0(t, c)$.

To reduce the system (34.1) to the system (34.4), it is necessary to choose an $(n + m) \times n$ matrix $P(t, c)$ such that the square matrix

$$\left\{ \frac{\partial y_0(t, c)}{\partial c}, \quad P(t, c) \right\} \tag{34.5}$$

is non-singular, sufficiently smooth and, together with its partial derivatives with respect to t and c, is bounded for $(t, c) \in (-\infty, \infty) \times D_1$, and then to make the substitution

$$y = y_0(t, c) + P(t, c)\, z. \tag{34.6}$$

In particular, if the matrix $\{\partial y_i^0 / \partial c_j\}$, $i, j = 1, \ldots, m$, is non-singular for $(t, c) \in (-\infty, \infty) \times D_1$ then the substitution (34.6) can be taken in the form

$$\begin{gathered} y_1 = y_1^0(t, c), \ldots, y_m = y_m^0(t, c), \\ y_{m+1} = y_{m+1}^0(t, c) + z_1, \ldots, y_{m+n} = y_{m+n}^0(t, c) + z_n. \end{gathered} \tag{34.7}$$

We further require that the linear system of equations

$$\frac{dz}{dt} = Q_2(t, c)\, z \tag{34.8}$$

is reduced by the non-singular substitution

$$z = B(t, c)\, x_1 \tag{34.9}$$

to the system

$$\frac{dx_1}{dt} = H(c)\, x, \tag{34.10}$$

and that the matrices $B(t, c)$ and $H(c)$ have the same properties as indicated above for (34.5). This enables us to make use of substitution (34.9) to transform the system (34.4) into the system

$$\frac{dc}{dt} = R_1(t, c, x_1), \qquad \frac{dx_1}{dt} = H(c)\, x_1 + F_1(t, c, x_1) \tag{34.11}$$

with the functions $R_1(t, c, x_1)$ and $F_1(t, c, x_1)$ sufficiently smooth on the domain

$$(t, c) \in (-\infty, \infty) \times D_1, \qquad \| x_1 \| \leqslant \eta_1 \quad (\eta < \varepsilon), \tag{34.12}$$

bounded together with their partial derivatives and satisfying the condition

$$R_1(t, c, 0) = F_1(t, c, 0) = \frac{\partial F_1(t, c, 0)}{\partial x_1} = 0. \tag{34.13}$$

We make the following remarks concerning the conditions, which ensure the indicated reductions : for $n = 1$ or the matrix $Q_2(t, c)$ periodic in t, their verification does not involve any particular difficulty; for in the first case equation (34.8) is solvable explicitly and in the second case the Floquet-Lyapunov theorem holds.

The final requirement is that the matrix $H(c)$ is reducible to $A(c)$, in real canonical form, by a matrix $C(c)$, sufficiently smooth and bounded together with its partial derivatives:

$$H(c) = C(c)\, A(c)\, C^{-1}(c) \qquad (c \in D_1). \tag{34.14}$$

This requirement enables us to transform the system (34.11) into the system

$$\frac{dc}{dt} = R(t, c, x), \qquad \frac{dx}{dt} = A(c)\, x + F(t, c, x) \tag{34.15}$$

with the function $R(t, c, x)$ and $F(t, c, x)$ having the properties similar to those of the functions $R_1(t, c, x)$ and $F_1(t, c, x)$.

Concerning the generality of the assumptions made, we may only comment that for conservative matrices $H(c)$, i.e. for matrices preserving the same Jordan (or real canonical) form for all c in D_1, the representation (34.14) is always realizable in a class of locally smooth matrices [9].

For these hypotheses to be satisfied with $c \in D_1$ in a class of smooth matrices, it is necessary to impose, besides conservativeness, some restrictions on the topology of the manifold D_1 stipulating, for example, that it should be homeomorphic to a ball, a torus or the direct product of a ball and a circle.

The transformations made on system (34.1) carry a neighbourhood of the manifold M_t into a neighbourhood of the trivial manifold $x = 0$ of system (34.15)

By way of investigating this neighbourhood, we shall show (under specific restriction on the eigenvalues of the matrix $A(c)$) the existence of a non-singular and sufficiently smooth substitution

$$\begin{aligned} c &= \theta + \Phi(t, \theta, \xi), \\ x &= \xi + Y(t, \theta, \xi), \end{aligned} \tag{34.16}$$

bounded for $t \in (-\infty, \infty)$, which reduces the system of equations (34.15) to the system

$$\begin{aligned} \frac{d\theta}{dt} &= 0, \\ \frac{d\xi}{dt} &= A(0)\,\xi; \end{aligned} \tag{34.17}$$

we demonstrate a method for constructing the transformation functions $\Phi(t, \theta, \xi)$ and $Y(t, \theta, \xi)$ and establish some of their properties.†

§ 35. Lemma on the Solutions of an Auxiliary System

In forming a transformation that reduces the system of equations (34.1) to system (34.17), an important role is played by solutions, bounded for $t \in (-\infty, \infty)$, of an auxiliary system of equations

$$\begin{aligned} \frac{\partial w}{\partial x} A_0(c)\, x + \frac{\partial w}{\partial t} &= R(t, c, x), \\ \frac{\partial u}{\partial x} A_0(c)\, x + \frac{\partial u}{\partial t} &= A_0(c)\, u + \frac{\partial A_0(c)\, x}{\partial c} w + F(t, c, x). \end{aligned} \tag{35.1}$$

The following lemma verifies the existence and some properties of such solutions.

Lemma 18. *Suppose that the right-hand side of the system* (35.1) *satisfies the conditions* :

(i) *the functions* $R(t, c, x)$, $F(t, c, x)$ *and the matrix* $A_0(c)$ *are defined for*

$$(t, c) \in (-\infty, \infty) \times D_1, \quad \| x \| \leqslant \eta, \tag{35.2}$$

†See Appendix 14.

are τ times continuously differentiable, $\tau \geqslant 3$, are bounded, together with their partial derivatives, and satisfy the condition

$$R(t, c, 0) = F(t, c, 0) = \frac{\partial F(t, c, 0)}{\partial x} = 0. \quad (35.3)$$

(ii) *The matrix $A_0(c)$ is in real canonical form, and its eigenvalues $\mu_1(c), \ldots, \mu_n(c)$ satisfy the inequality*

$$\operatorname{Re}[\mu_\alpha(c) + \mu_\beta(c) - \mu(c)] \leqslant -\gamma \quad (35.4)$$

for every $c \in D_1$, any $\alpha, \beta, j = 1, \ldots, n$ and some $\gamma > 0$.

Then, the system (35.1) *has a solution*

$$w(t, c, x) = -\int_0^\infty R(t + z, c, e^{A_0(c)z} x)\, dz,$$

$$u(t, c, x) = -\int_0^\infty \left[\frac{\partial A_0(c) x}{\partial c} w(t + z, c, e^{A_0(c)z} x) + \right. \quad (35.5)$$

$$\left. + e^{-A_0(c)z} F(t + z, c, e^{A_0(c)z} x)\right] dz,$$

($\tau - 2$)*-times continuously differentiable in the domain* (35.2) *and bounded, together with its partial derivatives, for which*

$$w(t, c, 0) = u(t, c, 0) = \frac{\partial u(t, c, 0)}{\partial x} = 0,$$

$$|w(t, c, x)| \leqslant q_2 |D_x R(t, c, x)|_0, \quad (35.6)$$

$$|u(t, c, x)| \leqslant q_2 (|D_c A_0(c)|_0 |D_x R(t, c, x)|_0 + |D_x^2 F(t, c, x)|_0),$$

where q_2 is a positive constant, depending only on γ and η.

Proof. The matrix $A_0(c)$ is assumed to be in real canonical form. Besides the eigenvalues, this form is characterized by the parameters $\varepsilon_\alpha(c)$ defined to within an arbitrary factor. These factors are held fixed and satisfy the inequalities

$$|\varepsilon_\alpha(c)| < \gamma \qquad (\alpha = 1, \ldots, k_0 < n,\ c \in D_1). \quad (35.7)$$

For the matrix $A_0(c)$ so chosen, the estimate of Lemma 13 of § 30 holds for the function $e^{A_0(c)z} x$ and, consequently, we have the inequality

$$\| e^{A_0(c)z} x \| \leqslant \| x \| \qquad \text{for } z \geqslant 0.$$

This inequality implies that the integrands of (35.5) are defined for all (t, c, x)

of the domain (35.2). By replacing $t + z$ by τ in (35.5) and differentiating the resulting expression term-by-term, it is easily verified that the integrals (35.5) satisfy the system of equations (35.1), whenever they exist and are continuously differentiable.

The convergence of the first integral in (35.5) follows from the estimate

$$| w(t, c, x) | \leqslant \int_0^\infty | R(t + z, c, e^{A_0(c)z} x) - R(t + z, c, 0) | \, dz \leqslant$$

$$\leqslant c \int_0^\infty | D_x R(t, c, x) |_0 \, | e^{A_0(c)z} x | \, dz \leqslant c_1 \, | D_x R(t, c, x) |_0 \, \| x \|, \tag{35.8}$$

implying also that $w(t, c, 0) = 0$ and giving the estimate (35.6) for $w(t, c, x)$. The differentiablity of the function $w(t, c, x)$ follows from the convergence of the integral in the expression

$$D_t^d D_x^r w(t, c, x) = - \int_0^\infty D_t^d D_x^r R(t + z, c, e^{A_0(c)z} x) \, dz,$$

which is established for $| r | = 0$ and $d \leqslant \tau - 1$ on the basis of estimates similar to (35.8), and for $| r | \geqslant 1$ by the inequality

$$| D_t^d D_x^r R(t + z, c, e^{A_0(c)z} x) | \leqslant c_2 \, | D_t^d D_x^r R(t, c, x) |_0 \, | e^{A_0(c)z} |,$$

with the notation

$$| e^{A_0(c)z} | = \max_{\alpha, \beta} | \{ e^{A_0(c)z} \}_{\alpha\beta} |.$$

Remembering the structure of matrix $e^{A_0(c)z}$ and the properties of the functions $w(t, c, x)$ and $F(t, c, x)$, we verify the convergence of the second integral in (35.5) by

$$| u(t, c, x) | \leqslant \int_0^\infty \left| \frac{\partial A_0(c) x}{\partial c} (w(t + z, c, e^{A_0(c)z} x) - w(t + z, c, 0)) + \right.$$

$$\left. + e^{-A_0(c)z} \sum_{\alpha,\beta=1}^{\infty} \frac{\partial^2 F(t + z, c, e^{A_0(c)z} x)}{\partial x_\alpha \, \partial x_\beta} \{ e^{A_0(c)z} x \}_\alpha \{ e^{A_0(c)z} x \}_\beta \right| dz \leqslant$$

$$\leqslant c_2 \left[| D_c A_0(c) |_0 \, | D_x w(t, c, x) |_0 + \right. \tag{35.9}$$

$$\left. + | D_x^2 F(t, c, x) |_0 \int_0^\infty P(z, \varepsilon) \, e^{\max_{\alpha,\beta,j} \mathrm{Re} \, [\mu_\alpha + \mu_\beta - \mu_j] z} \, dz \right] \| x \|^2 \leqslant$$

$$\leqslant \bar{c}_2 \, [| D_c A_0(c) |_0 \, | D_x R(t, c, x) |_0 + | D_x^2 F(t, c, x) |_0] \, \| x \|^2.$$

The inequality (35.9) implies the validity of the relation $u(t, c, 0) = (\partial u(t, c, 0)/\partial x) = 0$ and the estimate (35.6) for the function $u(t, c, x)$. The differentiability up to order $\tau - 2$ of the function $u(t, c, x)$ follows from the convergence of the integrals, given by the formal differentiation of expression (35.5) for the function $u(t, c, x)$. The convergence of these integrals is, however, proved by estimates analogous to those derived above for the functions $u(t, c, x)$ and $w(t, c, x)$.

§ 36. Inductive Theorem

We now investigate the character of the s-th step of the interation process used in the transformation of the system of equations (34.15). For this, starting from $Q_0 > 0$, we introduce the notation

$$N_s = Q_s^{\alpha}, \quad \delta_s = N_s^{-\beta}, \quad Q_s = Q_{s-1}^{\varkappa}, \quad (s = 1, 2, \ldots), \tag{36.1}$$

where the positive constants α, β, $\varkappa$ satisfy the system of inequalities

$$\beta > \max\left\{\frac{5\varkappa}{2-\varkappa}, \varkappa\left(4 + \frac{l_0+1}{\alpha}\right), \varkappa\left[s_0\left(1 + \frac{1}{\alpha}\right) + 3\right] + \frac{1}{\alpha}\right\},$$

$$l_0 > \frac{\varkappa}{\varkappa - 1}(\beta + 1), \quad l_1 > \alpha\beta, \quad 1 < \varkappa < 2, \quad s_0 \geqslant 2. \tag{36.2}$$

The theorem formulated and proved below characterizes the iteration process.

Inductive Theorem 31. *For the system of equations* (34.15), *suppose that the vector functions* $R(t, c, x)$ *and* $F(t, c, x)$ *are* l_0*-times continuously differentiable on the domain* (35.2) *and satisfy the condition* (35.3), *together with the inequalities*

$$|R(t, c, x)| + |F(t, c, x)| \leqslant \delta_{s-1}, \tag{36.3}$$

$$|D_t^d D_c^{\rho} D_x^r R(t, c, x)| + |D_t^d D_c^{\rho} D_x^r F(t, c, x)| \leqslant N_{s-1}^{l_0}$$

$$\text{for } |d| + |\rho| + |r| = l_0. \tag{36.4}$$

In addition, suppose that $A(c)$ *is a real canonical matrix,* l_1*-times continuously differentiable for* $c \in D$, *bounded, together with its partial derivatives, with respect to* c, *and such that its eigenvalues* $\mu_1(c), \ldots, \mu_n(c)$ *satisfy the inequalities* (35.4).

Then, for sufficiently small δ_0, *there exists the coordinate transformation*

$$\begin{aligned} c_0 &= \theta + \Phi(t, \theta\, \xi), \\ x &= \xi + Y(t, \theta, \xi), \end{aligned} \tag{36.5}$$

defined on the domain†

$$-\infty < t < \infty, \quad \theta \in D_1 - Q_s^{-1}, \quad \| \xi \| \leqslant \eta - N_s^{-1} \tag{36.6}$$

and satisfying the condition

$$\Phi(t, \theta, 0) = Y(t, \theta, 0) = \frac{\partial Y(t, \theta, 0)}{\partial \xi} = 0 \tag{36.7}$$

and the inequality

$$| \Phi(t, \theta, \xi) |_{s_0} + | Y(t, \theta, \xi) |_{s_0} \leqslant Q_{s-1}^{-1}, \tag{36.8}$$

such that the system (34.15) *in the coordinates* θ, ξ *assumes the form*

$$\begin{aligned} \frac{d\theta}{dt} &= R^{(t)}(t, \theta, \xi), \\ \frac{d\xi}{dt} &= A(\theta)\,\xi + F^{(1)}(t, \theta, \xi); \end{aligned} \tag{36.9}$$

here the functions $R^{(1)}(t, \theta, \xi)$ *and* $F^{(1)}(t, \theta, \xi)$ *are* l_0*-times continuously differentiable on the domain* (36.6) *and satisfy the condition*

$$R^{(1)}(t, \theta, 0) = F^{(1)}(t, \theta, 0) = \frac{\partial F^{(1)}(t, \theta, 0)}{\partial \xi} = 0 \tag{36.10}$$

and the inequalities

$$\begin{gathered} | R^{(1)}(t, \theta, \xi) | + | F^{(1)}(t, \theta, \xi) | \leqslant \delta_s, \\ | D_t^d D_\theta^\rho D_\xi^r R^{(1)}(t, \theta, \xi) | + | D_t^d D_\theta^\rho D_\xi^r F^{(1)}(t, \theta, \xi) | \leqslant N_s^{l_0} \\ \text{for } d + |\rho| + |r| = l_0. \end{gathered} \tag{36.11}$$

Proof. The substitution (36.5) made in the system (34.15) yields the system

$$\begin{aligned} \left(E + \frac{\partial \Phi}{\partial \theta} \right) \frac{d\theta}{dt} + \frac{\partial \Phi}{\partial \xi} \frac{d\xi}{dt} + \frac{\partial \Phi}{\partial t} &= R(t, \theta + \Phi, \xi + Y), \\ \frac{\partial Y}{\partial \theta} \frac{d\theta}{dt} + E \left(E + \frac{\partial Y}{\partial \xi} \right) \frac{d\xi}{dt} + \frac{\partial Y}{\partial t} &= A(\theta + \Phi)(\xi + Y) + \\ &\quad + F(t, \theta + \Phi, \xi + Y). \end{aligned} \tag{36.12}$$

Let us take the functions $\Phi(t, \theta, \xi)$ and $Y(t, \theta, \xi)$ as solutions, bounded for $t \in (-\infty, \infty)$, of the system

$$\begin{aligned} \frac{\partial \Phi}{\partial \xi} T A(\theta)\, \xi + \frac{\partial \Phi}{\partial t} &= T^0 R(t, \theta, \xi), \\ \frac{\partial Y}{\partial \xi} T A(\theta)\, \xi + \frac{\partial Y}{\partial t} &= T A(\theta)\, Y + \frac{\partial T A(\theta)\, \xi}{\partial \theta} \Phi + T^1 F(t, \theta, \xi), \end{aligned} \tag{36.13}$$

†Here and in what follows $D_1 - Q_s^{-1}$ denotes the point set contained in the manifold D_1 together with its Q-neighbourhood.

where $T = T_{Q_s}$, $T^0 = T^0_{N_s}$, $T^1 = T^1_{N_s}$ are the smoothing operators studied in § 13.

Substracting equations (36.12) from equations (36.13) and solving the resulting system in $\left\{ \frac{d\theta}{dt}, \frac{d\xi}{dt} - A(\theta)\,\xi \right\}$, we obtain the system

$$\frac{d\theta}{dt} = R^{(1)}(t, \theta, \xi),$$

$$\frac{d\xi}{dt} = A(\theta)\,\xi + F^{(1)}(t, \theta, \xi); \tag{36.14}$$

we have used here the notation

$$\{R^{(1)}, F^{(1)}\} = \begin{pmatrix} E + \frac{\partial \Phi}{\partial \theta} & \frac{\partial \Phi}{\partial \xi} \\ \frac{\partial Y}{\partial \theta} & E + \frac{\partial Y}{\partial \xi} \end{pmatrix} \{R(t, \theta + \Phi, \xi + Y)$$

$$- T^0 R(t, \theta, \xi) - \frac{\partial \Phi}{\partial \xi}(A(\theta) - TA(\theta))\,\xi,\ F(t, \theta + \Phi, \xi + Y)$$

$$- T^1 F(t, \theta, \xi) + (A(\theta + \Phi) - TA(\theta))(\xi + Y) - \frac{\partial TA(\theta)\,\xi}{\partial \theta}\Phi$$

$$- \left(E + \frac{\partial Y}{\partial \xi}\right)(A(\theta) - TA(\theta))\,\xi\Big\}. \tag{36.15}$$

We now consider equations (36.13). In view of the fact that

$$|\,A(\theta) - TA(\theta)\,| \leqslant cQ_s^{-l_1}\,|\,A(\theta)\,|_{l_1} \ll Q_s^{-1}$$

for sufficiently small δ_0, we see that the matrix $TA(\theta)$ has eigenvalues, close to $\mu_1(c), \ldots, \mu_n(c)$ and, consequently, such that the inequalities (35.4) hold for them. Since the form of a matrix is not affected by the smoothing process, the system (36.13) satisfies all the hypotheses of Lemma 18. By this lemma, the functions $\Phi(t, \theta, \xi)$ and $Y(t, \theta, \xi)$ can be represented by

$$\Phi(t, \theta, \xi) = -\int_0^\infty T^0 R(t + z, \theta, e^{TA(\theta)z}\,\xi)\,dz,$$

$$Y(t, \theta, \xi) = -\int_0^\infty \Big[\frac{\partial TA(\theta)\,\xi}{\partial \theta}\Phi(t + z, \theta, e^{TA(\theta)z}\,\xi) + \tag{36.16}$$

$$+ e^{-TA(\theta)z}\,T^1 F(t + z, \theta, e^{TA(\theta)z}\,x)\Big]\,dz,$$

and satisfy the estimate

$$|\Phi(t, \theta, \xi)| + |Y(t, \theta, \xi)| \leqslant q_2 [(1 + |D_c TA|_0) |D_\xi T^0 R|_0 + |D_\xi^2 T^1 F|_0] \leqslant$$
$$\leqslant \bar{q}_l (Q_s N_s + N_s^3) \delta_{s-1} \leqslant N_{s-1}^{-\varkappa} = N_s^{-1}. \quad (36.17)$$

Taking note of (36.16), we can evaluate the derivatives of the functions $\Phi(t, \theta, \xi)$ and $Y(t, \theta, \xi)$, so that

$$D_t^d D_\theta^\rho D_\xi^r \Phi(t, \theta, \xi) = -\int_0^\infty D_t^d D_\theta^\rho D_\xi^r T^0 R(t + z, \theta, e^{TA(\theta)z}\xi)\, dz. \quad (36.18)$$

The partial derivatives $D_y^\rho f(y, \eta(y))$ of the function $f(y, \eta(y)) = f(y_1, \ldots, y_{k_0}, \eta_1(y_1, \ldots, y_{k_0}), \ldots, \eta_l(y_1, \ldots, y_{k_0}))$ form a linear combination of the expressions

$$D_y^{i_1} D_\eta^{i_2} f(y, \eta)\, D_y^{j_0+1} \eta_{\beta_0} D_y^{j_1+1} \eta_{\beta_1} \ldots D_y^{j_{i_2}+1} \eta_{\beta_{i_2}},$$

i.e. $D_y^\rho f(y, \eta(y))$ is defined by

$$D_y^\rho f(y, \eta(y)) = L_{|\rho|}\left[D_y^{i_1} D_\eta^{i_2} f(y, \eta)\, D_y^{j_0+1} \eta_{\beta_0} \ldots D_y^{j_{i_2}+1} \eta_{\beta_{i_2}} \right], \quad (36.19)$$

where $i_1, i_2, j_1, \ldots, j_{i_2}$ are non-negative integers, $i_1 + i_2 = 1, \ldots, |\rho|$, $j_1 + \ldots + j_{i_2} = |\rho| - (i_1 + i_2)$; $\beta_1, \ldots, \beta_{i_2}$ are members of the series $1, \ldots, l$; $D_y^{j_0+1} \eta_{\beta_0} = 1$ and $L_{|\rho|}[\cdot]$ is a linear form with non-negative integral coefficients.

From (36.19), we find that

$$D_t^d D_\theta^\rho D_\xi^r T^0 R(t + z, \theta, e^{TA(\theta)z}\xi) = D_t^d D_\theta^\rho L_{|r|}\Big[D_\eta^{|r|} T^0 R(t + z, \theta, \eta) D_\xi \times$$
$$\times \{e^{TA(\theta)z}\xi\}_{\beta_1} \ldots D_\xi \{e^{TA(\theta)z}\xi\}_{\beta_{|r|}} \Big] =$$
$$= L_{|r|}\Big[D_\theta^\rho (D_t^d D_\eta^{|r|} T^0 R(t + z, \theta, \eta) \times$$
$$\times \{e^{TA(\theta)z}\}_{\alpha_1\beta_1} \ldots \{e^{TA(\theta)z}\}_{\alpha_{|r|}\beta_{|r|}}) \Big]$$
$$(36.20)$$

for $|r| > 0$, where $\eta = e^{TA(\theta)z}\xi$.

The derivatives $D_\theta^\rho (a(\theta)\, b(\theta))$, are given by

$$D_\theta^\rho (a(\theta)\, b(\theta)) = L_{|\rho|}\, [D_\theta^{i_1} a(\theta)\, D_\theta^{i_2} b(\theta)], \quad (36.21)$$

where i_1, i_2 are non-negative integers, so that $i_1 + i_2 = |\rho|$; hence,

$$D_\theta^\rho \left(D_t^d D_\eta^{|r|} T^0 R(t+z, \theta, \eta(\theta)) \{e^{TA(\theta)z}\}_{\alpha_1\beta_1} \dots \{e^{TA(\theta)z}\}_{\alpha_{|r|}\beta_{|r|}} \right) =$$

$$= L_{|\rho|} \left[D_\theta^{i_1} (D_t^d D_\eta^r T^0 R(t+z, \theta, \eta(\theta)))\, D_\theta^{i_2} (\{e^{TA(\theta)z}\}_{\alpha_1\beta_1} \dots \right.$$

$$\left. \dots \{e^{TA(\theta)z}\}_{\alpha_{|r|}\beta_{|r|}}) \right]. \tag{36.22}$$

The structure of the matrix $e^{TA(\theta)z}$ implies that

$$\{e^{TA(\theta)z}\}_{\alpha_1\beta_1} \dots \{e^{TA(\theta)z}\}_{\alpha_{|r|}\beta_{|r|}} = p_1(z, T\varepsilon, \cos T\beta_i z, \sin T\beta_i z) \times$$

$$\times\, e^{\mathrm{Re}\, T\mu(\theta)}\, z, \tag{36.23}$$

where

$$T\mu = T\mu_{i_1} + \dots + T\mu_{i_{|r|}}, \quad \beta_i = \mathrm{Im}\, \mu_i \text{ and } p_1(z, T\varepsilon, \cos T\beta_i z, \sin T\beta_i z$$

is a polynomial (in the arguments appearing there) of degree not higher than n. Furthermore,

$$|D_\theta^{i_2} (p_1(z, T\varepsilon, \cos T\beta_i z, \sin T\beta_i z)\, e^{\mathrm{Re}\, T\mu(\theta)z})|$$

$$\leqslant L_{i_1} [\,|D_\theta^{i_3} p_1|\, |D_\theta^{i_2 - i_3} e^{\mathrm{Re}\, T\mu(\theta)z}|\,] \leqslant$$

$$\leqslant q_3 p_2(z)\, Q_s^{i_2}\, |e^{\mathrm{Re}\, T\mu(\theta)z}| \leqslant q_3\, Q_s^{i_2} p_2(z)\, e^{-\gamma_1 z}, \tag{36.24}$$

where $p_2(z)$ is a polynomial in z and γ_1 is a positive constant. Let us estimate the expression

$$D_\theta^{i_1} (D_t^d D_\eta^{|r|} T^0 R(t+z, \theta, \eta(\theta))) = L_{i_1} \left[D_\theta^{i_1'} D_\eta^{i_2'} \left(D_t^d D_\eta^{|r|} T^0 \mathrm{R}(t+z, \theta, \eta) \times \right.\right.$$

$$\left.\left. \times\, D_\theta^{j_0+1} \{e^{TA(\theta)z}\xi\}_{\beta_0}\, D_\theta^{j_1+1} \{e^{TA(\theta)z}\xi\}_{\beta_1} \dots D_\theta^{j_{i_2'}+1} \{e^{TA(\theta)z}\}_{\beta_{i_2'}} \right) \right].$$

We have

$$\left| D_\theta^{j_{k_0}+1} \{e^{TA(\theta)z}\xi\}_{\beta_{k_0}} \right| \leqslant q_4\, Q^{j_{k_0}+1} \left| \{e^{TA(\theta)z}\xi\}_{\beta_{k_0}} \right| \leqslant \bar{q}_4\, Q_s^{j_{k_0}+1} \tag{36.25}$$

for $z \geqslant 0$, so that

$$| D_\theta^{i_1} (D_t^d D_\eta^{|r|} T^0 R(t+z, 0, \eta(\theta))) | \leqslant$$

$$\leqslant L_{i_1} \left[\left| D_t^d D_\theta^{i_1'} D_\eta^{|r| + i_2'} T^0 R(t, 0, \eta) \right| \bar{q}_4 Q_s^{j_{k_0} + \dots + j_{i_2'} + i_2'} \right] \leqslant$$

$$\leqslant L_{i_1} \left[\tilde{q}_4 N_s^{d + i_1' + |r| + i_2'} | R(t, \theta, \eta) |_0 Q^{i_1 - (i_2' + i_1')} Q_s^{i_2'} \right] \leqslant$$

$$\leqslant q_5 N_s^{d+|r|+i_1} Q_s^{i_1} \delta_{s-1}. \qquad (36.26)$$

From (36.22) and the estimates (36.24) and (36.25) above, we deduce that

$$\left| D_\theta^\rho (D_t^d D_\eta^{|r|} T^0 R(t+z, 0, \eta(\theta))) \{e^{TA(\theta)z}\}_{\alpha_1\beta_1} \dots \{e^{TA(\theta)z}\}_{\alpha_1\beta_1} \right| \{e^{TA(\theta)z}\}_{\alpha_{|r|}\beta_{|r|}} \Big|$$

$$\leqslant L_{|\rho|} [q_5 N_5^{d+|r|+i_1} Q_s^{i_1} \delta_{s-1} q_3 Q_s^{i_2} p_2(z) e^{-\gamma_1 z}] \leqslant$$

$$\leqslant q_6 p_2(z) e^{-\gamma_1 z} N_s^{d+|r|+|\rho|} Q_s^{|\rho|} \delta_{s-1},$$

whence we have

$$\left| D_t^d D_\theta^\rho D_\xi^r T^0 R(t+z, 0, e^{TA(\theta)z}\xi) \right| \leqslant q_7 p_2(z) e^{-\gamma_1 z} N^{d+|r|+|\rho|} Q_s^{|\rho|} \delta_{s-1}. \quad (36.27)$$

An estimate of the integral in (36.18) is given by

$$| D_t^d D_\theta^\rho D_\xi^r \Phi(t, 0, \xi) | \leqslant q_7 N^{d+|r|+|\rho|} \delta_{s-1} \int_0^\infty p_2(z) e^{-\gamma_1 z} dz =$$

$$= \bar{q}_7 N^{d+|r|+|\rho|} Q_s^{|\rho|} \delta_{s-1} \quad \text{for } |r| > 0. \qquad (36.28)$$

To evaluate the derivatives of the function $\Phi(t, 0, \xi)$ for $|r| = 0$, we set up the identity

$$D_t^d D_\theta^\rho \Phi(t, 0, \xi) = - \int_0^\infty [D_t^d D_\theta^\rho T^0 R(t+z, 0, e^{TA(\theta)z}\xi)$$

$$- D_t^d D_\theta^\rho T^0 R(t+z, 0, 0)] \, dz,$$

and estimate the difference

$$| D_t^d D_\theta^\rho T^0 R(t+z, \theta, e^{TA(\theta)z}\xi) - D_t^d D_\theta^\rho T^0 R(t+z, \theta, 0) | \leqslant$$

$$\leqslant q_8 | D_\xi D_t^d D_\theta^\rho T^0 R(t, \theta, e^{TA(\theta)z}\xi) |_0 \, | e^{TA(\theta)z} | \leqslant$$

$$\leqslant q_7 q_8 | p_2(z) e^{-\gamma_1 z} |_0 \, N_s^{d+|\rho|+1} Q_s^{|\rho|} \delta_{s-1} | e^{TA(\theta)z} |.$$

We obtain

$$| D_t^d D_\theta^\rho \Phi(t, \theta, \xi) | \leqslant \int_0^\infty q_7 q_8 | p_2(z) e^{-\gamma_1 z} |_0 N_s^{d+|\rho|+1} Q_s^{|\rho|} \delta_{s-1} | e^{TA(\theta)z} | \, dz \leqslant$$

$$\leqslant q_9 N_s^{d+|\rho|+1} Q^{|\rho|} \delta_{s-1}. \tag{36.29}$$

Combining the inequalities (36.28) and (36.29), we arrive at an estimate for the derivatives of function $\Phi(t, \theta, \xi)$:

$$| D_t^d D_\theta^\rho D_\xi^r \Phi(t, \theta, \xi) | \leqslant b N_s^{d+|\rho|+|\bar{r}|} Q_s^{|\rho|} \delta_{s-1}, \tag{36.30}$$

where

$$|\bar{r}| = \begin{cases} |r| & \text{for } |r| > 0, \\ 1 & \text{for } |r| = 0, \end{cases}$$

and b is a constant not depending on Q_s.

Similarly, we obtain an estimate for the derivatives of the function $Y(t, \theta, \xi)$ by the inequality

$$| D_t^d D_\theta^\rho D_\xi^r Y(t, \theta, \xi) | \leqslant b N_s^{d+|\rho|+|\bar{r}|+2} Q_s^{|\rho|} \delta_{s-1}. \tag{36.31}$$

The inequalities (36.30) and (36.31) imply the estimate (36.8) of Theorem 32, since

$$| \Phi(t, \theta, \xi) |_{s_0} + | Y(t, \theta, \xi) |_{s_0} \leqslant b [N_s^{s_0+1} Q_s^{s_0} \delta_{s-1} + N_s^{s_0+3} Q_s^{s_0} \delta_{s-1}] \leqslant$$

$$\leqslant 2b N_{s-1}^{\varkappa\left(s_0+3+\frac{s_0}{\alpha}\right)-\beta} < Q_{s-1}^{-1}. \tag{36.32}$$

We now consider the functions $R^{(1)}(t, \theta, \xi)$ and $F^{(1)}(t, \theta, \xi)$. In the first place, we remark that by the inequalities (36.32) the inverse

$$A_1^{-1} = \begin{pmatrix} E + \dfrac{\partial \Phi}{\partial \theta} & \dfrac{\partial \Phi}{\partial \xi} \\ \dfrac{\partial Y}{\partial \theta} & E + \dfrac{\partial Y}{\partial \xi} \end{pmatrix}^{-1} \tag{36.33}$$

exists and satisfies the inequality $| A^{-1} | < 2$. Taking note of the notation (36.15), we see that the estimates for the functions $R^{(1)}(t, \theta, \xi)$ and $F^{(1)}(t, \theta, \xi)$ are given by

$$\begin{aligned} &| R^{(1)}(t, \theta, \xi) | + | F^{(1)}(t, \theta, \xi) | \\ &\leqslant 2b_{11} \Big\{ | R(t, \theta + \Phi(t, \theta, \xi), \xi + Y(t, \theta, \xi)) - T^0 R(t, \theta, \xi) | + \\ &+ \left| \frac{\partial \Phi(t, \theta, \xi)}{\partial \xi} [A(\theta) - TA(\theta)] \xi \right| + \\ &+ | F(t, \theta + \Phi(t, \theta, \xi), \xi + Y(t, \theta, \xi)) - T^1 F(t, \theta, \xi) | + \\ &+ \left| [A(\theta + \Phi(t, \theta, \xi)) - TA(\theta)] \xi - \frac{\partial TA(\theta) \xi}{\partial \theta} \Phi(t, \theta, \xi) \right| + \\ &+ | [A(\theta + \Phi(t, \theta, \xi)) - TA(\theta)] Y(t, \theta, \xi) | + \\ &+ \left| \left(E + \frac{\partial Y(t, \theta, \xi)}{\partial \xi} \right) [A(\theta) - TA(\theta)] \xi \right| \Big\}. \end{aligned} \tag{36.34}$$

Making use of the properties of operators T, T^0 and T^1 established in § 14, and the inequalities (36.30) and (36.31) proved for the functions $\Phi(t, \theta, \xi)$ and $Y(t, \theta, \xi)$, we can evaluate the respective summands appearing on the right-hand side of (36.34). We have

$$\begin{aligned} &| R(t, \theta + \Phi(t, \theta, \xi), \xi + Y(t, \theta, \xi)) - T^0 R(t, \theta, \xi) | \leqslant \\ &\leqslant \bar{b}_1 [N_s^{-l_0} | D_t^d D_\theta^\rho D_\xi^r R(t, \theta, \xi) |_0 + D_\theta T^0 R(t, \theta, \xi) |_0 | \Phi(t, \theta, \xi) | + \\ &+ | D_\xi T^0 R(t, \theta, \xi) |_0 | Y(t, \theta \xi) |] \leqslant \tilde{b}_1 \left[\left(\frac{N_{s-1}}{N_s} \right)^{l_0} + N_s^4 \delta_{s-1}^2 \right], \end{aligned} \tag{36.35}$$

and

$$\begin{aligned} &| F(t, \theta + \Phi(t, \theta, \xi), \xi + Y(t, \theta, \xi)) - T^1 F(t, \theta, \xi) | \leqslant \\ &\leqslant \tilde{b}_1 \left| \left(\frac{N_{s-1}}{N_s} \right)^{l_0} + CN_s^2 \delta_{s-1} N_s \delta_{s-1} + CN_s^2 \delta_{s-1} b N_s^3 \delta_{s-1} \right| \leqslant \\ &\leqslant \tilde{b}_1 \left[\left(\frac{N_{s-1}}{N_s} \right)^{l_0} + N_s^5 \delta_{s-1}^2 \right]. \end{aligned} \tag{36.36}$$

Moreover, since

$$\left|\frac{\partial \Phi(t, \theta, \xi)}{\partial \xi}[A(\theta) - TA(\theta)]\,\xi\right| + \left|\left(E + \frac{\partial Y(t, \theta\,\xi)}{\partial \xi}\right)[A(\theta) - TA(\theta)]\,\xi\right| \leqslant$$

$$\leqslant b_2 \,|\, A(\theta) - TA(\theta)\,|_0 \leqslant \tilde{b}_1 Q_s^{-l_1},$$

$$\left|[\,A\,(\theta + \Phi(t, \theta, \xi)) - TA(\theta)]\,\xi - \frac{\partial TA(\theta)\,\xi}{\partial \theta}\,\Phi(t, \theta, \xi)\right| \leqslant$$

$$\leqslant b_3\,[Q_s^{-l_1} + |\,D_\theta^2\,TA(\theta)\,|_0\,|\,\Phi(t, \theta, \xi)\,|_0^2] \leqslant \tilde{b}_1\,[Q_s^{-l_1} + Q_s^2\,N_s^2\,\delta_{s-1}^2], \tag{36.37}$$

$$|\,[\,A\,(\theta + \Phi(t, \theta, \xi)) - TA(\theta)]\,Y(t, \theta, \xi)\,| \leqslant$$

$$\leqslant \tilde{b}_3\,[Q_s^{-l_1}\,|\,A(\theta)\,|_{l_1} + |\,D_\theta\,TA(\theta)\,|_0\,|\,\Phi(t, \theta, \xi)\,|]\,|\,Y(t, \theta, \xi\,)\,| \leqslant$$

$$\leqslant \tilde{b}_1\,[Q_s^{-l_1}\,N_s^3\,\delta_{s-1} + Q_s N_s^4\,\delta_{s-1}^2],$$

a comparison of the inequality (36.34) with the inequalities (36.35) — (36.37) finally gives

$$|\,R^{(1)}(t, \theta, \xi)\,| + |\,F^{(1)}(t, \theta, \xi)\,| \leqslant$$

$$\leqslant 4\tilde{b}_1\,(N_{s-1}^{\beta\varkappa - (\varkappa-1)l_0} + Y_{s-1}^{5\varkappa - \beta(2-\varkappa)} + Q_s^{-l_1+\beta\alpha})\,\delta_s \leqslant \delta_s, \tag{36.38}$$

proving the first inequality in (36.11). Now, in order to examine the l_0-th derivatives of the functions $F^{(1)}(t, \theta, \xi)$ and $R^{(1)}(t, \theta, \xi)$, we express the vector $\{R^{(1)}(t, \theta, \xi), F^{(1)}(t, \theta, \xi)\}$ as a sum of two vectors, by

$$\{R^{(1)}(t, \theta, \xi), F^{(1)}(t, \theta, \xi)\} = B_1(t, \theta, \xi) + B_2(t, \theta, \xi), \tag{36.39}$$

where

$$B_1(t, \theta, \xi) = A_1^{-1}(\theta)\,\{R\,(t, \theta + \Phi(t, \theta, \xi), \xi + Y(t, \theta, \xi)) - T^0 R\,(t, \theta, \xi),$$
$$F(t, \theta + \Phi(t, \theta, \xi), \xi + Y(t, \theta, \xi)) - T^1 F(t, \theta, \xi)\}; \tag{36.40}$$

$$B_2(t, \theta, \xi) = A_1^{-1}(\theta))\left\{\frac{\partial \Phi(t, \theta, \xi)}{\partial \xi}(TA(\theta) - A(\theta))\,\xi,\ [A\,(\theta + \Phi(t, \theta, \xi))\right.$$

$$- TA(\theta)]\,(\xi + Y(t, \theta, \xi)) - \frac{\partial TA(\theta)\,\xi}{\partial \theta}\,\Phi(t, \theta, \xi) +$$

$$\left. + \left(E + \frac{\partial Y(t, \theta, \xi)}{\partial \xi}\right)(TA(\theta) - A(\theta))\,\xi\right\}$$

$$= A_1^{-1}(\theta) B_3(t, \theta, \xi).$$

To evaluate the l_0-th derivatives of the function $B_1(t_1, \theta_1, \xi_1)$, we put

$$\begin{aligned}\Phi_1(t_1, \theta_1, \xi_1) &= N_s\Phi\left(\frac{t_1}{N_s}, \frac{\theta_1}{N_s}, \frac{\xi_1}{N_s}\right),\\ Y_1(t_1, \theta_1, \xi_1) &= N_sY\left(\frac{t_1}{N_s}, \frac{\theta_1}{N_s}, \frac{\xi_1}{N_s}\right),\\ R_1(t_1, \theta_1, \xi_1) &= N_sR\left(\frac{t_1}{N_s}, \frac{\theta_1}{N_s}, \frac{\xi_1}{N_s}\right),\\ F_1(t_1, \theta_1, \xi_1) &= N_sF\left(\frac{t_1}{N_s}, \frac{\theta_1}{N_s}, \frac{\xi_1}{N_s}\right),\\ B_1^{(1)}(t_1, \theta_1, \xi_1) &= N_sB_1\left(\frac{t_1}{N_s}, \frac{\theta_1}{N_s}, \frac{\xi_1}{N_s}\right).\end{aligned} \tag{36.41}$$

Differentiating and evaluating the functions $\Phi_1(t_1, \theta_1, \xi_1)$, $R_1(t_1, \theta_1, \xi_1)$ and $F_1(t_1, \theta_1, \xi_1)$, we obtain

$$\begin{aligned}&| D^d_{t_1} D^\rho_{\theta_1} D^r_{\xi_1} \Phi_1(t_1, \theta_1, \xi_1) | + | D^d_{t_1} D^\rho_{\theta_1} D^r_{\xi_1} Y_1(t_1, \theta_1, \xi_1) | \leqslant\\ &\leqslant bN_sN_s^{-d-|\rho|-|r|}(N_s^{d+|\rho|+|\bar{r}|}Q_s^{|\rho|}\delta_{s-1} + N_s^{d+|\rho|+|\bar{r}|+r}Q_s^{|\rho|}\delta_{s-1}) \leqslant\\ &\leqslant 2bN_s^{\varkappa\left(4+\frac{|\rho|}{\alpha}\right)-\beta} < 1\end{aligned}$$

for all $d + |\rho| + |r| \leqslant l_0 + 1$, and

$$\begin{aligned}&| D^d_{t_1} D^\rho_{\theta_1} D^r_{\xi_1} R_1(t_1, \theta_1, \xi_1) | + | D^d_{t_1} D^\rho_{\theta_1} D^r_{\xi_1} F_1(t_1, \theta_1, \xi_1) |\\ &\leqslant N_s\left(\frac{N_{s-1}}{N_s}\right)^{l_0} \leqslant N_{s-1}^{\varkappa-(\varkappa-1)l_0} < 1, \qquad d + |\rho| + |r| = l_0, \qquad (36.42)\\ &| R_1(t_1, \theta_1, \xi_1) | + | F_1(t_1, \theta_1, \xi_1) | \leqslant N_s(| R(t, \theta, \xi) |_0 + | F(t, \theta, \xi) |_0)\\ &\qquad \leqslant N_s\delta_{s-1} < 1.\end{aligned}$$

From the inequality (36.42) it follows that the functions $R_1(t_1, \theta_1, \xi_1)$ and $F_1(t_1, \theta_1, \xi_1)$ and their l_0-th derivatives with respect to $z = (t_1, \theta_1, \xi_1)$ are bounded by unity. Hence, the derivatives of $R_1(t_1, \theta_1, \xi_1)$ and $F_1(t_1, \theta_1, \xi_1)$ of orders less than l_0 are bounded by a constant, not depending on N_s.

Reverting to the notations (36.40) and (36.41), we can assert that $B_1^{(1)}(t_1, \theta_1, \xi_1)$ is expressed by means of functions, whose first l_0-derivatives are bounded by a constant not depending on N_s : consequently, we deduce that

$$| D^d_{t_1} D^\rho_{\theta_1} D^r_{\xi_1} B_1^{(1)}(t_1, \theta_1, \xi_1) | \leqslant b_4 \qquad \text{for } d + |\rho| + |r| \leqslant l_0.$$

The last inequality obviously implies, for the l_0-th derivatives of the function $B_1(t, \theta, \xi)$, the inequality

$$| D^d_t D^\rho_\theta D^r_\xi B_1(t, \theta, \xi) | \leqslant b_4 N_s^{-1}N_s^{l_0} \leqslant \frac{N_s^{l_0}}{4} \qquad \text{with } d + |\rho| + |r| = l_0 \tag{36.43}$$

for all (t, θ, ξ) in the domain (36.6).

For estimating the l_0-th derivatives of the function $B_2(t, \theta, \xi)$, we differentiate the second relation in (36.40), to obtain

$$D_z^{l_0} B_2(z) = L_{l_0} \{ D_z^{i_1} A_1^{-1}(z) D_z^{i_2} B_3(z) \mid \}; \tag{36.44}$$

here $i_1 + i_2 = l_0$, $z = (t, \theta, \xi)$.

It is to be shown that the derivatives of the matrix A^{-1} satisfy the inequality

$$| D_z^{i_1} A_1^{-1}(z) | \leqslant b_5 N_s^{i_1+4} Q_s^{t+i_1} \delta_{s-1} \qquad \text{for } 0 \leqslant i_1 \leqslant l_0. \tag{36.45}$$

Indeed, for $i_1 = 0$ this inequality holds by virtue of the structure of the matrix A_1^{-1} and the estimates (36.30) and (36.31). Suppose that it holds also for all $0 \leqslant i_1 \leqslant k_0 < l_0$. Then, we have the estimates

$$\begin{aligned} | D_z^{k_0+1} A_1^{-1}(z) | &\leqslant | D_z^{k_0} (- A_1^{-1}(z) D_z A_1(z) A_1^{-1}(z)) | \leqslant \\ &\leqslant L_{k_0} [| D_z^{j_1} A_1^{-1}(z) | \, | D_z^{j_2+1} A_1(z) | \, | D_z^{j_3} A_1^{-1}(z) |] \leqslant \\ &\leqslant L_{k_0} [b_5 N_s^{j_1+j_3+8} Q_s^{j_1+j_3+2} \delta_{s-1}^2 \, | D_z^{j_2+2} Y(z) |] \leqslant \\ &\leqslant b_6 N_s^{k_0+13} Q_s^{k_0+4} \delta_{s-1}^3 \leqslant b_5 N_s^{k_0+5} Q_s^{k_0+2} \delta_{s-1}, \end{aligned} \tag{36.46}$$

and the inequality (36.45) is proved.

It is now required to show that the derivatives of the function $B_3(z)$ satisfy the inequality

$$| D_z^{i_2} B_3(z) | \leqslant b_5 N_s^{i_2} Q_s^{l_2} \qquad \text{for } 0 \leqslant i_2 \leqslant l_0. \tag{36.47}$$

In terms of the notation (36.40), we have

$$\begin{aligned} | D_3^{i_2} B_3(z) | \leqslant q_2 \Big(& | D_{t,\theta,\xi}^{i_2} A(\theta + \Phi(t, \theta, \xi)) \xi | + | D_{t,\theta,\xi}^{i_2} TA(\theta) Y(t, \theta, \xi) | + \\ & + | D_{t,\theta,\xi}^{i_2} TA(\theta) \xi | + \left| D_{t,\theta,\xi}^{i_2} \frac{\partial TA(\theta) \xi}{\partial \theta} \Phi(t, \theta, \xi) \right| + \\ & + \left| D_{t,\theta,\xi}^{i_2} \frac{\partial Y(t, \theta, \xi)}{\partial \xi} TA(\theta) \xi \right| + \\ & + \left| D_{t,\theta,\xi}^{i_2} A(\theta + \Phi(t, \theta, \xi)) Y(t, \theta, \xi) \right| + \\ & + | D_{t,\theta,\xi}^{i_2} A(\theta) \xi | + \left| D_{t,\theta,\xi}^{i_2} \frac{\partial Y(t, \theta, \xi)}{\partial \xi} A(\theta) \xi \right| \Big). \end{aligned} \tag{36.48}$$

We now proceed to evaluate the summands appearing on the right-hand side of (36.48). For the first one, (36.30) yields

$$| D^{i_2}_{t,\theta,\xi} A(\theta + \Phi(t, \theta, \xi))\, \xi | \leqslant b_7 \, | D^{l_1}_{t,\theta,\xi} A(\theta + \Phi(t, \theta, \xi)) |$$

$$\leqslant b_7 L_{i_2} \Big[| D^{\nu}_{\eta} A(\eta) | \, | D^{j_0+1}_{t,\theta,\xi} (\theta + \Phi(t, \theta, \xi))_{\beta_0} | \ldots$$

$$\ldots | D^{j_\nu+1}_{t,\theta,\xi} A(\theta + \Phi(t, \theta, \xi))_{\beta_\nu} | \Big]$$

$$\leqslant \bar{b}_7 L_{i_2} \Big[(e_1 + N_s^{j_1+2} Q_s^{j_1+1} \delta_{s-1}) \ldots$$

$$\ldots (e_\nu + N_s^{j_\nu+2} Q_s^{j_\nu+1} \delta_{s-1}) \Big], \tag{36.49}$$

where the constants $e_1, \ldots, e_\nu$ are either zero or 1; $\nu = 1, 2, \ldots, i_2$ and $j_1 + j_2 + \ldots + j_\nu = i_2 - \nu$.

The inequality (36.49) implies the estimate

$$| D^{i_2}_z A(\theta + \Phi(t, \theta, \xi))\, \xi | \leqslant \bar{b}_7 L_{i_2} [N_s^{i_2-\nu} Q_s^{i_2-\nu} (e_1 N_s^{-j_1} Q_s^{-j_1} + N_s^2 Q_s \delta_{s-1}) \ldots$$

$$\ldots (e_\nu N_s^{-j_\nu} Q_s^{-j_\nu} + N_s^2 Q_s \delta_{s-1})] \leqslant b_8 N_s^{i_2-1} Q_s^{i_2-1}. \tag{36.50}$$

Manipulating the inequality (36.50), we find that

$$| D^{i_2}_{t,\theta,\xi} A(\theta + \Phi(t, \theta, \xi))\, Y(t, \theta, \xi) | \leqslant$$

$$\leqslant b_7 L_{2_1} [| D^{\nu}_{t,\theta,\xi} A(\theta + \Phi(t, \theta, \xi)) | \, | D^{i_2-\nu}_{t,\theta,\xi} Y(t, \theta, \xi) | \leqslant$$

$$\leqslant \bar{b}_8 L_{i_2} [| N_s^{\nu-1} Q_s^{\nu-1} N_s^{i_2-\nu+3} Q_s^{i_2-\nu} \delta_{s-1}] \leqslant b_9 N_s^{i_2} Q_s^{i_2}. \tag{36.51}$$

The estimates for the remaining summands in (36.48) are obtained in a similar manner:

$$| D^{i_2}_{t,\theta,\xi} TA(\theta)\, \xi | \leqslant b_{10} Q_s^{i_2}, \quad | D^{i_2}_{t,\theta,\xi} TA(\theta)\, Y(t, \theta, \xi) | \leqslant b_{10} N_s^{i_2+3} Q_s^{i_2} \delta_{s-1},$$

$$\left| D^{i_2}_{t,\theta,\xi} \frac{\partial TA(\theta)\, \xi}{\partial \theta} \Phi(t, \theta, \xi) \right| \leqslant b_{10} N_s^{i_2+1} Q_s^{i_2+1} \delta_{s-1},$$

$$\left| D^{i_2}_{t,\theta,\xi} \frac{\partial Y(t, \theta, \xi)}{\partial \xi} TA(\theta)\, \xi \right| \leqslant b_{10} N_s^{i_2+4} Q_s^{i_2+1} \delta_{s-1},$$

$$| D^{i_2}_{t,\theta,\xi} A(\theta)\, \xi | \leqslant b_{10}, \tag{36.52}$$

$$\left| D^{i_2}_{t,\theta,\xi} \frac{\partial Y(t, \theta, \xi)}{\partial \xi} A(\theta)\, \xi \right| \leqslant b_{10} N_s^{i_2+4} Q_s^{i_2+1} \delta_{s-1}$$

for $0 \leqslant i_2 \leqslant l_0$. Comparing the inequalities (36.48) – (36.52), we arrive at the desired estimate

$$| D_z^{i_2} B_3 (z) | \leqslant b_5 N_s^{i_2} Q_s^{i_2} .$$

For the l_0-th derivatives of the function $B_2 (t, \theta, \xi)$, we can now make use of the inequalities (36.45) and (36.48) to obtain

$$| D_t^d D_\theta^\rho D_\xi^r B_2 (t, \theta, \xi) | \leqslant L_{l_0} [b_5 N_s^{i_2+i_1} Q_s^{i_1+i_2} N_s^4 Q_s \delta_{s-1}] \leqslant$$

$$\leqslant b_{11} N_s^4 Q_s^{i_0+1} \delta_{s-1} N_s^{i_0} \leqslant \frac{N_s^{l_0}}{4}$$

$$\text{for } d + | \rho | + | r | = l_0. \tag{36.53}$$

This, by the estimate (36.43), implies the inequality

$$| D_t^d D_\theta^\rho D_\xi^r R^{(1)} (t, \theta, \xi) | + | D_t^d D_\theta^\rho D_\xi^r F^{(1)} (t, \theta, \xi) | \leqslant N_s^{l_0}$$

for $d + | \rho | + | r | = l_0$, i. e. the second inequality in (36.11) of the theorem under consideration is proved.

Since the operators T^0 and T^1 preserve the zeros of the functions $R (t, \theta, \xi)$ and $F (t, \theta, \xi)$, we deduce from (36.15) the relations (36.10), which completes the proof of the present inductive theorem.

§ 37. Neighbourhood of an Invariant Manifold

In investigating the behaviour of the solutions of the system (34.1) in a neighbourhood of the manifold M_t, the solutions of the system (34.15) in a neighbourhood of the manifold $x = 0$ had also claimed our attention. Solving this problem, we intend to study the equations (34.17) in a neighbourhood of the manifold $x = 0$ and establish the following theorem.

Theorem 32. *For given positive constants a_0, a, γ, $\bar{\eta}$ and a given integer s_0 ($s_0 \geqslant 2$), there exist a positive number $\delta_0 = \delta_0 (a_0, \gamma, \bar{\eta}, s_0$ and integers $l_0 = l_0 (s_0)$ and $l_1 = l_1 (s_0)$ such that if for*

$$(t, c) \in (-\infty, \infty) \times D_1 \quad \| x \| \leqslant \eta \tag{37.1}$$

the functions $R (t, c, x)$ and $F (t, c, x)$ are l_0-times continuously differentiable and satisfy the condition

$$R (t, c, 0) = F (t, c, 0) = \frac{\partial F (t, c, 0)}{\partial x} = 0 \tag{37.2}$$

together with the inequalities

$$| R(t, c, x) | + | F(t, c, x) | \leqslant \delta_0,$$
$$| R(t, c, x) |_{l_0} + | F(t, c, x) |_{l_0} \leqslant a_0, \tag{37.3}$$

and if $A(c)$ is a real canonical matrix, l_1-times continuously differentiable, bounded, together with its partial derivatives,

$$| A(c) |_{l_1} \leqslant a, \tag{37.4}$$

and is such that its eigenvalues $\mu_1(c), \ldots, \mu_n(c)$ satisfy the inequalities†

$$\mathrm{Re}\,[\mu_\alpha(c) + \mu_\beta(c) - \mu_j(c)] \leqslant -\gamma \tag{37.5}$$

for every $c \in D_1$ and any $\alpha, \beta, j = 1, 2, \ldots, n$, then there exist functions $\Phi(t, \theta, \xi)$ and $Y(t, \theta, \xi)$, $(s_0 - 1)$-times continuously differentiable, bounded for

$$(t, c) \in (-\infty, \infty)\, D_1 - \bar{\eta}, \qquad \| \xi \| \leqslant \eta - \bar{\eta}, \tag{37.6}$$

and satisfying the relations

$$\Phi(t, \theta, 0) = Y(t, \theta, 0) = \frac{\partial Y(t, \theta, 0)}{\partial \xi} = 0 \tag{37.7}$$

and the inequality

$$| \Phi(t, \theta, \xi) |_{s_0-1} + | Y(t, \theta, \xi) |_{s_0-1} \leqslant \bar{\eta}, \tag{37.8}$$

so that the system of equations

$$\frac{dc}{dt} = R(t, c, x),$$
$$\frac{dx}{dt} = A(c)\, x + F(t, c, x) \tag{37.9}$$

is reduced by the substitution

$$c = \theta + \Phi(t, \theta, \xi),$$
$$x = \xi + Y(t, \theta, \xi) \tag{37.10}$$

to the system

$$\frac{d\theta}{dt} = 0,$$
$$\frac{d\xi}{dt} = A(\theta)\, \xi. \tag{37.11}$$

†See Appendix 15.

This theorem implies that the general solution of the system (37.9) in a neighbourhood of the manifold $x = 0$ has the form

$$c = c_0 + \Phi(t, c_0, e^{A(c_0)t} y_0),$$
$$x = e^{A(c_0)t} y_0 + Y(t, c_0, e^{A(c_0)t} y_0), \tag{37.12}$$

where c_0, y_0 are arbitrary constants in the domain $c_0 \in D_1 - \bar{\eta}$, $\| y_0 \| \leqslant \eta - \bar{\eta}$. The general solution of equations (37.12) is, consequently, defined for $t \geqslant 0$, and asymptotically attracted to the trivial manifold $c = c_0$, $x = 0$. It shall be seen from the proof below that the solution (37.12) can be taken in the first approximation, expressed by

$$c_t = c - \int_0^\infty R(t + \tau, c_0, e^{A(c_0)\tau} y_0)\, d\tau,$$
$$x_t = e^{A(c_0)t} y_0 - \int_0^\infty \left[\frac{\partial A(c_0) y_0}{\partial c} (c_{t+\tau} - c_0) + \right. \tag{37.13}$$
$$\left. + e^{A(c_0)\tau} F(t + \tau, c_0, e^{A(c_0)\tau} y_0) \right] d\tau.$$

Proof of Theorem 32. Let l_0, l_1 and δ_0 be so chosen that for $s = 1$ Theorem 31 is applicable to the system (37.9). The assertions of this theorem imply the existence of the substitution

$$c = c^{(1)} + \Phi^{(1)}(t, c^{(1)}, x^{(1)}),$$
$$x = x^{(1)} + Y^{(1)}(t, c^{(1)}, x^{(1)}), \tag{37.14}$$

as a result of which a system similar to (37.9) is obtained, though with the functions $R^{(1)}(t, c^{(1)}, x^{(1)})$ and $F^{(1)}(t, c^{(1)}, x^{(1)})$ satisfying the hypotheses of Theorem 32.

Now, apply successively the inductive Theorem 31 to the new system, to obtain at the s-th step the change of variable

$$c^{(s-1)} = c^{(s)} + \Phi^{(s)}(t, c^{(s)}, x^{(s)}),$$
$$x^{(s-1)} = x^{(s)} + Y^{(s)}(t, c^{(s)}, x^{(s)}), \tag{37.15}$$

reducing the system of equations for $c^{(s-1)}$, $x^{(s-1)}$ to the system

$$\frac{dc^{(s)}}{dt} = R^{(s)}(t, c^{(s)}, x^{(s)}),$$
$$\frac{dx^{(s)}}{dt} = A(c^{(s)})\, x^{(s)} + F^{(s)}(t, c^{(s)}, x^{(s)}).$$

In addition, the functions $\Phi^{(s)}(t, c^{(s)}, x^{(s)})$, $Y^{(s)}(t, c^{(s)}, x^{(s)})$, $R^{(s)}(t, c^{(s)}, x^{(s)})$ and $F^{(s)}(t, c^{(s)}, x^{(s)})$ are defined on the domain

$$(t, c^{(s)}) \in (-\infty, \infty)\, D_1 - \sum_{\nu=1}^{s} Q_\nu^{-1}, \quad \| x^{(s)} \| \leqslant \eta - \sum_{\nu=1}^{s} N_\nu^{-1}, \tag{37.16}$$

are l_0-times continuously differentiable and satisfy the relations

$$\Phi^{(s)}(t, c^{(s)}, 0) = Y^{(s)}(t, c^{(s)}, 0) = \left.\frac{\partial Y^{(s)}(t, c^{(s)}, x)}{\partial x}\right|_{x=0} = $$
$$= R^{(s)}(t, c^{(s)}, 0) = F^{(s)}(t, c^{(s)}, 0) = \left.\frac{\partial F^{(s)}(t, c^{(s)}, x)}{\partial x}\right|_{x=0} = 0 \tag{37.17}$$

and the inequalities

$$| \Phi^{(s)}(t, c^{(s)}, x^{(s)}) |_{s_0} + | Y^{(s)}(t, c^{(s)}, x^{(s)}) |_{s_0} \leqslant Q_{s-1}^{-1},$$
$$| R^{(s)}(t, c^{(s)}, x^{(s)}) | + | F^{(s)}(t, c^{(s)}, x^{(s)}) | \leqslant \delta_s, \tag{37.18}$$
$$| D_t^d D_c^\rho D_x^r R^{(s)}(t, c^{(s)}, x^{(s)}) | + | D_t^d D_c^\rho D_x^r F^{(s)}(t, c,^{(s)}, x^{(s)}) | \leqslant N_s^{l_0}$$
$$\text{for } d + |\rho| + |r| = l_0.$$

As in the preceding chapter (see § 32, p. 227), by using the inequality (37.18) we can prove the convergence of the iteration process without any difficulty. For this, we express c and x in terms of $c^{(s)}$ and $x^{(s)}$ by

$$c = c^{(s)} + \Psi^{(s)}(t, c^{(s)}, x^{(s)}), \quad x = x^{(s)} + X^{(s)}(t, c^{(s)}, x^{(s)}). \tag{37.19}$$

The functions $\Psi^{(s)}(t, c^{(s)}, x^{(s)})$ and $X^{(s)}(t, c^{(s)}, x^{(s)})$ are defined on the domain (37.16), are l_0-times continuously differentiable and satisfy the conditions (37.17).

A series of calculations, similar to those carried out in § 32, yields for these functions the limits

$$\lim_{s\to\infty} \Psi^{(s)}(t, \theta, \xi) = \Psi^{(\infty)}(t, \theta, \xi),$$
$$\lim_{s\to\infty} X^{(s)}(t, \theta, \xi) = X^{(\infty)}(t, \theta, \xi), \tag{37.20}$$

and the inequality

$$| \Psi^{(\infty)}(t, \theta, \xi) | + | X^{(\infty)}(t, \theta, \xi) | \leqslant 2\left[(1 + B_1) \sum_{\nu=0}^{\infty} Q_{1+\nu}^{-1} + Q_0^{-1}\right] < \bar{\eta}. \tag{37.21}$$

Conditions, similar to (37.17), imply that

$$\Psi^{(\infty)}(t, \theta, 0) = X^{(\infty)}(t, \theta, 0) = 0. \tag{37.22}$$

Since by the estimates (37.18), $R^{(s)}(t, \theta, \xi) \to 0$ and $F^{(s)}(t, \theta, \xi) \to 0$ uniformly in t, θ, ξ as $s \to \infty$, it follows that

$$\Phi(t, \theta, \xi) = \Psi^{(\infty)}(t, \theta, \xi), \qquad Y(t, \theta, \xi) = X^{(\infty)}(t, \theta, \xi). \tag{37.23}$$

To complete the proof of Theorem 32, it is now required to show that the functions $\Psi^{(\infty)}(t, \theta, \xi)$ and $X^{(\infty)}(t, \theta, \xi)$ are (s_{0-1})-times continuously differentiable and that the Jacobian

$$\det \left| E + \begin{pmatrix} \dfrac{\partial \Phi}{\partial \theta}, & \dfrac{\partial \Phi}{\partial \xi} \\ \dfrac{\partial Y}{\partial \theta}, & \dfrac{\partial Y}{\partial \xi} \end{pmatrix} \right| \tag{37.24}$$

is non-singular.

Evaluating the s_0-th derivatives of the functions $\Psi^{(s)}(t, \theta, \xi)$ and $X^{(s)}(t, \theta, \xi)$ and carrying out a series of calculations (for simplicity we consider only the partial derivatives with respect to t), we obtain the estimate

$$\left| \frac{\partial \Psi^{(s+1)}(t, \theta, \xi)}{\partial t} - \frac{\partial \Psi^{(s)}(t, \theta, \xi)}{\partial t} \right| \leqslant (1 + B_1 + B_2)\, Q_s^{-1}, \tag{37.25}$$

implying that the criterion for the uniform convergence of the function $\partial \Psi^{(s)}(t, \theta, \xi)/\partial t$ is satisfied:

$$\left| \frac{\partial \Psi^{(s+k)}(t, \theta, \xi)}{\partial t} - \frac{\partial \Psi^{(s)}(t, \theta, \xi)}{\partial t} \right| \leqslant (1 + B_1 + B_2) \sum_{\nu=0}^{k-1} Q_{s+\nu}^{-1}. \tag{37.26}$$

Hence it follows that the function $\Psi^{(\infty)}(t, \theta, \xi)$ has a continuous first derivative with respect to t and also that the relations

$$\frac{\partial \Psi^{(s)}(t, \theta, \xi)}{\partial t} \underset{s \to \infty}{\to} \frac{\partial \Psi^{(\infty)}(t, \theta, \xi)}{\partial t}, \qquad \left| \frac{\partial \Psi^{(\infty)}(t, \theta, \xi)}{\partial t} \right| < \frac{\bar{\eta}}{2} \tag{37.27}$$

hold.

By similar reasoning we verify that the functions $\Psi^{(\infty)}(t, \theta, \xi)$, $X^{(\infty)}(t, \theta, \xi)$ are differentiable with respect to θ, ξ, and t, θ, ξ, respectively. The relations (37.27), for $\bar{\eta}$ small, directly imply that the Jocobian (37.24) is non-singular.

The dependence of l_0 and l_1 on s_0, referred to in the theorem, is given implicitly by the system of inequalities (36.2). It implies, in particular, that to reduce the system (37.9) to the form (37.11) by the 5-times continuously differentiable substitution (37.10) it is sufficient that the functions $R(t, c, x)$ and $F(t, c, x)$ be 50-times and $A(c)$ be 125-times continuously differentiable.

It should be noticed that the system of inequalities (36.2) does not uniquely define the dependence of l_0 and l_1 on s_0. Giving a specific meaning to the parameters α and $\varkappa$, a concrete relationship between l_0, l_1 and s_0 can be determined. Thus, in particular, if the matrix $A(c)$ is infinitely differentiable, then we can take α to be sufficiently large, $\varkappa$ positive and equal to $(2s_0 + 1)/(s_0 + 3)$ and $l_0 = l_0(s)$ defined by

$$l_0 = \left[\frac{2(s_0 + 1)(2s_0 + 1)}{s_0 - 2}\right] + 1 \qquad (s_0 \geqslant 5), \tag{37.28}$$

where [. . .] denotes an integral part of the number $[2(s_0 + 1)(2s_0 + 1)]/(s_0 - 2)$. Hence it follows that for reducing the system (37.9) to the system (37.11), when A is a constant, by the 4-times continuously differentiable substitutions (37.10), it is sufficient that the right-hand side of system (37.9) be 45-times continuously differentiable.

§ 38. Behaviour of Solutions of a System of Two Equations in the Neighbourhood of Equilibrium Positions

We consider a system of equations

$$\begin{aligned} \frac{dx_1}{dt} &= y, \\ \frac{dy}{dt} &= f(t, x_1, y)\, y, \end{aligned} \tag{38.1}$$

having the invariant manifold

$$x_1 = c, \qquad y = 0, \qquad (c \in (-\infty, \infty)), \tag{38.2}$$

filled by equilibrium positions. For the manifold D_1, we take the segment $c_0 \leqslant c \leqslant c_1$, which is either of finite length or is the union of a finite number of such segments. The manifold M_t

$$x_1 = c, \qquad y = 0 \qquad (c \in D_1) \tag{38.3}$$

is an infinitely differentiable compact manifold. In its neighbourhood, the system of equations (38.1) assumes, for small y, the form

$$\begin{aligned} \frac{dc}{dt} &= z, \\ \frac{dz}{dt} &= f(t, c, 0)\, z + f_1(t, c, z), \end{aligned} \tag{38.4}$$

with the notation

$$x_1 = c, \qquad y = z, \qquad f_1(t, c, z) = [f(t, c, z) - f(t, c, 0)]\, z.$$

The change of variables†

$$z = e^{\int^t [f(t, c, 0) - f_0(c)]\, dt}\, x, \qquad f_0(c) = \frac{1}{2\pi}\int_0^{2\pi} f(t, c, 0)\, dt \tag{38.5}$$

reduces the system (38.4) to the system

$$\frac{dc}{dt} = e^{\int^t [f(t, c, 0) - f_0(c)]\, dt}\, x,$$

$$\frac{dx}{dt} = f_0(c)\, x + e^{-\int^t [f(t, c, 0) - f_0(c)]\, dt} f_1\left(t, c, e^{\int^t [f(t, c, 0) - f_0(c)]\, dt}\right). \tag{38.6}$$

Suppose that $x_1, x_2, \ldots, x_n$ denote the zeros of the function $f_0(c)$ ($x_{i+1} > x_i$, $i = 1, \ldots, n$). For every segment $[c_i, c_{i+1}]$, contained in the interval (x_i, x_{i+1}) ($i = 0, 1, \ldots, n, n+1$; $x_0 = -\infty$, $x_{n+1} = +\infty$), the system (38.6) satisfies the hypotheses of Theorem 32 either for $t > 0$, if in the segment considered $f_0(c) < 0$, or for $t < 0$, if $f_0(c) > 0$ in that segment. Hence it follows that outside a neighbourhood of the zeros of function $f_0(c)$, the substitution

$$\begin{aligned} c &= \theta + \Phi(t, \theta, \xi), \\ x &= \xi + Y(t, \theta, \xi) \end{aligned} \tag{38.7}$$

2π-periodic in t, reduces the system (38.6) to the system

$$\begin{aligned} \frac{d\theta}{dt} &= 0, \\ \frac{d\xi}{dt} &= f_0(\theta)\, \xi. \end{aligned} \tag{38.8}$$

Solving the system of equations (38.8) and substituting the result into (38.7), we obtain the general solution of equations (38.1) in a neighbourhood of the manifold $D_1 \cup D_1'$ in the form

$$\begin{aligned} c_t &= c_0 + \Phi(t, c_0, e^{f_0(c_0)t} y_0), \\ x_t &= e^{f_0(c_0)t} y_0 + Y(t, c_0, e^{f_0(c_0)t} y_0). \end{aligned} \tag{38.9}$$

† $\int^t f\, dx$ denotes an anti-derivative of the function f.

It is seen from (38.9) that the solutions of system (38.1) originating in a neighbourhood of the manifold D_1, are damped and oscillatory, and also that each solution tends to an equilibrium position on the manifold D_1. The rate of damping is not uniform; it depends on the function $f_0(c)$ which plays the role of the logarithmic 'decrement' of damping in its own way. Furthermore, the rate of the damping process is observed to be fastest when the absolute value of the function $f_0(c)$ is the largest. In a neighbourhood of the zeros of the function $f_0(c)$ the damping slows down.

The situation for $T < 0$ is, however, similar for the solutions originating from a neighbourhood of D_1'.

APPENDICES

Appendix 1 (cf. p. 47)

While investigating the problem of the convergent power series expansion of conditionally periodic motions, Moser subsequently obtained the existence conditions, which are more general than those derived in Theorem 4. The principal results of Moser's paper, "Convergent series expansions for quasi-periodic motions", *Math. Ann.*, **169** (1967), 136-176, are listed in the next theorem.

Theorem. *Suppose that the system*

$$\frac{dh}{dt} = Hh + Mh + \mu + \varepsilon F(h, \varphi, \varepsilon), \qquad \frac{d\varphi}{dt} = \omega + \Delta + \varepsilon f(h, \varphi, \varepsilon) \tag{1}$$

satisfies the following conditions :

(i) *the functions $F(h, \varphi, \varepsilon)$ and $f(h, \varphi, \varepsilon)$ are analytic on the domain*

$$|h| < \eta, \quad |\operatorname{Im} \varphi| > \rho, \quad |\varepsilon| < \varepsilon_0, \tag{2}$$

2π-periodic in φ and real for real arguments,

(ii) *the characteristic numbers, i.e. the frequencies $\omega = (\omega_1, \ldots, \omega_m)$ and the eigenvalues $\lambda = (\lambda_1, \ldots, \lambda_n)$ of the diagonal matrix H satisfy the inequalities*

$$|(l, \lambda) + i(k, \omega)| \geqslant K(|k|^d + 1)^{-1} \qquad (0 < K \leqslant 1) \tag{3}$$

for every integral-valued vector $k = (k_1, \ldots, k_m)$ and all integral-valued $l = (l_1, \ldots, l_n)$, so that

$$\left|\sum_{\alpha=1}^{n} l_\alpha\right| \leqslant 1, \qquad \sum_{\alpha=1}^{n} |l_\alpha| \leqslant 2, \tag{4}$$

except for the finite number $l, k = l, 0$, for which the left-hand side of (3) *vanishes.*

Then, there exist the uniquely defined power series $\mu(\varepsilon)$, $\Delta(\varepsilon)$ and $M(\varepsilon)$, satisfying the conditions

$$H\mu = 0, \qquad HM = MH \tag{5}$$

such that (1) *has a quasi-periodic solution with characteristic numbers ω, λ.*

Furthermore, there is a coordinate transformation of the form

$$h = g + \varepsilon U(g, \theta, \varepsilon),$$
$$\varphi = \theta + \varepsilon \Phi(\theta, \varepsilon), \tag{6}$$

with the functions $U(g, \theta, \varepsilon)$ and $\Phi(\theta, \varepsilon)$, analytic in g, θ, ε, 2π-periodic in θ and real for real arguments, which reduces the system of equations (1) *to the system*

$$\dot{g} = Hg + O(g^2),$$
$$\dot{\theta} = \omega + O(g). \tag{7}$$

Here $O(g^d)$ denotes an analytic function of g, θ, ε which, at $g = 0$, vanishes together with its derivatives with respect to g up to order $\alpha - 1 \geqslant 0$.

In particular,

$$h = \varepsilon U(0, \omega t + \theta_0, \varepsilon),$$
$$(\theta_0 = \text{const.}) \tag{8}$$
$$\varphi = \omega t + \theta_0 + \varepsilon \Phi(\omega t + \theta_0, \varepsilon)$$

is a quasi-periodic solution of (1) *with the characteristic numbers $\omega_1, \ldots, \omega_m$, $\lambda_1, \ldots, \lambda_n$. Each of the series μ, Δ, M, U, Φ has a non-zero radius of convergence which depends on ε, μ, Δ, M and which vanishes for $\varepsilon = 0$.*

Appendix II (cf. p. 55)

The case when some of the coordinates of the vector $a^{(0)}$ are zero was considered by V. T. Yatsuk in his paper, "On the existence of a quasi-periodic solution of a system of second order differential equations", *Ukr. Math. Zh.*, **24** (5), (1972).

As is known, in this case the amplitude-phase conversion of coordinates gives rise to additional difficulties connected with investigating the solutions of equations for amplitude and phase in the neighbourhood of a pole. These can be circumvented by applying, instead of the amplitude-phase change of the coordinates q_{k_j} $(j = 1, 2, \ldots, r)$ satisfying $a_{k_j}^{(0)} = 0$, the more extensively used Van-der-Pol transformation given by

$$q_{k_j} = \xi_j \sin \lambda_{k_j} t + \eta_j \cos \lambda_{k_j} t,$$
$$(j = 1, 2, \ldots, r) \tag{9}$$
$$\dot{q}_{k_j} = \xi_j \lambda_{k_j} \cos \lambda_{k_j} t - \eta_j \lambda_{k_j} \sin \lambda_{k_j} t.$$

This enables us to obtain the existence conditions for and the properties of quasi-periodic solutions of the system (6.1) (see p. 54) :

$$\frac{d^2 q_k}{dt} + \lambda_k^2 q_k = \varepsilon f_k \left(q_1, \ldots, q_n, \frac{dq_1}{dt}, \ldots, \frac{dq_n}{dt} \right),$$

$$(k = 1, 2, \ldots, n) \qquad (6.1)$$

with the frequency basis $\lambda_{k_1}, \ldots, \lambda_{k_r}, \omega_{k_{r+1}}, \ldots, \omega_{k_{n-r}}$.

Appendix III (cf. p. 55)

Theorem 6 elucidates the existence conditions for the quasi-periodic solutions of system (6.1) with the frequency basis $\omega_1, \ldots, \omega_n$. In their paper, "Asymptotic expansions of quasi-periodic solutions in quasi-linear systems of second order", in : *Asymptotic and Qualitative Methods in the Theory of Non-linear Oscillations"*, Inst. Mat., Acad. Nauk USSR, Kiev (1971), Mitropoliskii and Samoilenko have examined the quasi-periodic solutions of system (6.1) with the frequency basis

$$\omega_1, \ldots, \omega_r \qquad (r \leqslant n), \qquad (10)$$

i.e. quasi-periodic solutions with 'partial' frequency basis.

In the construction of such solutions, the right-hand sides of differential equations (6.1) are assumed to be the polynomial functions of their arguments. The initial frequencies $\lambda_1, \ldots, \lambda_n$ are expressed in the form

$$\lambda_i^2 = \omega_i^2 - \varepsilon \Delta_i (\varepsilon), \qquad (i = 1, 2, \ldots, r)$$

$$(11)$$

$$\lambda_j^2 = (m^j, \omega)^2 - \varepsilon \Delta_j (\varepsilon), \qquad (j = r + 1, \ldots, n),$$

where $m^j = (m_1^{(j)}, \ldots, m_r^{(j)})$ is a vector with integral coordinates and $\Delta_i (\varepsilon)$ is some function of the parameter ε.

The quasi-periodic solutions of the original system are sought in the form of a power series in the parameter ε :

$$q_i = a_i^0 \cos \psi_i + \varepsilon u_i^{(1)} (\psi) + \varepsilon^2 u_i^{(2)} (\psi) + \ldots$$

$$(12)$$

$$q_j = a_j^0 \cos (m^j, \psi) + \varepsilon u_j^{(1)} (\psi) + \varepsilon^2 u_j^{(2)} (\psi) + \ldots,$$

where $\psi = \omega t + \psi^0$. The functions $\Delta (\varepsilon)$ obey the condition $u_1^{(l)}, \ldots, u_n^{(l)}$, are defined for every $l = 1, 2, \ldots$ and are 2π-periodic in ψ.

Writing

$$A_i^0(a, \theta) = \frac{1}{(2\pi)^r} \int_0^{2\pi} \ldots \int_0^{2\pi} f_i\{a_1 \cos \psi_1, \ldots, a_r \cos \psi_r, a_{r+1} \cos \times$$

$$\times [(m^{r+1}, \psi) - \theta_1], \ldots, a_n \cos [(m^n, \psi) - \theta_{n-r}]; -a_1\omega_1 \times$$

$$\times \sin \psi_1, \ldots, - a_n (m^n, \omega) \sin [(m^n, \psi) - \theta_{n-r}]\} \sin \psi_i d\psi_1 \ldots d\psi_r, \tag{13}$$

$$A_j^0(a, \theta) = \frac{1}{(2\pi)^r} \int_0^{2\pi} \ldots \int_0^{2\pi} f_j \{a_1 \cos \psi_1, \ldots, - a_n (m^n, \omega) \sin \times$$

$$\times [(m^n, \psi) - \theta_{n-r}]\} \sin [(m^j, \psi) - \theta_{j-r}] \, d\psi_1 \ldots d\psi_r$$

$$(i = 1, \ldots, r; j = r + 1, \ldots, n),$$

$$A^0 = (A_1^0, A_2^0, \ldots, A_n^0),$$

the existence problem for the formal power series expansion (12) is solved in the following theorem.

Theorem. *Suppose that the system* (6.1) *under consideration is such that the equation*

$$A^0(a, 0) = 0 \tag{14}$$

has a solution $a = a^0$, $a^0 = (a_1^0, \ldots, a_n^0)$, $a_\alpha^0 \neq 0$ for $\alpha = 1, 2, \ldots, n$ *and*

$$\det \left\| \frac{\partial A^0(a^0, 0)}{\partial a} \right\| \neq 0. \tag{15}$$

Then for a sequence $\{\Delta^{(\nu)}\}$, *prescribed for each value of* $\nu = 0, 1, \ldots$ *with* $\Delta(\varepsilon)$ *defined by*

$$\Delta(\varepsilon) = \sum_{\nu=0}^{\infty} \varepsilon^\nu \Delta^{(\nu)}, \tag{16}$$

the system of equations (6.1) *has the formal quasi-periodic solution* (12) *with the frequency basis* (10).

Let

$$B_i^0(a, \theta) = \frac{1}{(2\pi)^r} \int_0^{2\pi} \ldots \int_0^{2\pi} \frac{f_i\{a_1 \cos\psi_1, \ldots, -a_n (m^n, \omega) \sin [(m^n, \psi) - \theta_{n-r}]\}}{a_i\omega_i} \times$$

$$\times \cos\psi_i d\psi_1 \ldots d\psi_n, \qquad (i = 1, 2, \ldots, r) \tag{17}$$

$$B_j^0(a, \theta) = \frac{1}{(2\pi)^r} \int_0^{2\pi} \cdots \int_0^{2\pi} \frac{f_j\{a_1 \cos\psi_1, \ldots, -a_n(m^n, \omega) \sin[(m^n, \psi) - \theta_{n-r}]\}}{a_j(m^j, \omega)} \times$$

$$\times \cos[(m^j, \psi) - \theta_{j-r}]\, d\psi_1 \ldots d\psi_n, \qquad (j = r + 1, \ldots, n)$$

$$B^0 = (B_1^0, B_2^0, \ldots, B_r^0), \quad B' = (B_{r+1}^0, \ldots, B_n^0),$$

$$\mathfrak{M} = \text{colon}\,\{m^{r+1}, \ldots, m^n\}.$$

The next proposition holds in analogy to Theorem 6 (see p. 55).

Theorem. *Suppose that for the system of differential equations (6.1) the following conditions are fulfilled :*

(i) *The functions $f_k(q, q')$ are analytic in the variables q, q' in the neighbourhood of the torus*

$$\begin{aligned} q_i &= a_i^0 \cos\psi_i, & q_j &= a_j^0 \cos(m^j, \psi), \\ q_i' &= -a_i^0 \omega_i \sin\psi_i, & q_j' &= -a_j^0 (m^j, \omega) \sin(m^j, \psi). \\ (i &= 1, \ldots, r; & j &= r + 1, \ldots, n). \end{aligned} \tag{18}$$

(ii) *The frequencies $\omega = (\omega_1, \ldots, \omega_r)$ satisfy the inequalities*

$$|(k, \omega)| \geqslant K |k|^{-(r+1)} \qquad (|k| \neq 0) \tag{19}$$

for every integral-valued vector $k = (k_1, \ldots, k_r)$.

(iii) *For $\alpha = 1, \ldots, n$, equation (14) has a solution $a = a^0$, $a^0 = (a_1^0, \ldots, a_n^0)$, $a_\alpha^0 \neq 0$, so that the inequality (15) is satisfied and the real parts of all the eigenvalues of the matrix*

$$\left\| \begin{matrix} \dfrac{\partial A^0(a^0, 0)}{\partial a}, & \dfrac{\partial A^0(a^0, 0)}{\partial \theta} \\ \dfrac{\partial [B'(a^0, 0) - \mathfrak{M} B^0(a^0, 0)]}{\partial a}, & \dfrac{\partial [B'(a^0, 0) - \mathfrak{M} B^0(a^0, 0)]}{\partial \theta} \end{matrix} \right\| \tag{20}$$

are negative.

Then there exists a uniquely defined $\Delta(\varepsilon)$ for $0 < \varepsilon < \varepsilon_0$ such that the system of equations (6.1) with λ defined as a function of $\omega_1, \ldots, \omega_r$ and $\Delta_1(\varepsilon), \ldots, \Delta_n(\varepsilon)$ according to (11), has a quasi-periodic solution with the frequency $\omega_1, \ldots, \omega_r$:

$$q = u(\omega t + \psi_0, \varepsilon), \tag{21}$$

and every solution, sufficiently close to the quasi-periodic solution, tends to it as $t \to \infty$.

Moreover, the formal power series expansion in ε *of the functions* $u(\psi, \varepsilon)$, $\Delta(\varepsilon)$ *is defined by expressions* (12), (16) *and is asymptotically convergent.*

Appendix IV (cf. p. 59)

The conditions for the existence of quasi-periodic solutions, deduced in Chap. 1, indicate the possibility of selecting that frequency $\lambda(\varepsilon)$ which keeps the given frequency basis ω of the quasi-periodic solutions fixed for every ε.

These conditions, however, fail to provide sufficiently detailed information on either the values of the original frequencies $\lambda(\varepsilon)$ or their arithmetical properties. The main results of the first chapter are statistical in character and their application to a concrete system or to systems with fixed initial frequencies λ is not altogether convenient.

For proving the existence of a quasi-periodic solution without the *tour de force* of explicitly determining its frequency basis, the phase-averaging method developed by Mitropoliskii and Samoilenko in their papers, "On quasi-periodic oscillations in non-linear systems" *Ukr. Math. Zh.* **24**(2), (1972), and "Conditionally periodic oscillations in systems of non-linear mechanics", *Mathematical Physics*, Kiev, **2** (1972), has been found convenient for a system of the form (6.4) :

$$\frac{da}{dt} = \varepsilon A(a, \varphi, \varepsilon),$$
$$\frac{d\varphi}{dt} = \lambda + \varepsilon B(a, \varphi, \varepsilon). \tag{22}$$

Using the convenient change of variables

$$a = b + \sum_{l=1}^{p} \varepsilon^l u_l(b, \theta), \qquad \varphi = \theta + \sum_{l=1}^{p} \varepsilon^l v_l(b, \theta), \tag{23}$$

this method permits the reduction of system (22), with sufficiently smooth right-hand sides in which it is assumed that A and B are polynomials in a, ε and trigonometric polynomials in φ, to the system

$$\frac{db}{dt} = \varepsilon \bar{A}(b, \theta, 0) + \sum_{l=2}^{p} \varepsilon^l \bar{A}_l(b, \theta) + \varepsilon^{p+1} \ldots,$$
$$\frac{d\theta}{dt} = \lambda + \varepsilon \bar{B}(b, \theta, 0) + \sum_{l=2}^{p} \varepsilon^l \bar{B}_l(b, \theta) + \varepsilon^{p+1} \ldots, \tag{24}$$

where the terms $\bar{A}_l(b, \theta)$ and $\bar{B}_l(b, \theta)$ $(l = 2, \ldots, p)$ contain only 'resonance' harmonics :

$$\bar{A}_l(b, \theta) = \sum_{(k, \lambda)=0} A_l^k(b)\, e^{i(k,\theta)}, \qquad \left(\begin{array}{l} l = 1, 2, \ldots, p; \\ \bar{A}_1 = \bar{A};\ \ \bar{B}_1 = \bar{B} \end{array} \right) \tag{25}$$

$$\bar{B}_l(b, \theta) = \sum_{(k, \lambda)=0} B_l^k(b)\, e^{i(k,\theta)}.$$

Equations (24) on rejection of terms of order ε^{p+1} represent p-th approximation equations. Their solutions substituted in (23) yield the p-th best approximations. The presence of only resonance harmonics in the p-th approximation equations leads to the decomposition of these equations : in place of the angular variables $\theta'' = (\theta_{\alpha_1}, \theta_{\alpha_2}, \ldots, \theta_{\alpha\rho})$, we can introduce the variables $\psi = (\psi_1, \ldots, \psi_\rho)$ through the linear transformation

$$\psi = K_0\theta, \quad \theta'' = K^0\psi + L\theta', \quad K_0\lambda = 0 \tag{26}$$

K_0, K^0, L are integral matrices, $\theta' = (\theta_{i_1}, \ldots, \theta_{i_{m-\rho}})$, $i_1, \ldots, i_{m-\rho} \neq \alpha_1, \ldots, \alpha_\rho)$, so that the system (24) takes the form

$$\frac{db}{dt} = \sum_{l=1}^{p} \varepsilon^l \hat{A}_l(b, \psi) + \varepsilon^{p+1} \ldots,$$

$$\frac{d\psi}{dt} = K_0 \sum_{l=1}^{p} \varepsilon^l \hat{B}_l(b, \psi) + \varepsilon^{p+1} \ldots, \tag{27}$$

$$\frac{d\theta'}{dt} = \lambda' + \sum_{l=1}^{p} \varepsilon^l \hat{B}'_l(b, \psi) + \varepsilon^{p+1} \ldots .$$

Thus, the p-th approximation equations decompose into two subsystems, which are solved successively. In addition, the smaller the ρ (ρ is the rank of the spectral basis of the right-hand side of the p-th approximation equation), the larger is the number of independent equations in the second subsystem and, consequently, the greater is the ease with which the p-th approximation equations can be solved.

It is obvious that the quasi-static equilibrium positions

$$b = b_0(\varepsilon) \simeq \sum_{l=0}^{p-1} \varepsilon^l b_l = b^0, \qquad \psi = \psi_0 \simeq \sum_{l=0}^{p-1} \varepsilon^l \psi_l = \psi^0 \tag{28}$$

the decomposed system of p-th approximation equations define approximate quasi-periodic solutions of the original system

$$a = b^0 + \sum_{l=1}^{p} \varepsilon^l u_l (b^0, \lambda' (\varepsilon) t + \theta_0', K^0\psi^0 + L (\lambda' (\varepsilon) t + \theta_0')),$$

$$\varphi' = \lambda' (\varepsilon) t + \theta_0' + \sum_{l=1}^{p} \varepsilon^l v_l' (b^0, \lambda' (\varepsilon) t + \theta_0', K^0\psi^0 + L (\lambda' (\varepsilon) t + \theta_0')), \quad (29)$$

$$\varphi'' = K^0\psi^0 + L (\lambda' (\varepsilon) t + \theta_0') + \sum_{l=1}^{p} \varepsilon^l v_l'' (b^0, \lambda' (\varepsilon) t + \theta_0', K^0\psi^0 + + L (\lambda' (\varepsilon) t + \theta_0')),$$

where $\varphi = (\varphi', \varphi'') = (\varphi_{i_1}, \ldots, \varphi_{i_{m-\rho}}, \varphi_{\alpha_1}, \ldots, {}_{\alpha_\rho}), \lambda = (\lambda', \lambda''), v_l = (v_l', v_l'')$, and θ_0' are arbitrary constants.

The existence problem for the quasi-periodic solutions of system (22), some of their properties and their relationship with the approximate solutions (29) are dealt with in the next theorem, formulated for $p = 1$.

Theorem. *Suppose that the decomposed system of first approximation equation has a quasi-static equilibrium position* $b = b^0$, $\psi = \psi^0$:

$$\hat{A} (b^0, \psi^0, 0) = 0, \qquad K_0 \hat{B} (b^0, \psi^0, 0). \quad (30)$$

Assume further that the eigenvalues of the matrix

$$H = \frac{\partial (\hat{A}, K_0 \hat{B})}{\partial (b, \psi)} = \begin{pmatrix} \frac{\partial \hat{A}}{\partial b} & \frac{\partial \hat{A}}{\partial \psi} \\ K_0 \frac{\partial \hat{B}}{\partial b} & K_0 \frac{\partial \hat{B}}{\partial \psi} \end{pmatrix}_{\substack{b=b^0 \\ \psi=\psi^0}} \quad (31)$$

have non-zero real parts.

Then $\varepsilon_0 > 0$, *and* $\rho_0 > 0$ *can be found such that for every* $\varepsilon \in [0, \varepsilon_0]$:

(i) *in the* $\varepsilon\rho_0$*-neighbourhood of the position* $a = b^0$, $K_0\theta = \psi^0$, *the system equations* (22) *has an* $(m - \rho)$*-dimensional smooth invariant toroidal manifold* :

$$a = b^0 + \varepsilon U (\varphi', \varepsilon), \qquad \varphi'' = K^0\psi^0 + L\varphi' + \varepsilon V (\varphi', \varepsilon) \quad (3$$

(*the functions U and V are* 2π*-periodic in* φ' *and sufficiently smooth with respect* φ', ε);

(ii) *the quasi-periodic motions of system* (22), *characterized by the equation*

$$\frac{d\varphi'}{dt} = \lambda' + \varepsilon B' (b^0 + \varepsilon U (\varphi', \varepsilon), \varphi', K^0\psi^0 + L\varphi' + \varepsilon V (\varphi', \varepsilon)) \quad (3$$

are satisfied on the manifold (32) ;

(iii) *the manifold* (32) *has stability of the same character as the quasi-static equilibrium position* b^0, ψ^0 *of the decomposed first approximation equation* ;

(iv) *for* $p = \infty$, *the formulas* (29) *are asymptotically convergent expansions of the quasi-periodic solutions.*

Appendix V (cf. p. 82)

It may be remarked that Likova and Bogatirev [37] have successfully carried over to equations in a Banach space most of the results of § 11 on the reducibility of non-linear equations (11.1) and, in the particular, those of Theorem 9 (p. 82).

Appendix VI (cf. p. 120)

The arguments given in this chapter show that if $f(\varphi)$ is sufficiently small, then for most $\lambda = \mu + \Delta$ the motions on the torus, defined by the system of equations (17.3), are conditionally periodic with strongly non-commensurable frequencies. The representability of the set $\mathfrak{M}$ for those values of λ for which reduction to absolute rotation is either impossible or not demonstrable, and also the determination of the character of trajectories on the torus $\mathcal{T}_m$ for $\lambda \in \mathfrak{M}$ are important but still open questions except for a two-dimensional torus.

We shall deduce here two results in this direction.

Theorem [64]. *Suppose that the function* $f(\varphi) = (f_1, \ldots, f_m)$ *is* 2π*-periodic in* φ *and that* $\mu = (\mu_1, \ldots, \mu_m)$ *is a vector with commensurable components such that*

$$\mu_1 : \mu_2 : \ldots : \mu_m = q_1 : q_2 : \ldots : q_m \tag{34}$$

for some integers $q_1, q_2, \ldots, q_m$.

Suppose also that the right-hand side of the system

$$\frac{d\varphi}{dt} = \mu + \Delta + f(\varphi) \tag{35}$$

satisfies the inequalities

$$|f_i| \leqslant M_i, \quad |f_i(\varphi_1', \ldots, \varphi_m') - f_i(\varphi_1'', \ldots, \varphi_m'')| \leqslant \frac{M_i}{4\pi r m} \sum_{i=1}^{m} |\varphi_i' - \varphi_i''|, \tag{36}$$

where r *is the least common multiple of numbers* $q_1, \ldots, q_m$; $i = 1, 2, \ldots, m$.

Then, for arbitrary $\varphi_0 = (\varphi_1^0, \ldots, \varphi_m^0)$ *there is a unique vector* $\Delta = \Delta(\varphi_0) =$

$= (\Delta_1, \ldots, \Delta_m)$, *satisfying the inequality*

$$| \Delta_i | \leqslant M_i, \qquad (i = 1, 2, \ldots, m), \tag{37}$$

such that the trajectory of system (35), *passing through the point* φ_0, *is closed on* $\mathcal{T}_m$ *and the solutions originating on this trajectory are periodic with period* $T = 2\pi(r/\mu_1)$,

$$\varphi = \mu t + \varphi_0 + \varphi(t), \qquad \Phi(t) = \varphi(t + T) \tag{38}$$

and satisfy the inequalities

$$| \Phi_i(t) | \leqslant F_1 \frac{r}{\mu_1} M_i \qquad (i = 1, \ldots, n). \tag{39}$$

Moreover, $\Delta(\varphi_0)$ *is a continuous,* 2π*-periodic function of the argument* φ_0 *and its values lie on some surface*

$$\psi(\Delta_1, \ldots, \Delta_m) = 0 \tag{40}$$

of the Euclidean space E_m.

The second result relates to the system (17.3), among whose frequencies $\mu_1, \ldots, \mu_m$, only r are non-commensurable. It is convenient to express such a system in the form

$$\begin{aligned} \frac{d\varphi}{dt} &= \mu + \Delta + f(\varphi, \theta), \\ \frac{d\theta}{dt} &= \mu' + \Delta' + f'(\varphi, \theta), \end{aligned} \tag{41}$$

regarding φ, μ, Δ and f as r-dimensional and θ, μ', Δ' and f' as $(m - r)$-dimensional vectors, and assuming the frequencies $\mu = (\mu_1, \ldots, \mu_r)$ to be non-commensurable as well as $\mu' = (\mu_{r+1}, \ldots, \mu_m)$ to be expressible in terms of μ by the formula

$$\mu' = K_0 \mu, \tag{42}$$

where K_0 is an integral-valued matrix.

Put

$$\psi = K_0 \varphi - \theta, \tag{43}$$

so that system (41) has the form

$$\begin{aligned} \frac{d\psi}{dt} &= \delta + F(\varphi, K_0\varphi - \psi), \\ \frac{d\varphi}{dt} &= \mu + \Delta + f(\varphi, K_0\varphi - \psi), \end{aligned} \tag{44}$$

where $\delta = K_0\Delta - \Delta'$, $F(\varphi, \theta) = K_0 f(\varphi, \theta) - f'(\varphi, \theta)$.

Assuming δ, F, Δ and f to be sufficiently small, the phase-averaging method mentioned above can be used to investigate the trajectories of the system (44), and the presence of quasi-periodic solutions with frequency basis $\mu_1, \ldots, \mu_r$ in the original system (41) can be established. This leads to the next theorem.

Theorem. *Suppose that the functions $f(\varphi, \theta)$, $F'(\varphi, \theta)$ are sufficiently smooth, 2π-periodic in φ, θ and sufficiently small, and let $\mu = (\mu_1, \ldots, \mu_r)$ be a vector with non-commensurable components, satisfying the inequality*

$$|(\mu, k)| \geqslant K |k|^{-m} \qquad (|k| \neq 0) \tag{45}$$

for every integral-valued $k = (k_1, \ldots, k_r)$,

$$F_0(\psi) = \frac{1}{(2\pi)^r} \int_0^{2\pi} \cdots \int_0^{2\pi} F(\varphi, K_0\varphi - \psi)\, d\varphi_1 \ldots d\varphi_r. \tag{46}$$

Then, for any $\psi_0 = (\overset{0}{\psi}_1, \ldots, \overset{0}{\psi}_{m-r})$ for which the real parts of the eigenvalues of matrix $\partial F_0(\psi_0)/\partial\psi$ are non-zero, a sufficiently small $\Delta = \Delta(\psi_0)$ can be found such that with $\Delta = \Delta(\psi_0)$ and $\Delta' = K_0\Delta - F_0(\psi_0)$ the system of equations (41) *has quasi-periodic solutions $\varphi = \varphi_t$, $\theta = \theta_t$ with frequencies $\mu_1, \ldots, \mu_r$, having the form*

$$\begin{aligned} \varphi_t &= \mu t + \varphi_0 + \Phi(\mu t + \varphi_0, \psi_0), \\ \theta_t &= K_0\varphi_t - \psi_0 + \Phi'(\mu t + \varphi_0, \psi_0), \end{aligned} \tag{47}$$

where the functions $\Phi(\varphi, \psi_0)$, $\Phi'(\varphi, \psi_0)$ are sufficiently small, smooth and 2π-periodic in φ, and φ_0 is an arbitrary constant.

As $t \to \infty$, the quasi-periodic solutions are attracted to each other, the solutions getting closer every time the real parts of all the eigenvalues of matrix $\partial F_0(\psi_0)/\partial\psi$ are negative.

This theorem prescribes conditions under which there exist, on a torus, the invariant tori of a dimension less than that of the original one. The values of Δ, Δ' for which such a situation occurs lie on an r-dimensional hyperplane

$$\Delta' = K_0\Delta - F_0(\psi) \tag{48}$$

of the space E_m.

Appendix VII (cf. p. 121)

When the requirement of strong non-commensurability of frequencies ω is not involed, the existence of non-reducibile systems is established by trivial examples. For frequencies, satisfying the condition of strong non-commensurability (20.6), examples of non-reducible systems have been constructed by A.M. Gleason for the non-smooth matrix $P(\varphi)$ and by I. N. Blinov for an analytic matrix $P(\varphi)$. The example due to Blinov, derived in his paper, "On the reducibility problem of a system of linear differential equations with quasi-periodic-coefficients", *Izv. Akad. Nauk SSSR, ser., mat.* **31** (2), (1967), has the form

$$\begin{aligned} \frac{dx_1}{dt} &= i\alpha_1 x_1, \\ \frac{dx_2}{dt} &= P(\omega_1 t, \omega_2 t)\, x_1 + i\alpha_2 x_2 \end{aligned} \tag{49}$$

(α_1 and α_2 real numbers). This example involves complex-valued coefficients. Incidentally, an example of a non-reducible system with real-valued coefficients and with the form of its fundamental matrix made explicit would have been of more interest.

Appendix VIII (cf. p. 187)

Mitropoliskii and Belan [46] have applied the method of accelerated convergence for reducing to the diagonal form a system whose right-hand side differs slightly from the diagonal. The basic results of this paper are listed in the following theorem.

Theorem [46]. *Suppose that* $A(t) = \operatorname{diag}\{a_1(t), \ldots, a_n(t)\}$ *is a diagonal matrix and that* $B(t)$ *is an* $n \times n$ *square matrix continuous for* $t \in (-\infty, \infty)$. *Assume that*

$$\inf_{i \neq j} |\, a_i(t) - a_j(t) \,| \geqslant K_0 > 0 \qquad (i, j = 1, 2, \ldots, n). \tag{50}$$

Then, a sufficiently small $\bar{d} = \bar{d}(k_0, \varkappa)$, $1 < \varkappa < 2$, *can be found such that for*

$$\|\, B(t) \,\| \leqslant \bar{d} \qquad (t \in (-\infty, \infty)), \tag{51}$$

the system of equations

$$\frac{dx}{dt} = (A(t) + B(t))\, x \tag{52}$$

reduces, under the non-singular linear transformation

$$x = \Phi(t)\, y, \tag{53}$$

to the form

$$\frac{dy}{dt} = A^0(t)\, y, \tag{54}$$

where $A^0(t) = A(t) + \operatorname{diag} B(t) + \sum_{j=1}^{\infty} \operatorname{diag} B_j(t)$ *is a diagonal matrix and the matrix* $\Phi(t) = \prod_{j=1}^{\infty}(E + Y_j(t))$ *is bounded for* $t \in (-\infty, \infty)$:

$$\left\| \Phi(t) - \prod_{j=s}^{\infty}(E + Y_j(t)) \right\| \leqslant 2d^{-(\varkappa-1)\varkappa^{s-1}},$$

$$\left\| \sum_{j=s}^{\infty} \operatorname{diag} B_j(t) \right\| \leqslant 2d^{-\varkappa^{s-1}} \qquad (s = 1, 2, \ldots). \tag{55}$$

Further, if $A(t)$ *and* $B(t)$ *are almost-periodic matrices, then so also are* $\Phi(t)$ *and* $A^0(t)$ *with the same frequency basis. The condition* (50) *governing the reducibility of system* (52) *to the diagonal form* (54) *assumes the mutual separation of the diagonal elements of matrix* $A(t)$.

If the condition (50) is supposed to be fulfilled for only some and not all the i, then we can prove the reducibility of system (52) to the form (54) with the matrix $A^0(t)$ of a special but not diagonal form, characterized by the corresponding rows of $A^0(t)$ being filled by zeros everywhere except the diagonal entry. For $i = 1$, this result has been obtained by E. P. Belan in his paper, "On the stability of almost diagonal systems of linear differential equations", *Ukr. Math. Zh.* **20** (4), (1968), and has been used, as we also did in Theorem 14 of § 23, for elucidating the stability of the trivial solution of system (52), when one of the Lyapunov characteristic exponents of the generated system (of matrix $A(t)$), is zero.

Appendix IX (cf. p. 191)

The metric postulates of this chapter (Theorems 19 and 22) throw definite light on Arnold's problem ([4], problem IV) in its later form: to determine whether every system

$$\frac{dx}{dt} = (A + \mu P(\varphi))\,x,$$

$$\frac{d\varphi}{dt} = \omega \tag{56}$$

is reducible for almost all matrices A and almost all values of the parameter μ without the requirement that μ be small ($P(\varphi)$ is assumed analytic and ω strongly non-commensurable), if it is solved completely for small values of μ.

Appendix X (cf. p. 192)

Of the recent studies relating to this problem, it is worthwhile to mention the result due to V. K. Melinikov. In his paper, "On the behaviour of the trajectory of a class of dynamical systems", *Mat. Sb.* **73** (4), (1967), he has considered the problem of reducing the following dynamical system in a three-dimensional Euclidean space

$$\frac{dx}{dt} = f(x) \qquad (x = (x_1, x_2, x_3)) \tag{57}$$

in a neighbourhood of the periodic solution $x = x_0(t)$ to the form

$$\dot{u} = -\lambda_1 (u^2 + v^2)\, v, \quad \dot{v} = \lambda_1 (u^2 + v^2)\, u, \quad \dot{\sigma} = \lambda_2 (u^2 + v^2), \tag{58}$$

where $\lambda_1 = \lambda_1 (u^2 + v^2)$ and $\lambda_2 = \lambda_2 (u^2 + v^2)$ are functions of only the combination $u^2 + v^2$. The basic consequences of this paper consist in determining the necessary and sufficient conditions, subject to whose fulfilment, it is possible to transform the system (57) to (58), and in constructing suitable substitutions, which can be found via the Newton type iteration methods.

Appendix XI (cf. p. 227)

In his paper, "On reduction of a dynamical system to canonical form in the neighbourhood of a smooth invariant torus", *Izv. Akad. Nauk SSSR, ser. mat.* **36** (1), (1972), Samoilenko has dispensed with both the hypotheses on the reducibitity of a linear system as well as on the parallel flow of trajectories on a torus.

This paper deals with a system of the form

$$\begin{aligned} \frac{dy}{dt} &= A(\varphi, y)\, y, \\ \frac{d\varphi}{dt} &= f(\varphi, y), \end{aligned} \tag{59}$$

where $A(\varphi, y)$ and $f(\varphi, y)$ are, respectively, a matrix and a vector function, defined and sufficiently smooth in a certain domain $D_\mu = \left\{\varphi, y;\ \| y \| = \sqrt{\sum_{i=1} y_i^2} \leqslant \mu\right\}$, and 2π-periodic in each of the variables φ_ν $(\nu = 1, 2, \ldots, m)$.

Side by side with the system (59), there is a discussion on its 'canonical' part

$$\frac{dx}{dt} = A_0(\theta)\,x,$$
$$\frac{d\theta}{dt} = f_0(\theta), \tag{60}$$

where $A_0(\theta) = A(\theta, 0), f_0(\theta) = f(\theta, 0)$, and conditions are determined governing the transformation of the system (59) into the system (60) in some neighbourhood of the torus $y = 0$; the conditions required for reducibility are given, and the iteration process is described for the construction of the corresponding transformation

$$y, \varphi \to x, \theta. \tag{61}$$

By $\theta_t = \theta_t(\theta), \theta_0(\theta) = \theta$ is denoted the flow of trajectories on the torus $y=0$, i.e. the general solution of the second equation of system (60) and by $\Omega_\tau^t(B) = \Omega_\tau^t(B;\theta)$ the fundamental solution matrix of the linear system

$$\frac{dx}{dt} = B(\theta_t)\,x, \tag{62}$$

assuming B to be either of the matrices $A_0(\theta)$, or $\dfrac{\partial f_0(\theta)}{\partial\theta}$; $\Omega_\tau^\tau(B) = E$ the identity matrix.

Conditions sufficient for ensuring the reduction of system (59) to the canonical form (60) in D_μ are set forth in the next theorem.

Theorem. *Assume that the functions $A(\varphi, y)$ and $f(\varphi, y)$ are three times continuously differentiable with respect to φ, y in the domain D_μ and 2π-periodic in φ. Suppose that the matrices $\partial f_0(\theta)/\partial(\theta)$ and $A_0(\theta)$ are such that the inequalities*

$$\|\Omega_0^t(A_0)\| \leqslant C_1 \quad \text{for} \quad t \geqslant 0,$$

$$\int_0^\infty \max\Big\{\Big\|\Omega_\tau^0\Big(\frac{\partial f_0}{\partial\theta}\Big)\Big\|\,\|\Omega_0^\tau(A_0)\|;\ \Big\|\frac{\partial}{\partial\theta_j}\Omega_\tau^0\Big(\frac{\partial f_0}{\partial\theta}\Big)\Big\|\,\|\Omega_0^\tau(A_0)\|;$$
$$\Big\|\Omega_\tau^0\Big(\frac{\partial f_0}{\partial\theta}\Big)\Big\|\,\Big\|\frac{\partial}{\partial\theta_j}\Omega_0^\tau(A_0)\Big\|;\ \Big\|\Omega_\tau^0\Big(\frac{\partial f_0}{\partial\theta}\Big)\Big\|\,\Big\|\Omega_0^\tau\Big(\frac{\partial f_0}{\partial\theta}\Big)\Big\|\,\|\Omega_0^\tau(A_0)\|\Big\}\,d\tau \leqslant c_2,$$
$$\tag{63}$$
$$\int_0^\infty \max\Big\{\|\Omega_\tau^0(A_0)\|\,\|\Omega_0^\tau(A_0)\|^2;\ \Big\|\frac{\partial}{\partial\theta_j}\Omega_\tau^0(A_0)\Big\|\,\|\Omega_0^\tau(A_0)\|^2;$$
$$\|\Omega_\tau^0(A_0)\|\,\Big\|\Omega_0^\tau\Big(\frac{\partial f_0}{\partial\theta}\Big)\Big\|\,\|\Omega_0^\tau(A_0)\|^2;\ \|\Omega_\tau^0(A_0)\|\,\Big\|\frac{\partial}{\partial\theta_j}\Omega_0^\tau(A_0)\Big\|\,\|\Omega_0^\tau(A_0)\|\Big\}\times$$
$$\times\, d\tau \leqslant c_2$$

are satisfied, for every $j = 1, 2, \ldots, m$ *and some positive constants* c_1, c_2.

Then we can find h_0 $(0 < h_0 \leqslant \mu)$ *and the functions* $Y(\theta, x)$, $\Phi(\theta, x)$, *continuously differentiable with respect to* θ, x *for* $\| x \| \leqslant h_0$, 2π*-periodic in* θ, *and satisfying the conditions*

$$Y(\theta, 0) = \frac{\partial Y(\theta, 0)}{\partial x} = \Phi(\theta, 0) = 0, \tag{64}$$

such that the system of differential equations (59) *reduces, by the change of variables*

$$y = x + Y(\theta, x), \qquad \varphi = \theta + \Phi(\theta, x), \tag{65}$$

to the canonical form (60).

Among the necessary conditions for the inequalities (63), we mention :

$$\max_{\theta} \Big\{ \beta^0(\theta) \,;\, \beta^0(\theta) - \alpha_0(\theta) \,;\, \beta^0(\theta) - \alpha_0(\theta) + \alpha^0(\theta) \,;\, 2\beta_0(\theta) - \beta^0(\theta) \,;$$
$$2\beta^0(\theta) - \beta_0(\theta) + \alpha^0(\theta) \Big\} < 0, \tag{66}$$

where $\alpha_0(\theta)$ and $\alpha^0(\theta)$, and $\beta_0(\theta)$ $\beta^0(\theta)$ are equal to the least and greatest eigenvalues of the matrices

$$\frac{1}{2}\left[\frac{\partial f_0(\theta)}{\partial \theta} + \left(\frac{\partial f_0(\theta)}{\partial \theta}\right)^*\right] \quad \text{and} \quad \frac{1}{2}\,[A_0(\theta) + A_0^*(\theta)]$$

respectively, the asterisk sign $*$ denoting the adjoint of a matrix.

Thus, the assertions derived in the above theorem remain valid, if in place of the inequalities (63), the simpler inequalities (66) have to be satisfied.

Appendix XII cf. p. (234)

The dependence of l_0 on s_0, given by the inequality (32.40), is considerably overstated. In the paper cited in the preceding appendix, it has been refined to the form

$$l_0 = s_0 + 2. \tag{67}$$

Appendix XIII (cf. p. 242)

In the present section, the theory of perturbations of an invariant torus has been sketched only briefly with the main object to make explicit the conditions that ensure the preservation of an invariant torus under a small perturbation of a dynamical system. However, there is a large and growing literature devoted to this theory. Among more recent contributions in this field, a mention may be made of Samoilenko's works, "On preservation of the invariant torus under perturbation", *Izv. Akad. Nauk SSSR, ser. mat.* **34** (6), (1970), and "On the theory of perturbations of invariant manifolds of dynamical system" in : *Proceedings of International Conference on Non-linear Oscillations,* Vol. I ; and *Analytic Methods in the Theory of Nonlinear Oscillations*, Izd. Inst. Math. Akad. Nauk USSR, Kiev (1970). These contributions lend a new approach to the theory of perturbations of invariant tori of dynamical systems, involving Green's functions for linearized problems. The treatment given develops, from a unified and general view point, the theory of perturbations for both smooth and non-differentiable invariant toroidal manifolds of dynamical systems and obtains conditions close to those necessary for the existence of such manifolds. One of the results is presented below.

Let $C^r(D)$ be the Banach space of the functions $u(\varphi, y)$, defined on a compact set D, 2π-periodic in φ and having on D partial derivatives up to order r ; $|\cdot|_r$ is the norm in $C^r(D)$. Assume that, in the neighbourhood of a toroidal manifold M of the unperturbed dynamical system, it is possible to introduce angular coordinates $\varphi = (\varphi_1, \ldots, \varphi_m)$ and normal coordinates $y = (y_1, \ldots, y_n)$ such that the equation of the manifold has the form $y = 0$, and the perturbed system of equations takes the form

$$\frac{dy}{dt} = -b(\varphi, y, \mu)\, y + c(\varphi, \mu),$$
$$\frac{d\varphi}{dt} = a(\varphi, y, \mu), \tag{68}$$

with $b, c, a \in C^r(\| y \| \leqslant d)$, for every fixed $\mu \in [0, \mu_0]$, $\mu_0 > 0$, $c(\varphi, 0) = 0$.

Theorem. *Assume that the right-hand side of system* (68) *satisfies the following conditions* :

(i) max $\{| a(\varphi, y, \mu) - a(\varphi. 0, 0)|_r, | b(\varphi, y, \mu) - b(\varphi, 0, 0)|_r, | c(\varphi, \mu)|_r\}$ $\leqslant L_r(d, \mu_0)$, *where* $L_r(d, \mu_0) \to 0$ *as* $d \to 0$, $\mu_0 \to 0$.

(ii) *for every space* C^a, *sufficiently small in norm, of the functions* $a_1(\varphi)$, $b_1(\varphi)$ *and* $c_1(\varphi)$ *the perturbed linearized equation*

$$\frac{dy}{dt} = (-b(\varphi, 0, 0) + b_1(\varphi))\, y + c_1(\varphi),$$
$$\frac{d\varphi}{dt} = a(\varphi, 0, 0) + a_1(\varphi) \tag{69}$$

has an invariant torus $\mathcal{T} : y = u(\varphi)$, *satisfying*

$$| u(\varphi)|_a \leqslant K | c_1(\varphi)|_r, \tag{70}$$

with K *a positive constant.*

Then, $\mu^0 \leqslant \mu_0$ can be found such that for every $\mu \in [0, \mu^0]$ the system of equations (68) *has, in the space C^{r-1}_{Lip}, for $r \geqslant 1$, an invariant torus $\mathcal{T}(\mu) : y = u(\varphi, \mu)$ satisfying*

$$\lim_{\mu \to 0} | u(\varphi, \mu) |_{r-1} = 0. \tag{71}$$

This theorem defines conditions under which the perturbed system has an invariant torus, continuously deforming under a decreasing perturbation. The condition (ii), which is quite restrictive, is equivalent to the existence of the Green's function $G_0(\tau, \varphi)$ for the problem on an invariant torus of the linearized system (69), satisfying the inequality

$$| G_0(\tau, \varphi) c_1(\varphi_\tau(\varphi)) |_r \leqslant K e^{-\gamma |\tau|} | c_1(\varphi) |_r \tag{72}$$

for every $\tau \in (-\infty, \infty)$ and some positive K, γ.

For the existence of the function $G_0(\tau, \varphi)$, satisfying (72), it is sufficient, in particular, that the inequalities

$$\beta(\varphi) + r\alpha(\varphi) > 0, \qquad \min_{\varphi} \beta(\varphi) = \beta > 0 \tag{73}$$

are satisfied, for the least eigenvalues $\beta(\varphi)$ and $\alpha(\varphi)$ of the matrices

$$\frac{1}{2}[b(\varphi, 0, 0) + b^*(\varphi, 0, 0)] \quad \text{and} \quad \frac{1}{2}\left[\frac{\partial a(\varphi, 0, 0)}{\partial \varphi} + \left(\frac{\partial a(\varphi, 0, 0)}{\partial \varphi}\right)^*\right].$$

Thus, subject to inequalities (73) and the hypothesis (i) being satisfied, the system of equations (68) has in C^{r-1}_{Lip} $(r \geqslant 1)$ an invariant torus satisfying (71). This last assertion represents an important conclusion proved by Sacker in [59] for $r \leqslant 2$.

Appendix XIV (cf. p. 246)

Making use of the ideas set forth in Appendix XI, we can get rid of the restrictions of the reducibilitty of the linear system (34.8) to the form (34.10) and the representability of the matrix $H(c)$ by the formula(34.14). This enables us to consider a dynamical system in the neighbourhood of a manifold with the only assumption that it has the form

$$\frac{dc}{dt} = R(t, c, x),$$

$$\frac{dx}{dt} = A(t, c)x + F(t, c, x) \tag{74}$$

$$\left(R(t, c, 0) = F(t, c, 0) = \frac{\partial F(t, c, 0)}{\partial x} = 0\right),$$

and to determine the conditions under which the coordinate transformation of the form (34.16) reduces it to the canonical form

$$\frac{d\theta}{dt} = 0,$$
$$\frac{d\xi}{dt} = A(t, 0)\,\xi. \qquad (75)$$

Appendix XV (cf. p. 262)

Theorem 32 remains valid if the condition (37.5) on the eigenvalues of matrix $A(c)$ is replaced by the conditions analogous to (32.9) :

$$\begin{aligned} &\operatorname{Re}[(r, \mu(c)) - \mu_j(c)] \leqslant -\gamma \qquad \text{for } |r| = \tau_0 + 1, \\ &|\operatorname{Re}[(r, \mu(c)) - \mu_j(c)]| \geqslant \varepsilon \qquad \text{for } 2 \leqslant |r| \leqslant \tau_0 \end{aligned} \qquad (76)$$

for every $c \in D_1$, any $j = 1, 2, \ldots, n$, non-negative integral-valued $r = (r_1, \ldots$ $\ldots, r_n)$ and some positive γ and ε.

It is not difficult to arrive at this conclusion if, in the iteration process applied in the proof of Theorem 32, we use Taylor's formula when solving the auxiliary system (35.1) : we represent the right-hand sides of (35.1) at the point $x = 0$ by Taylor's formula with the remainder term defined by the τ_0-th derivatives with respect to x, and seek the solutions $w(t, c, x)$ and $u(t, c, x)$ in a similar form.

REFERENCES

[1] Andrianova, L. Ya. (1962). On the reducibility of a system of linear differentiai equations with quasiperiodic coefficients, *Vestnik Leningrad Gos. Univ., Ser. Mat. Mekh. Astronom.* **7**, 14-24.

[2] Arnol'd, V. L. (1961), Small divisors. I. On the mapping of a circle into itself, *Izv. Akad. Nauk SSSR, Ser. Mat.*, **25** (1), 21-86.

[3] Arnol'd, V. L. (1963), Small divisors. II. Proof of A. N. Kolmogorov's theorem on conservation of conditionally-periodic motion under small perturbations of the Hamiltonian functions, *Usp. Mat. Nauk.* **18** (5), 13-40.

[4] Arnol'd, V. L. (1963), Small divisor problems in classical and celestical mechanics, *Usp. Mat. Nauk* **18** (6), 91-192.

[5] Artem'ev, N. A. (1944), A method for the determination of a characteristic exponent and its application to two problems in celestial mechanics, *Izv. Akad. Nauk SSSR* **8**(2), 61-100.

[6] Belaga, E. G. (1962), On the reducibility of a system of ordinary differential equations in the neighbourhood of conditionally-periodic motion, *Dokl. Akad. Nauk SSSR* **143** (2), 255-58.

[7] Blinov, I. N. (1965), Analytic representation of solutions of systems of linear differential equations with almost periodic coefficients depending on parameters, *Diff. Uravneniya* **1** (8), 1042-53.

[8] Bogatirev, B. M. (1968), On the construction of solutions of differential equations with quasiperiodic coefficients in the space L_2 [0, 2π], *Ukr. Mat. Zh.* **20** (4).

[9] Bogdanov, J. C. and Chebotarev, G. N. (1959), On matrices commuting with their products, *Izv. Vuzov. Matematika* **4**, 27-37.

[10] Bogoljubov, N. N. (1945), *On some Statistical Methods in Methematical Physics*, Izv. vo Akad. Nauk Ukr. SSR, Kiev.

[11] Bogoljubov, N. N. (1964), On quasiperiodic solutions in the problems of nonlinear Mechanics, *Tr. Pervoi Letnei Mat. Shkol.* **1**, Kiev.

[12] Bogoljubov, N. N. and Mitropoliskii, Yu. A. (1963), *Asymptotic Methods in the Theory of Nonlinear Oscillations*, Fizmatgiz ; (English Tr. Hindustan Publishing Corpn. Delhi (1961)).

[13] Bogoljubov, N. N. and Mitropoliskii, Yu. A. (1965), On a study of quasiperiodic régimes in nonlinear oscillatory systems, *Les Vibrations forces dans les Systems nonlineares*, Paris, 181-92.

[14] Bolyai, P. G. (1961), *Selected Works*, Izd-vo Latv. Akad. Nauk, Riga.

[15] Cherry, T. M. (1938), Analytic quasi-periodic curves of discontinuous type on the torus, *Proc. London Math. Soc.* **44**.

[16] Coddington E. A. and Levinson, N. L. (1955), *Theory of Ordinary Differential Equations*, McGraw-Hill, New York.

[17] Denjoy, A. (1932), Sur les courbes définies par les équations differentialles á la surface du tore, *J. Math. Pures Appl.* **11** (9), 333-375.

[18] Erugin, N. P. (1946), Reducible systems, *Trudy Mat. Inst. Steklov*, **13**, 95.

[19] Erugin, N. P. (1963), *Linear Systems of Ordinary Differential Equations*, Izd-vo Akad, Nauk BSSR.

[20] Finzi, A. (1950, 1952), Sur Les problèmes de la genèration d'une transformation donnée d'une courbe fermée par une transformation infinitesimale, *Ann. Soç. I'Ec. Norm. Sup.* **67** (3), 243-305 ; **69** (4), 371-430.

[21] Gantmacher, F. R. (1967), *The Theory of Matrices*, Vols. I and II, Chelsea, New York.

[22] Gel'man A. E. (1954), On the reducibility of a class of systems of differential equations with quasiperiodic coefficients, *Dokl. Akad. Nauk SSSR* **116** (4), 535-37.

[23] Gel'man, A. E. (1965), On the reducibility of a system with quasiperiodic matrices, *Diff. Uravnehiya* **1** (3), 283-94.

[24] Harasahal, V. (1964), *The Theory of Almost Periodic Solutions of Ordinary Differential Equations*, Doctoral thesis, Inst. Mat. Akad. Nauk Ukr. SSR, Kiev.

[25] Hardy, G. H., Littlewood, J. E. and Polya, G. (1964), *Inequalities*, Cambridge University Press, London.

[26] Kneser, H. (1924), Regulare Kurvenscharen auf Ringflachen, *Math. Ann.* **91**, 135-54.

[27] Kolmogorov, A. N. (1953), On dynamical systems with integral invariant on the torus, *Dokl. Akad. Nauk SSSR* **93** (5), 763-66.

[28] Kolmogorov, A. N. (1954), On the conservation of conditionally periodic motion under small perturbations of the Hamiltonian functions, *Dokl. Akad. Nauk SSSR* **98** (4), 527-30.

[29] Kolmogorov, A. N. (1961), A general theory of dynamical systems and classical mechanics, *Proceedings of the International Mathematical Congress*, Amsterdam.

[30] Krylov, N. M. and Bogoljubov, N. N. (1934), Sur quelques developments formulas en series dans la mécanique non-lineaire, *Ukranin. Akad. Nauk Kiev* **4**, 56.

[31] Krylov, N. M. and Bogoljubov, N. N. (1934), *Applications of the Methods of Nonlinear Mechanics to the Theory of Stationary Oscillations*, Ukrain. Akad. Nauk, Kiev.

[32] Krylov, N. M. and Bogoljubov, N. N. (1934), *New Methods in Nonlinear Mechanics GTTI*, Kiev.

[33] Krylov, N. M. and Bogoljubov, N. N. (1937), *Introduction to Nonlinear Mechanics*, Izd-vo Akad Nauk Ukr. USSR, Kiev; (a free translation by S. Lefshetz) *Ann. of Math. Studies* **11**, 106 (1947).

[34] Kupka, I. (1964), Stabilitè des varièties invariates d'un champ de vecteurs pour les petites perturbations, *C.R. Acad. Sci. Paris* **258** (17).

[35] Landau, E. (1913), Einigeungleichungen für zweimal differentiberbare Funktionen, *Proc. London Math. Soc.* **13** (2), 43-49.

[36] Likova, O. B. (1964), On quasiperiodic solutions of nearly canonical systems, *Ukr. Mat. Zh.* **16** (6), 752-68.

[37] Likova, O. B. and Bogatirev, B. M. (1968), On the reducibility of some differential equations in Banach space, *Ukr. Mat. Zh.* **20** (5).

[38] Lojasiewicz, S. (1961), Sur le problème de la division, *Rozprawy Matematyczne* **22**.

[39] Lyashenko, N. Ya. (1961), An analogue of Floquet's theorem for a special case of a linear homogeneous system of differential equations with quasiperiodic coefficients, *Dokl. Akad. Nauk SSSR* **111** (2), 295-98.

[40] Malkin, I. G. (1966), *Theory of Stability of Motions*, Nauka, Moscow (German Tr. : Oldenbourg, Munchen) ; (English tr. : Atomic Energy Commission, Washington, D. C.).

[41] Mitropoliskii, Yu. A. (1964), On the construction of general solutions of nonlinear differential equations by the method of 'accelerated convergence' *Ukr. Mat. Zh.* **16** (4), 475-501.

[42] Mitropoliskii, Yu. A. 1(966), The method of accelerated convergence in problems of nonlinear mechanics, *Funkcialaj Ekvacioj* (*Serio Internacia*), **9** (1-3), 27-42.

[43] Mitropoliskii, Yu. A. (1967), The method of accelerated covergence in problems of nonlinear mechanics, *Proceedings of the Third Scientific Congress of Young Ukranian Mathematicians*, Kiev.

[44] Mitropoliskii, Yu. A. (1968), On the construction of solutions and the reducibility of differential equations with quasiperiodic coefficients, *Proccedings of the Third All-Union Conference of Theoretical and Applied Mechanics*, Moscow, 212.

[45] Mitropoliskii, Yu. A. (1968), On the construction of solutions and the reducibility of differential equations with quasiperiodic coefficients, *Proceedings of the Fourth Congress on Nonlinear Oscillations*, Prague.

[46] Mitropoliskii, Ju. A. and Belan, E. P. (1968), On the construction of solutions of an almost diagonal system of linear differential equations by a method which assures accelerated convergence, *Ukr. Mat. Zh.* **20** (2), 166-75.

[47] Mitropoliskii, Ju. A. and Likova, O. B. (1965), On the existence of quasiperiodic solutions of perturbed canonical systems, *Les Vibrations forcèes dans les Systèms non-lineáres*, Paris, 407-14.

[48] Mitropoliskii, Ju. A. and Samoilenko, A. M. (1964), The structure of trajectories on a toroidal manifold, *Dokl. Akad. Nauk USSR* **8**, 984-85.

[49] Mitropoliskii, Ju. A. and Samoilenko, A. M. (1965), On construction of solutions of linear differential equations with quasiperiodic coefficients by the method of accelerated convergence, *Ukr. Mat. Zh.* **17** (6), 42-59.

[50] Mitropoliskii, Ju. A. and Samoilenko, A. M. (1967), The problem of the construction of solutions of linear differential equations with quasiperiodic coefficients, in : *Mathematical Physics*, Kiev, 185-98.

[51] Moser, J. (1961), A new technique for the construction of solutions of nonlinear differential equa tions, *Proc. Nat. Acad. Sci. USA* **47** (11), 1824-31.

[52] Moser, J. (1962), On invariant curves of area-preserving mappings of an annulus, *Nachr. Akad, Wiss. Göttingen. Math-Phys* : Kl. N 1 (Russian tr. *Sb. Mat.* **6** (4), 1962, 3-10).

[53] Moser, J. (1965), On invariant surfaces and almost periodic solutions of ordinary differential equations, *Notices Amer. Math. Soc.* **12** (1).

[54] Moser, J. (1966), A rapidly convergent iteration method and nonlinear differential equations, II, *Ann. Scoula Norm. Sup. Pisa* **20** (3), 499-536.

[55] Nach, J. (1956), The imbedding problem for Riemannian manifolds, *Ann. Math.* **63** (1), 20-63.

[56] Pliss, V. A. (1964), *Nonlocal Problems in the Theory of Oscillations*, 'Nauka' M-L.

[57] Poincaré, A. L. (1881), Sur les courbes definies par les èquations differentielles, *J. de. Math.* **7** (3), 375-422 ; (Russian Tr. : (1947), OGIZ, M-L).

[58] Poincaré, A. L. (1879), Sur les propriétés des fonctions définies par les équations aux differénces partielles; In : *Thesis de Mathematiques*, Gauthier-Villers, Paris.

[59] Sacker, R. (1965), A new approach to the perturbation theory of invariant surfaces, *Comm. Pure Appl. Math.* **18** (4).

[60] Samoilenko, A. M. (1964), The problem of *the structure* of trajectories on the torus, *Ukr. Mat. Zh.* **16** (6), 769-82.

[61] Samoilenko, A. M. (1966), On the reducibility of a system of ordinary differential equations in the neighbourhood of a smooth toroidal manifold, *Izv. Akad. Nauk SSSR, ser. mat.* **30** (5), 1047-72.

[62] Samoilenko, A. M. (1966), On the reducibility of a system of ordinary differential equations in the neighbourhood of an invariant smooth manifold (a brief scientific report on a thesis), *Proceedings of the International Congress of Mathematicians*, Moscow, Section, 6, 47.

[63] Samoilenko, A. M. (1966), On the reducibility of a system of ordinary differential equations in the neighbourhood of a smooth integral manifold, *Ukr. Math. Zh.* **18** (6), 41-64.

[64] Samoilenko, A. M. (1967), Trajectories on the Tori, *Proceedings of the Third Scientific Congress of Young Ukranian Mathematicians*, Kiev.

[65] Samoilenko, A. M. (1967), *Some Problems in the Theory of Periodic and Quasiperiodic Systems*, Doctoral thesis, Inst. Mat. Akad. Nauk Ukr. SSSR, Kiev.

[66] Samoilenko, A. M. (1968), On the reducibility of a system of linear differential equations with quasiperiodic coefficients, *Ukr. Mat. Zh.* **20** (2).

[67] Shtokalo, I. J. (1960), *Linear Differential Equations with Variable Coefficients*, Izd-vo Akad. Nauk Ukr. SSR, Kiev; (English tr. : Hindustan Publishing Corporation (India), Delhi).

[68] Siegel, C. L. (1945), Note on differential equations on the torus, *Ann. of Math.* **46** (2), 423-426.

[69] Siegel, C. L. (1932), Über die Normalform analytischer Differentialgleichungen in der Nähe einer Gleichgewichtslösung, *Nachr. Akad. Wiss., Göttingen, Math. Phys.*, 21-30.
[70] Sobolev, S. L. (1950), *Some Applications of Functional Analysis in Mathematical Physics*, Leningrad University Press.
[71] Sobolev, S. L. (1966), *Equations of Mathematical Physics*, Nauka, Moscow.
[72] Zadiraka, K. V. (1965), *The Study of an Irregularly Perturbed Differential System by the Method of Integral Manifolds*, Doctoral dissertation, Akad. Nauk Ukr. SSR, Kiev.